CAMBRIDGE TRACTS IN MATHEMATICS

General Editors

B. BOLLOBÁS, W. FULTON, A. KATOK, F. KIRWAN,
P. SARNAK, B. SIMON, B. TOTARO

CAMBRIDGE TRACTS IN MATHEMATICS

General Editors:

B. BOLLOBÁS, W. FULTON, A. KATOK, F. KIRWAN, P. SARNAK, B. SIMON, B. TOTARO

A complete list of books in the series can be found at www.cambridge.org/mathematics.
Recent titles include the following:

Jordan Structures in Geometry and Analysis

CHO-HO CHU
Queen Mary, University of London

CAMBRIDGE UNIVERSITY PRESS
Cambridge, New York, Melbourne, Madrid, Cape Town,
Singapore, São Paulo, Delhi, Tokyo, Mexico City

Cambridge University Press
The Edinburgh Building, Cambridge CB2 8RU, UK

Published in the United States of America by Cambridge University Press, New York

www.cambridge.org
Information on this title: www.cambridge.org/9781107016170

First published 2012

Printed in the United Kingdom at the University Press, Cambridge

A catalogue record for this publication is available from the British Library

ISBN 978-1-107-01617-0 Hardback

To my family

Contents

vii

Preface

Despite the rapid advances in Jordan theory and its diverse applications in the last two decades, there are few convenient references in book form for beginners and researchers in the field. This book is a modest attempt to fill part of this gap.

The aim of the book is to introduce to a wide readership, including research students, the close connections between Jordan algebras, geometry, and analysis. In particular, we give a self-contained and systematic exposition of a Jordan algebraic approach to symmetric manifolds which may be infinite-dimensional, and some fundamental results of Jordan theory in complex and functional analysis. In short, this book is about *Jordan geometric analysis.*

Although the concept of a Jordan algebra was introduced originally for quantum formalism, by P. Jordan, J. von Neumann and E. Wigner [64], unexpected and fruitful connections with Lie algebras, geometry and analysis were soon discovered. In the last three decades, many more applications of Jordan algebraic structures have been found. We expose some of these applications in this book. Needless to say, the choice of topics is influenced by the author's predilections, and regrettable omissions are inevitable if the length of the book is to be kept manageable. Nevertheless, an effort has been made to cover sufficient basic results and Jordan techniques to provide a handy reference.

We begin by discussing the basic structures of Jordan algebras and Jordan triple systems in Chapter 1, and the connections of these Jordan structures to Lie theory. An important link is the Tits–Kantor–Koecher construction, which establishes the correspondence between Jordan triple systems and a class of graded Lie algebras. We discuss some details of classical matrix Lie groups and their Lie algebras and use them as examples to illustrate these connections, as well as preparation for the introduction of Banach Lie groups in the following chapter.

Since É. Cartan's seminal work, Lie theory has been an important tool in the study of Riemannian symmetric spaces and their classification. It was found relatively recently that Jordan algebras and Jordan triple systems can be used to give an algebraic description of a large class of symmetric spaces which is also accessible in infinite dimension. This is the subject of Chapter 2. We give a concise introduction to Banach manifolds and Banach Lie groups. We show the connections between Jordan algebras and symmetric cones, and the correspondence of Jordan triple systems and Riemannian symmetric spaces in the infinite-dimensional setting. We complete the discussion by showing that the bounded symmetric domains in complex Banach spaces are exactly the open unit balls of JB*-triples which are complex Banach spaces equipped with a Jordan triple structure.

A large part of Chapter 3 is devoted to the study of JB*-triples. They play an important role in geometry and analysis, as informed by the previous result. The open unit balls of JB*-triples can be regarded as an infinite-dimensional generalisation of the open unit disc in the complex plane and provide a natural setting for complex function theory. As examples, we discuss distortion theorems and iterations of holomorphic maps on these open balls, where Jordan techniques come into play. In a functional-analytic vista, JB*-triples form an important class of Banach spaces, which includes C*-algebras, spaces of operators between Hilbert spaces and some exceptional Jordan algebras. We present a sufficient number of basic properties of JB*-triples as research tools, but a complete treatment would lengthen the book to excess. From the viewpoint of JB*-triples, many results in C*-algebras, for example, those on contractive projections and isometries, can be explained simply in a geometric perspective. Finally, we discuss Jordan structures in Hilbert spaces, which are important in the geometry of infinite-dimensional Riemannian symmetric spaces; for instance, the curvature tensor is related to the Jordan triple product. The last chapter contains some new results.

It is a great pleasure to thank many colleagues and friends for valuable conversations concerning the subject matter of this book. I have benefited especially from inspiring discussions with Wilhelm Kaup and the late Issac Kantor on many occasions. I thank Pauline Mellon for reading part of the manuscript and for her useful comments. I much appreciate the sabbatical leave from Queen Mary College in 2010, which enabled me to complete the manuscript. I would also like to thank my wife, Yen, and my daughter, Clio, for their constant support and encouragement.

1

Jordan and Lie theory

1.1 Jordan algebras

We begin by discussing the basic structures and some examples of Jordan algebras which are relevant in later chapters. One important feature is that multiplication in these algebras need not be associative.

By an *algebra* we mean a vector space $\mathcal{A}$ over a field, equipped with a bilinear product $(a, b) \in \mathcal{A}^2 \mapsto ab \in \mathcal{A}$. We do not assume associativity of the product. If the product is associative, we call $\mathcal{A}$ *associative*.

Homomorphisms and isomorphisms between two algebras are defined as in the case of associative algebras. An *antiautomorphism* of an algebra $\mathcal{A}$ is a linear bijection $\varphi : \mathcal{A} \longrightarrow \mathcal{A}$ such that $\varphi(ab) = \varphi(b)\varphi(a)$ for all $a, b \in \mathcal{A}$.

We call an algebra $\mathcal{A}$ *unital* if it contains an identity, which will always be denoted by **1**, unless stated otherwise. As usual, one can adjoin an identity **1** to a nonunital algebra $\mathcal{A}$ to form a unital algebra $\mathcal{A}_1$, called the *unit extension* of $\mathcal{A}$.

A *Jordan algebra* is a commutative algebra over a field $\mathbb{F}$, and satisfies the Jordan identity

$$(ab)a^2 = a(ba^2) \qquad (a, b \in \mathcal{A}).$$

We always assume that $\mathbb{F}$ is not of characteristic 2; however, in later sections, $\mathbb{F}$ is usually either $\mathbb{R}$ or $\mathbb{C}$.

The concept of a Jordan algebra was introduced by P. Jordan, J. von Neumann, and E. Wigner [64] to formulate an algebraic model for quantum mechanics. They introduced the notion of an *r-number system* which is, in modern terminology, a finite-dimensional, formally real Jordan algebra. In fact, the term *Jordan algebra* first appeared in an article by A. A. Albert [3]. It denotes an

1

algebra of linear transformations closed in the product

$$A \cdot B = \frac{1}{2}(AB + BA).$$

Although Jordan algebras were motivated by quantum formalism, unexpected and important applications in algebra, geometry and analysis have been discovered. Some of these discoveries are the subject of discussions in ensuing chapters.

On any associative algebra $\mathcal{A}$, a product $\circ$ can be defined by

$$a \circ b = \frac{1}{2}(ab + ba) \qquad (a, b \in \mathcal{A}),$$

where the product on the right-hand side is the original product of $\mathcal{A}$. The algebra $\mathcal{A}$ becomes a Jordan algebra with the product $\circ$. We call this product *special*. A Jordan algebra is called *special* if it is isomorphic to, and hence identified with, a Jordan subalgebra of an associative algebra $\mathcal{A}$ with respect to the special Jordan product $\circ$. Otherwise, it is called *exceptional*.

It is often convenient to express the Jordan identity as an operator identity. Given an algebra $\mathcal{A}$ and $a \in \mathcal{A}$, we define a linear map $L_a : \mathcal{A} \longrightarrow \mathcal{A}$, called *left multiplication by a*, as follows:

$$L_a(x) = ax \qquad (x \in \mathcal{A}).$$

The Jordan identity can be expressed as

$$[L_a, L_{a^2}] = 0 \qquad (a \in \mathcal{A}),$$

where $[\cdot, \cdot]$ is the usual Lie bracket product of linear maps. Given $a, b \in \mathcal{A}$, we define the *quadratic operator* $Q_a : \mathcal{A} \longrightarrow \mathcal{A}$ and *box operator* $a \,\square\, b : \mathcal{A} \longrightarrow \mathcal{A}$ by

$$Q_a = 2L_a^2 - L_{a^2}, \quad a \,\square\, b = L_{ab} + [L_a, L_b]. \tag{1.1}$$

These operators are fundamental in Jordan theory, as is the linearization of the quadratic operator:

$$Q_{a,b} = Q_{a+b} - Q_a - Q_b.$$

Let $\mathcal{A}$ be an algebra and let $a \in \mathcal{A}$. We define $a^0 = \mathbf{1}$ if $\mathcal{A}$ is unital,

$$a^1 = a, \quad a^{n+1} = aa^n \qquad (n = 1, 2, \ldots).$$

The following power associative property depends on the assumption that the scalar field $\mathbb{F}$ for $\mathcal{A}$ is not of characteristic 2.

Theorem 1.1.1 *A Jordan algebra $\mathcal{A}$ is power associative; that is,*

$$a^m a^n = a^{m+n} \qquad (a \in \mathcal{A}; m, n = 1, 2, \ldots).$$

In fact, we have $[L_{a^m}, L_{a^n}] = 0$.

Proof For any α, β in the underlying field $\mathbb{F}$, we have

$$[L_{a+\alpha b+\beta c}, L_{(a+\alpha b+\beta c)^2}] = 0$$

for all $a, b, c \in \mathcal{A}$. Expanding the product, we find that the coefficient of the term $\alpha\beta$ is

$$2[L_a, L_{bc}] + 2[L_b, L_{ca}] + 2[L_c, L_{ab}],$$

which must be 0. Since $\mathbb{F}$ is not of characteristic 2, we have

$$[L_a, L_{bc}] + [L_b, L_{ca}] + [L_c, L_{ab}] = 0.$$

Applying this operator identity to an element $x \in \mathcal{A}$ and using commutativity of the Jordan product yields

$$
\begin{aligned}
&(L_a L_{bc} + L_b L_{ca} + L_c L_{ab})(x) \\
&= (L_{bc} L_a + L_{ca} L_b + L_{ab} L_c)(x) \\
&= L_{bc} L_x(a) + L_{bx} L_c(a) + L_{xc} L_b(a) \\
&= (L_{bc} L_x + L_{cx} L_b + L_{xb} L_c)(a) \\
&= (L_x L_{bc} + L_b L_{cx} + L_c L_{xb})(a) \\
&= (L_{((bc)a)} + L_b L_a L_c + L_c L_a L_b)(x).
\end{aligned}
$$

Putting $b = a^n$ and $c = a$ in this identity, we obtain a recursive formula,

$$L_{a^{n+2}} = 2L_a L_{a^{n+1}} + L_{a^n} L_{a^2} - L_{a^n} L_a^2, -L_a^2 L_{a^n},$$

which implies that each L_{a^n} is a polynomial in L_a and L_{a^2} which commute. It follows that L_{a^n} commutes with L_{a^m} for all $m, n \in \mathbb{N}$. In particular, we have

$$L_{a^n} L_a(a^m) = L_a L_{a^n}(a^m),$$

and power associativity follows from induction. $\qquad\square$

Corollary 1.1.2 *Let $\mathcal{A}$ be a Jordan algebra and let $a \in \mathcal{A}$. The subalgebra $\mathcal{A}(a)$ generated by a in $\mathcal{A}$ is associative.*

In fact, we have the following deeper result. It can be derived from Macdonald's theorem, which states that if an identity in 3 variables is linear in 1 variable and holds in all special Jordan algebras, then it holds in all Jordan algebras. We omit the proof, which can be found, for instance, in the books by Jacobson [62], McCrimmon [88], and Zhevlakov *et al.* [123]. We remark, however, that for Jordan algebras over a field of characteristic 2, a Jordan algebra with a single generator need not be special.

Shirshov–Cohn Theorem *Let $\mathcal{A}$ be a Jordan algebra and let $a, b \in \mathcal{A}$. Then the Jordan subalgebra $\mathcal{B}$ generated by a, b (and $\mathbf{1}$, if $\mathcal{A}$ is unital) is special.*

One can use the Shirshov–Cohn theorem to establish various identities in Jordan algebras. For instance, in any Jordan algebra $\mathcal{A}$, we have the identity

$$2L_a^3 - 3L_{a^2}L_a + L_{a^3} = 0 \tag{1.2}$$

for each $a \in \mathcal{A}$. In other words, we have

$$2a(a(ab)) - 3a^2(ab) + a^3b = 0$$

for $a, b \in \mathcal{A}$. To see this, let $\mathcal{B}$ be the Jordan subalgebra of $\mathcal{A}$ generated by a and b. Then it is special and hence embeds in some associative algebra $(\mathcal{A}', \times)$ with

$$ab = \frac{1}{2}(a \times b + b \times a).$$

In $\mathcal{B}$, we have

$$2a(a(ab)) = \frac{1}{4}(a^3 \times b + 3a^2 \times b \times a + 3a \times b \times a^2 + b \times a^3)$$

$$3a^2(ab) = \frac{1}{4}(3a^3 \times b + 3a^2 \times b \times a + 3a \times b \times a^2 + 3b \times a^3),$$

which, together with $2a^3b = a^3 \times b + b \times a^3$, verifies the identity.

Definition 1.1.3 Two elements a and b in a Jordan algebra $\mathcal{A}$ are said to *operator commute* if the left multiplications L_a and L_b commute. The *centre* of $\mathcal{A}$ is the set $Z(\mathcal{A}) = \{z \in \mathcal{A} : L_zL_a = L_aL_z, \forall a \in \mathcal{A}\}$.

We observe that $L_aL_b = L_bL_a$ if, and only if, $(ax)b = a(xb)$ for all $x \in \mathcal{A}$. Evidently, the centre $Z(\mathcal{A}) = \{z \in \mathcal{A} : (za)b = z(ab), \forall a, b \in \mathcal{A}\}$ is an associative subalgebra of $\mathcal{A}$.

Example 1.1.4 The Cayley algebra $\mathcal{O}$, known as the *octonions*, is a complex nonassociative algebra with a basis $\{e_0, e_1, \ldots, e_7\}$ and satisfies

$$a^2b = a(ab), \qquad ab^2 = (ab)b \qquad (a, b \in \mathcal{O}), \tag{1.3}$$

where e_0 is the identity of $\mathcal{O}$, $e_j^2 = -e_0$ for $j \neq 0$, and the multiplication is encoded in the following Fano plane, consisting of seven points and seven lines. The points are the basis elements but e_0, and the lines are the sides of the triangle, together with the circle. Each line has a cyclic ordering shown by the arrow. If e_i, e_j and e_k are cyclically ordered, then $e_i e_j = -e_j e_i = e_k$. For instance, $e_6 e_2 = (-e_4 e_2)e_2 = -e_4 e_2^2 = e_4$.

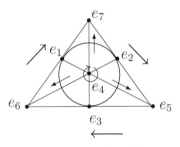

Octonion multiplication

The algebra $\mathcal{O}$ is an *alternative algebra* in the sense that the *associator*

$$[x, y, z] = (xy)z - x(yz)$$

is an alternating function of x, y, z: exchanging any two variables entails a sign change of the function. This condition is a reformulation of the multiplication rules in (1.3).

We will denote by $\mathbb{O}$ the real Caylay algebra, which is the real subalgebra of $\mathcal{O}$ with basis $\{e_0, \ldots, e_7\}$. Historically, octonions were discovered by a process of duplicating the real numbers $\mathbb{R}$. Indeed, the complex numbers arise from $\mathbb{R}$ as the product $\mathbb{R} \times \mathbb{R}$ with the multiplication

$$(a, b)(c, d) = (ac - db, bc + da) \qquad (a, b, c, d \in \mathbb{R}).$$

The real associative *quaternion* algebra $\mathbb{H}$ can be constructed by an analogous duplication process. One can define $\mathbb{H}$ as $\mathbb{C} \times \mathbb{C}$ with the multiplication

$$(a, b)(c, d) = (ac - \bar{d}b, b\bar{c} + da) \qquad (a, b, c, d \in \mathbb{C}),$$

which is isomorphic to the following real non-commutative algebra of 2×2 matrices:

$$\left\{ \begin{pmatrix} a & b \\ -\bar{b} & \bar{a} \end{pmatrix} : (a, b) \in \mathbb{C} \times \mathbb{C} \right\}. \tag{1.4}$$

In the identification with this algebra, $\mathbb{H}$ has a basis

$$\mathbf{1} = \begin{pmatrix} 1 & 0 \\ 0 & 1 \end{pmatrix}, \quad \mathbf{i} = \begin{pmatrix} i & 0 \\ 0 & -i \end{pmatrix}, \quad \mathbf{j} = \begin{pmatrix} 0 & 1 \\ -1 & 0 \end{pmatrix}, \quad \mathbf{k} = \begin{pmatrix} 0 & i \\ i & 0 \end{pmatrix},$$

satisfying

$$\mathbf{i}^2 = \mathbf{j}^2 = \mathbf{k}^2 = \mathbf{ijk} = -\mathbf{1}, \qquad \mathbf{ij} = -\mathbf{ji} = \mathbf{k}.$$

Likewise, $\mathbb{O}$ can be defined as the product $\mathbb{H} \times \mathbb{H}$ with the multiplication

$$(a, b)(c, d) = (ac - \bar{d}b, b\bar{c} + da) \qquad (a, b, c, d \in \mathbb{H}),$$

where the *conjugate* $\bar{c}$ of a quaternion $c = \alpha\mathbf{1} + x\mathbf{i} + y\mathbf{j} + z\mathbf{k}$ is defined by

$$\bar{c} = \alpha\mathbf{1} - x\mathbf{i} - y\mathbf{j} - z\mathbf{k},$$

so that the *real part* of c is $\mathrm{Re}\, c = \frac{1}{2}(c + \bar{c}) = \alpha\mathbf{1}$. A *positive* quaternion is one of the form $\alpha\mathbf{1}$ for some $\alpha > 0$. The basis elements of $\mathbb{H} \times \mathbb{H}$ are

$$e_0 = (\mathbf{1}, 0), \quad e_1 = (\mathbf{i}, 0), \quad e_2 = (\mathbf{j}, 0), \quad e_3 = (\mathbf{k}, 0),$$
$$e_4 = (0, \mathbf{1}), \quad e_5 = (0, \mathbf{i}), \quad e_6 = (0, \mathbf{j}), \quad e_7 = (0, \mathbf{k}).$$

The algebras $\mathbb{C}$, $\mathbb{H}$ and $\mathbb{O}$ are *quadratic*; that is, each element x satisfies the equation $x^2 = \alpha x + \beta\mathbf{1}$ for some $\alpha, \beta \in \mathbb{R}$, where $\mathbf{1}$ denotes the identity of the algebra. If $x = (a_1, a_2) \in \mathbb{H} \times \mathbb{H}$ with

$$a_n = \alpha_n\mathbf{1} + x_n\mathbf{i} + y_n\mathbf{j} + z_n\mathbf{k} \qquad (n = 1, 2),$$

then we have

$$x^2 = 2\alpha_1 x - (x_1^2 + y_1^2 + z_1^2 + x_2^2 + y_2^2 + z_2^2)e_0.$$

Example 1.1.5 A well-known example in Albert [2] of an exceptional Jordan algebra is the 27-dimensional real algebra

$$H_3(\mathbb{O}) = \{(a_{ij})_{1 \le i, j \le 3} : (a_{ij}) = (\tilde{a}_{ji}), a_{ij} \in \mathbb{O}\}$$

of 3×3 matrices over $\mathbb{O}$, Hermitian with respect to the usual involution $\sim$ in $\mathbb{O}$ defined by

$$(\alpha_0 e_0 + \cdots + \alpha_7 e_7)\widetilde{} = \alpha_0 e_0 - \cdots - \alpha_7 e_7.$$

The Jordan product is given by

$$A \circ B = \frac{1}{2}(AB + BA) \qquad (A, B \in H_3(\mathbb{O})),$$

where the multiplication on the right is the usual matrix multiplication. We refer to Jacobson [62] and McCrimmon [88] for a more detailed analysis of $H_3(\mathbb{O})$. The exceptionality of $H_3(\mathbb{O})$ involves the so-called *s-identities*, which are valid in all *special* Jordan algebras but not all Jordan algebras. One such identity was first found by Glennie [43]:

$$2Q_x(z)Q_{y,x}Q_z(y^2) - Q_xQ_zQ_{x,y}Q_y(z)$$
$$= 2Q_y(z)Q_{x,y}Q_z(x^2) - Q_yQ_zQ_{y,x}Q_x(z),$$

which does not hold in $H_3(\mathbb{O})$. An alternative proof of exceptionality bypassing s-identities can be found in Hanche-Olsen and Størmer [47].

Definition 1.1.6 An element e in an algebra $\mathcal{A}$ is called an *idempotent* if $e^2 = e$. Two idempotents e and u are said to be *orthogonal* if $eu = ue = 0$. An element $a \in \mathcal{A}$ is called *nilpotent* if $a^n = 0$ for some positive integer n.

Lemma 1.1.7 *Let $\mathcal{A}$ be a unital Jordan algebra with an idempotent e. Let $a \in \mathcal{A}$. The following conditions are equivalent:*

(i) *a and e operator commute.*
(ii) *$Q_e(a) = L_e a$.*
(iii) *a and e generate an associative subalgebra of $\mathcal{A}$.*

Proof (i) $\Rightarrow$ (ii). We have

$$Q_e(a) = 2(L_e^2 - L_e)(a) = 2e(ea) - ea = 2e^2a - ea = ea.$$

(ii) $\Rightarrow$ (iii). Let $\mathcal{B}$ be the subalgebra generated by a and e. By the Shirshov–Cohn theorem, $\mathcal{B}$ is isomorphic to a Jordan subalgebra $\mathcal{B}'$ of an associative algebra $(\mathcal{A}', \times)$ with respect to the special Jordan product. Identify a and e as elements in $\mathcal{B}'$. Then

$$L_e a = \frac{1}{2}(e \times a + a \times e) = Q_e(a) = e \times a \times e,$$

since $e = e^2 = e \times e$. Multiplying the above identity on the left by e, we get $e \times a = e \times a \times e$. Multiplying the identity on the right by e gives $a \times e = e \times a \times e$. Hence $e \times a = a \times e$ and $ea = e \times a$. Hence $(\mathcal{B}', \times)$ is a commutative subalgebra of $(\mathcal{A}, \times)$ and the special Jordan product in $\mathcal{B}'$ is just the product $\times$ and is, in particular, associative.

(iii) $\Rightarrow$ (i). In the proof of Theorem 1.1.1, we have the operator identity

$$[L_e, L_{bc}] + [L_b, L_{ce}] + [L_c, L_{eb}] = 0$$

for all $b, c \in \mathcal{A}$. Putting $c = e$, we have

$$[L_e, L_{be}] + [L_b, L_e] + [L_e, L_{eb}] = 0,$$

which gives

$$2[L_e, L_{be}] = [L_e, L_b]. \tag{1.5}$$

Since $e^2 = e$, in the special Jordan algebra $\mathcal{A}(a, e, \mathbf{1})$ generated by a, e and $\mathbf{1}$, it can be verified easily that

$$a = Q_e(a) + Q_{1-e}(a)$$

and $Q_{1-e}(a)e = 0$, as well as $Q_e(a)e = Q_e(a)$. Substituting $Q_{1-e}(a)$ for b in (1.5), we get $[L_e, L_{Q_{1-e}(a)}] = 0$. Putting $b = Q_e(a)$ in (1.5) gives $[L_e, L_{Q_e(a)}] = 0$. It follows that

$$[L_e, L_a] = [L_e, L_{Q_e(a)}] + [L_e, L_{Q_{1-e}(a)}] = 0.$$

$\square$

Lemma 1.1.8 *Let $\mathcal{A}$ be a finite-dimensional associative algebra containing an element a which is not nilpotent and not an identity. Then $\mathcal{A}$ contains a nonzero idempotent, which is a polynomial in a, without constant term.*

Proof We may assume that $\mathcal{A}$ has an identity $\mathbf{1}$. Finite dimensionality implies that there is a nonzero polynomial p of least degree and without constant term, such that $p(a) = 0$. Write $p(x) = x^k q(x)$, where $k \geq 1$ and q is a polynomial, such that $q(0) \neq 0$. The degree $\deg q$ of q is strictly positive, since a is not nilpotent. There are then polynomials q_1 and q_2 with $\deg q_1 < \deg q$ and

$$x^k q_1(x) + q_2(x)q(x) = \mathbf{1},$$

where the nonzero polynomial $g(x) = x^k q_1(x)$ has no constant term and $\deg g < \deg p$. Hence $e = g(a) \neq 0$. We have $e^2 = e$, since $a^{2k}q_1(a) + a^k q_2(a)q(a) = a^k$ and

$$g(a)^2 - g(a) = a^{2k}q_1(a)^2 - a^k q_1(a) = a^k q_2(a)q(a)q_1(a) = 0.$$

$\square$

Lemma 1.1.9 *Let $\mathcal{A}$ be a Jordan algebra. Then an element $a \in \mathcal{A}$ is nilpotent if and only if the left multiplication $L_a : \mathcal{A} \longrightarrow \mathcal{A}$ is nilpotent.*

Proof If L_a is nilpotent, then $a^{n+1} = L_a^n(a)$ implies that a is nilpotent. Conversely, for any $a \in \mathcal{A}$ with $a^n = 0$, we show that L_a is nilpotent by induction on the exponent n. The assertion is trivially true if $n = 1$. Given that the assertion is true for n, we consider $a^{n+1} = 0$. We have $(a^2)^n = 0 = (a^3)^n$, and therefore L_{a^2} and L_{a^3} are nilpotent, by the inductive hypothesis. It follows from the identity

$$2L_a^3 = 3L_{a^2}L_a - L_{a^3}$$

that L_a is nilpotent.

$\square$

Given an idempotent e in a Jordan algebra $\mathcal{A}$, the left multiplication $L_e : \mathcal{A} \longrightarrow \mathcal{A}$ satisfies the equation

$$2L_e^3 - 3L_e^2 + L_e = 0 \tag{1.6}$$

by the identity (1.2). Hence an eigenvalue α of L_e is a root of

$$2\alpha^3 - 3\alpha^2 + \alpha = 0$$

and is 0, $\frac{1}{2}$ or 1. If $\mathcal{A}$ is associative, then $L_e^2 = L_e$ and $\frac{1}{2}$ is not an eigenvalue of L_e. Nevertheless, we denote the eigenspaces of $2L_e$ by

$$\mathcal{A}_k(e) = \{x \in \mathcal{A} : 2ex = kx\} \qquad (k = 0,\, 1,\, 2)$$

and call $\mathcal{A}_k(e)$ the *Peirce k-space* of e. The earlier remark implies that $\mathcal{A}_1(e) = \{0\}$ if $\mathcal{A}$ is associative.

We define two linear operators, $Q_e : \mathcal{A} \longrightarrow \mathcal{A}$, and $Q_e^\perp : \mathcal{A} \longrightarrow \mathcal{A}$, by

$$Q_e = 2L_e^2 - L_e, \quad Q_e^\perp = 4(L_e - L_e^2). \tag{1.7}$$

Evidently, L_e commutes with both Q_e and $Q_e^\perp$. Using the equation (1.6), one can easily establish

$$L_e Q_e = Q_e = Q_e^2, \quad L_e Q_e^\perp = \frac{1}{2} Q_e^\perp = \frac{1}{2}(Q_e^\perp)^2, \quad L_e(I - Q_e - Q_e^\perp) = 0,$$

where I is the identity operator on $\mathcal{A}$ and Q_e and $Q_e^\perp$ are mutually orthogonal. It follows that

$$\mathcal{A}_2(e) = Q_e(\mathcal{A}), \quad \mathcal{A}_1(e) = Q_e^\perp(\mathcal{A}), \quad \mathcal{A}_0(e) = (I - Q_e - Q_e^\perp)(\mathcal{A}), \tag{1.8}$$

which gives rise to the following *Peirce decomposition* of $\mathcal{A}$:

$$\mathcal{A} = \mathcal{A}_0(e) \oplus \mathcal{A}_1(e) \oplus \mathcal{A}_2(e).$$

We will return to the Peirce decomposition with more details in the more general setting of Jordan triple systems. We note for the time being that the Peirce spaces $\mathcal{A}_0(e)$ and $\mathcal{A}_2(e)$ are Jordan subalgebras of $\mathcal{A}$, as shown below. We also note that $\mathcal{A}_2(e)$ never vanishes.

Lemma 1.1.10 *The Peirce spaces of an idempotent e in a Jordan algebra $\mathcal{A}$ satisfy*

$$\mathcal{A}_0(e)\mathcal{A}_0(e) \subset \mathcal{A}_0(e), \quad \mathcal{A}_1(e)\mathcal{A}_1(e) \subset \mathcal{A}_0(e) \oplus \mathcal{A}_2(e), \quad \mathcal{A}_2(e)\mathcal{A}_2(e) \subset \mathcal{A}_2(e).$$

Proof We first prove the second inclusion. Let $x, y \in \mathcal{A}_1(e)$ and let $xy = a_0 + a_1 + a_2$ be the Peirce decomposition of xy. We have

$$0 = [L_y, L_{ex}] + [L_e, L_{xy}] + [L_x, L_{ye}]$$
$$= \frac{1}{2}[L_y, L_x] + [L_e, L_{xy}] + \frac{1}{2}[L_x, L_y]$$
$$= [L_e, L_{xy}].$$

In particular, $[L_e, L_{xy}](e) = 0$ gives

$$0 = e(xy) - e(e(xy))$$
$$= a_2 + \frac{1}{2}a_1 - e\left(a_2 + \frac{1}{2}a_1\right)$$
$$= a_2 + \frac{1}{2}a_1 - a_2 - \frac{1}{4}a_1 = \frac{1}{4}a_1.$$

Hence $xy \in \mathcal{A}_0(e) \oplus \mathcal{A}_2(e)$.

Let $x, y \in \mathcal{A}_j(e)$, where $j = 0, 2$. Then we have $[L_e, L_{x^2}] = 0$ by the first equation earlier in the proof. Using this and expanding $((x + e)y)(x + e)^2 = (x + e)(y(x + e)^2)$, we obtain

$$2(xy)(xe) + (xy)e + 2(ey)(xe) = 2x(y(xe)) + x(ye) + 2(ey)(xe),$$

which gives $(xy)e = \frac{j}{2}xy$. This proves the first and the last inclusion. $\square$

Definition 1.1.11 An idempotent e in a Jordan algebra $\mathcal{A}$ is called *maximal* if the Peirce 0-space $\mathcal{A}_0(e)$ is $\{0\}$. A nonzero idempotent e is called *primitive* if there are no nonzero orthogonal idempotents u and v satisfying $e = u + v$.

Lemma 1.1.12 *Let $\mathcal{A}$ be a finite-dimensional Jordan algebra which contains no nonzero nilpotent element. Then $\mathcal{A}$ contains a maximal idempotent.*

Proof Ignore the trivial case $\mathcal{A} = \{0\}$. Applying Lemma 1.1.8 to an associative subalgebra of $\mathcal{A}$ generated by a nonzero element, one finds a nonzero idempotent e. If $\mathcal{A}_0(e) \neq \{0\}$, then again one can pick a nonzero idempotent $u \in \mathcal{A}_0(e)$. Then $e' = e + u$ is an idempotent and $\mathcal{A}_0(e) \subset \mathcal{A}_0(e')$. Since $u \in \mathcal{A}_0(e')\backslash\mathcal{A}_0(e)$, we have $\dim \mathcal{A}_0(e) < \dim \mathcal{A}_0(e')$. By finite dimensionality of $\mathcal{A}$, this process of increasing dimension must stop, yielding a maximal idempotent. $\square$

Proposition 1.1.13 *Let $\mathcal{A}$ be a finite-dimensional Jordan algebra which contains no nonzero nilpotent element. Then $\mathcal{A}$ has an identity.*

Proof By Lemma 1.1.12, $\mathcal{A}$ contains a maximal idempotent e such that

$$\mathcal{A} = \mathcal{A}_1(e) \oplus \mathcal{A}_2(e).$$

We show that $\mathcal{A}_1(e) = \{0\}$. Let $x \in \mathcal{A}_1(e)$. Then $x^2 \in \mathcal{A}_0(e) \oplus \mathcal{A}_2(e) = \mathcal{A}_2(e)$, and we have

$$\frac{1}{2}x^3 = (xe)x^2 = x(ex^2) = x^3,$$

which implies that $x = 0$. Hence $\mathcal{A} = \mathcal{A}_2(e)$, in which e is the identity. $\quad\square$

A real Jordan algebra $\mathcal{A}$ is called *formally real* if $a_1^2 + \cdots + a_k^2 = 0$ implies $a_1 = \cdots = a_k = 0$ for $a_1, \ldots, a_k \in \mathcal{A}$. By Proposition 1.1.13, a finite-dimensional formally real Jordan algebra has an identity. The exceptional Jordan algebra $H_3(\mathbb{O})$ is formally real.

Theorem 1.1.14 *Let $\mathcal{A}$ be a finite-dimensional real Jordan algebra. The following conditions are equivalent:*

(i) *$\mathcal{A}$ is formally real.*
(ii) *$a^2 + b^2 = 0 \Rightarrow a = b = 0$ for any $a, b \in \mathcal{A}$.*
(iii) *Each $x \in \mathcal{A}$ admits a decomposition $x = \alpha_1 e_1 + \cdots + \alpha_n e_n$, where $\alpha_1, \ldots \alpha_n \in \mathbb{R}$ and $e_1, \ldots, e_n$ are mutually orthogonal idempotents in $\mathcal{A}$.*
(iv) *The bilinear form $(x, y) \in \mathcal{A}^2 \mapsto \mathrm{Trace}\,(x \,\square\, y) \in \mathbb{R}$ is positive definite.*

Proof (ii) $\Rightarrow$ (iii). Let $x \in \mathcal{A}$. Then the subalgebra $\mathcal{A}(x)$ generated by x is associative and has an identity e by Proposition 1.1.13. Finite dimensionality implies that

$$e = e_1 + \cdots + e_n$$

for some mutually orthogonal primitive idempotents in $\mathcal{A}(x)$. Hence we have, by associativity of $\mathcal{A}(x)$,

$$x = e_1 x e_1 + \cdots + e_n x e_n.$$

We show that the associative algebra $e_j \mathcal{A}(x) e_j$ reduces to $\mathbb{R}e_j$. Indeed, by primitivity, there is no nonzero idempotent in $e_j \mathcal{A}(x) e_j$ other than e_j, and it follows from Lemma 1.1.8 that each $a \in e_j \mathcal{A}(x) e_j \backslash \{0, e_j\}$ gives rise to a polynomial $g(a)$ that has no constant term and satisfies $g(a) = e_j$, which implies that a is invertible in $e_j \mathcal{A}(x) e_j$. Hence $e_j \mathcal{A}(x) e_j$ is a field over $\mathbb{R}$ and must be either $\mathbb{R}$ or $\mathbb{C}$. Since $\mathbb{C}$ is not formally real, we conclude that $e_j \mathcal{A}(x) e_j = \mathbb{R}e_j$ and therefore

$$x = \sum_j \alpha_j e_j$$

for some $\alpha_1, \ldots, \alpha_n \in \mathbb{R}$.

(iii) $\Rightarrow$ (iv). Given $x = \alpha_1 e_1 + \cdots + \alpha_n e_n$ for some mutually orthogonal idempotents $e_1, \ldots, e_n$, we have

$$\text{Trace}\,(x \,\square\, x) = \sum_j \alpha_j^2 \text{Trace}\,(e_j \,\square\, e_j) \geq 0.$$

If $\text{Trace}\,(x \,\square\, x) = 0$, then $\text{Trace}\,(e_j \,\square\, e_j) = 0$, and $e_j = 0$, since $e_j \,\square\, e_j = L_{e_j}$ has eigenvalues $0, \frac{1}{2}$ or 1, for all j.

(iv) $\Rightarrow$ (i). Let $a_1^2 + \cdots + a_k^2 = 0$. Then $\sum_j \text{Trace}\,(a_j \,\square\, a_j) = \text{Trace}\,(a_1 \,\square\, a_1 + \cdots + a_k \,\square\, a_k) = 0$. Hence $\text{Trace}\,(a_j \,\square\, a_j) = 0$ implies $a_j = 0$ for all j. $\qquad\square$

A subspace I of a Jordan algebra $\mathcal{A}$ is called an *ideal* if $a \in \mathcal{A}$ and $x \in I$ imply $ax \in I$, in which case I is also an ideal in the unit extension $\mathcal{A}_1$ of $\mathcal{A}$ and the quotient space $\mathcal{A}/I$ is a Jordan algebra with the natural product

$$(a + I)(b + I) = ab + I \qquad (a, b \in \mathcal{A}).$$

A Jordan algebra $\mathcal{A}$ with nontrivial multiplication (i.e., $ab \neq 0$ for some $a, b \in \mathcal{A}$), is called *simple* if the only ideals of $\mathcal{A}$ are $\{0\}$ and $\mathcal{A}$ itself.

To avoid confusion, homomorphisms and isomorphisms between Jordan algebras are sometimes called *Jordan homomorphisms* and *Jordan isomorphisms* to distinguish them from the ones for other algebraic structures. The kernel $\varphi^{-1}(0)$ of a Jordan homomorphism $\varphi : \mathcal{A} \longrightarrow \mathcal{B}$ is an ideal of $\mathcal{A}$. Given an ideal I of a Jordan algebra $\mathcal{A}$, the quotient map $q : \mathcal{A} \longrightarrow \mathcal{A}/I$ is a homomorphism with kernel $q^{-1}(0) = I$.

Finite-dimensional formally real Jordan algebras have been classified in the seminal article by Jordan *et al.* [64]. Jordan algebras of any dimension were classified by Zelmanov in [120] and [121]. There are three different types of simple Jordan algebras: namely, Hermitian, Clifford and Albert.

Example 1.1.15 Jordan algebras of the Hermitian type are special Jordan algebras obtained from associative algebras. They are of the form

$$H(\mathcal{A}, *) = \{a \in \mathcal{A} : a^* = a\},$$

where $\mathcal{A}$ is an associative algebra with involution $*$ and $H(\mathcal{A}, *)$ is equipped with the special Jordan product

$$a \circ b = \frac{1}{2}(ab + ba).$$

It is simple if $\mathcal{A}$ is simple. The special Jordan algebra $(\mathcal{A}, \circ)$ constructed at the beginning of the section from an associative algebra $\mathcal{A}$ is of this type. Indeed, let $\mathcal{A}^{op}$ denote the opposite algebra (i.e., $\mathcal{A}^{op}$ has the same vector space structure as $\mathcal{A}$, but the multiplication is reversed) and define an involution $*$ on $\mathcal{A} \oplus \mathcal{A}^{op}$ by $(a \oplus b)^* = b \oplus a$. Then $(\mathcal{A}, \circ)$ is isomorphic to $H(\mathcal{A} \oplus \mathcal{A}^{op}, *)$.

Example 1.1.16 Of the Clifford type are Jordan algebras of a symmetric bilinear form. They are special Jordan algebras of the form $V \oplus \mathbb{F}$, where V is a vector space over a field $\mathbb{F}$, equipped with a symmetric bilinear form $f : V \times V \longrightarrow \mathbb{F}$. The direct sum $V \oplus \mathbb{F}$ forms a Jordan algebra in the product

$$(x \oplus \alpha)(y \oplus \beta) = (\alpha x + \beta y) \oplus (f(x, y) + \alpha \beta).$$

We call $V \oplus \mathbb{F}$ a *spin factor*. It is simple if f is non-degenerate and dim $V \geq 2$.

Example 1.1.17 The Albert type Jordan algebras are the exceptional Jordan algebras which are 27-dimensional over the centre. For example, $H_3(\mathbb{O})$.

1.2 Jordan triple systems

A real or complex vector space V equipped with a triple product $\{\cdot, \cdot, \cdot\}$: $V^3 \longrightarrow V$ is called a *Jordan triple* if the triple product is trilinear and symmetric in the outer variables and satisfies the following identity:

$$\{a, b, \{x, y, z\}\} = \{\{a, b, x\}, y, z\} - \{x, \{b, a, y\}, z\} + \{x, y, \{a, b, z\}\}. \tag{1.9}$$

A complex vector space V with a triple product $\{\cdot, \cdot, \cdot\}$ which is linear and symmetric in the outer variables, but conjugate linear in the middle variable, and satisfies (1.9), is called a *Hermitian Jordan triple*. By a *Jordan triple system* V, we mean a Jordan triple or a Hermitian Jordan triple V. By restricting to the real scalar field, a complex Jordan triple or a Hermitian Jordan triple can, and will, be regarded as a real Jordan triple.

A vector subspace W of a Jordan triple system V is called a *subtriple* if $x, y, z \in W$ implies $\{x, y, z\} \in W$. Given subsets A, B and C of V, we define

$$\{A, B, C\} = \{\{a, b, c\} : a \in A, b \in B, c \in C\}.$$

A linear map $f : V \longrightarrow W$ between two Jordan triple systems V and W is called a *(Jordan) triple homomorphism* if it preserves the triple product:

$$f\{a, b, c\} = \{f(a), f(b), f(c)\} \qquad (a, b, c \in V).$$

A triple homomorphism is called a *(Jordan) triple monomorphism* if it is injective. A bijective triple homomorphism is called a *(Jordan) triple isomorphism*.

We call the identity (1.9) the *main triple identity* of a Jordan triple system. It is often written as

$$\{\{a, b, x\}, y, z\} - \{x, \{b, a, y\}, z\} = \{a, b, \{x, y, z\}\} - \{x, y, \{a, b, z\}\}.$$
(1.10)

Example 1.2.1 The complex space $\mathbb{C}^n$ can be given various Jordan triple structures; for instance, one can define a Jordan triple product by

$$\{x, y, z\} = xyz \qquad (x, y, z \in \mathbb{C}^n),$$

which turns $\mathbb{C}^n$ into a complex Jordan triple, where the product on the right is the coordinatewise product. On the other hand, the triple product

$$\{x, y, z\} = x\overline{y}z \qquad (x, y, z \in \mathbb{C}^n)$$

makes $\mathbb{C}^n$ into a Hermitian Jordan triple. We can also equip $\mathbb{C}^n$ with the Hermitian Jordan triple product

$$\{x, y, z\} = \langle x, y \rangle z + \langle z, y \rangle x,$$

where $\langle \cdot, \cdot \rangle$ denotes the usual inner product.

The concept of a Jordan triple system was originally derived from the generalization of Jordan algebras relating to Lie algebras and differential geometry.

In a real or complex Jordan algebra $\mathcal{A}$, we define a canonical (*Jordan*) *triple product* by

$$\{a, b, c\} = (ab)c + a(bc) - b(ac) \tag{1.11}$$

for $a, b, c \in \mathcal{A}$. If $\mathcal{A}$ is a complex Jordan algebra equipped with a conjugate linear algebra involution $*$, the canonical Hermitian (Jordan) triple product is defined by

$$\{a, b, c\} = (ab^*)c + a(b^*c) - b^*(ac). \tag{1.12}$$

Equipped with one of these triple products, a real or complex Jordan algebra is a Jordan triple system.

We first prove three basic identities in a Jordan triple system.

Lemma 1.2.2 *Given x, y, z in a Jordan triple system, we have*

$$\{\{x, y, x\}, y, z\}\} = \{x, \{y, x, y\}, z\}. \tag{1.13}$$

Proof This follows by putting $a = x$ and $b = y$ in the main triple identity. $\square$

Lemma 1.2.3 *Given x, y, z is a Jordan triple system, we have*

$$\{x, y, \{x, z, x\}\} = \{x, \{y, x, z\}, x\}. \tag{1.14}$$

Proof Applying the triple identity repeatedly yields

$$
\begin{aligned}
\{x, y, \{x, z, x\}\} &= \{\{x, y, x\}, z, x\}\} - \{x, \{y, x, z\}, x\} + \{x, z, \{x, y, x\}\} \\
&= 2\{x, z, \{x, y, x\}\} - \{x, \{y, x, z\}, x\} \\
&= 2\{\{x, z, x\}, y, x\} - 2\{x, \{z, x, y\}, x\} \\
&\quad + 2\{x, y, \{x, z, x\}\} - \{x, \{y, x, z\}, x\} \\
&= 4\{x, y, \{x, z, x\}\} - 3\{x, \{y, x, z\}, x\} \\
&= \{x, \{y, x, z\}, x\}.
\end{aligned}
$$ □

This proof also yields the identity

$$\{x, y, \{x, z, x\}\} = \{\{x, y, x\}, z, x\}. \tag{1.15}$$

Lemma 1.2.4 *Given that x, y, z is a Jordan triple system, we have*

$$\{\{x, y, x\}, z, \{x, y, x\}\} = \{x, \{y, \{x, z, x\}, y\}, x\}. \tag{1.16}$$

Proof Adding the two triple identities

$$
\begin{aligned}
\{y, x, \{y, x, z\}\} &= \{\{y, x, y\}, x, z\} - \{y, \{x, y, x\}, z\} + \{y, x, \{y, x, z\}\} \\
\{z, x, \{y, x, y\}\} &= \{\{z, x, y\}, x, y\} - \{y, \{x, z, x\}, y\} + \{y, x, \{z, x, y\}\},
\end{aligned}
$$

we obtain

$$\{y, \{x, y, x\}, z\} = 2\{y, x, \{y, x, z\}\} - \{y, \{x, z, x\}, y\}.$$

It follows that

$$
\begin{aligned}
\{\{x, y, x\}, z, \{x, y, x\}\} &= 2\{\{\{x, y, x\}, z, x\}, y, x\} - \{x, \{z, \{x, y, x\}, y\}, x\} \\
&= 2\{\{\{x, y, x\}, z, x\}, y, x\} - 2\{x, \{y, x, \{y, x, z\}\}, x\} \\
&\quad + \{x, \{y, \{x, z, x\}, y\}, x\} \\
&= \{x, \{y, \{x, z, x\}, y\}, x\},
\end{aligned}
$$

where the last identity follows from repeated applications of (1.13):

$$
\begin{aligned}
\{\{\{x, y, x\}, z, x\}, y, x\} &= \{\{x, \{y, x, z\}, x\}, y, x\} \\
&= \{x, \{\{y, x, z\}, x, y\}, x\} \\
&= \{x, \{y, \{x, z, x\}, y\}, x\}.
\end{aligned}
$$ □

In place of the triple identity (1.9), one can also use the basic identities (1.13), (1.14) and (1.16) as the defining identities for a Jordan triple system, as in Loos [84], in which case the triple identity can be derived.

The arguments in the proof of Lemma 1.2.4 can be repeated to yield the following identities.

Lemma 1.2.5 *Given a, x, y, z in a Jordan triple system, we have*

$$2\{x, a, \{y, a, z\}\} = \{x, \{a, y, a\}, z\} + \{x, \{a, z, a\}, y\} \qquad (1.17)$$

$$2\{a, x, \{a, y, z\}\} = \{\{a, x, a\}, y, z\} + \{a, \{x, z, y\}, a\} \qquad (1.18)$$

$$2\{a, \{x, a, y\}, z\} = \{\{a, x, a\}, y, z\} + \{\{a, y, a\}, x, z\}. \qquad (1.19)$$

Proof The first identity follows from adding the two triple identities

$$\{y, a, \{x, a, z\}\} = \{\{y, a, x\}, a, z\} - \{x, \{a, y, a\}, z\} + \{x, a, \{y, a, z\}\}$$

$$\{z, a, \{x, a, y\}\} = \{\{z, a, x\}, a, y\} - \{x, \{a, z, a\}, y\} + \{x, a, \{z, a, y\}\}.$$

We obtain the third identity by adding the two triple identities

$$\{a, x, \{a, y, z\}\} = \{\{a, x, a\}, y, z\} - \{a, \{x, a, y\}, z\} + \{a, y, \{a, x, z\}\}$$

$$\{a, y, \{a, x, z\}\} = \{\{a, y, a\}, x, z\} - \{a, \{y, a, x\}, z\} + \{a, x, \{a, y, z\}\}.$$

The second identity follows from the triple identity

$$\{\{a, x, a\}, y, z\} = \{a, x, \{a, y, z\}\} - \{a, \{x, z, y\}, a\} + \{\{a, y, z\}, x, a\}. \qquad \square$$

In a Jordan triple system V, we define the odd powers of an element x by induction as follows:

$$x^1 = x, \quad x^3 = \{x, x, x\}, \quad x^{2n+1} = \{x, x^{2n-1}, x\} \quad (n = 2, 3, \ldots).$$

By (1.14) and induction, we have

$$x^{2n+1} = \{x^{2n-1}, x, x\}. \qquad (1.20)$$

One often makes use of the following *polarization* in a Jordan triple system V:

$$2\{x, y, z\} = \{x + z, y, x + z\} - \{x, y, x\} - \{z, y, z\}. \qquad (1.21)$$

It follows that the triple product in a Jordan triple system is completely determined by the *symmetrized* product $\{x, y, x\}$.

If V is a Hermitian Jordan triple, we also have the polarization identity

$$4\{z, y, z\} = (y + z)^3 + (y - z)^3 - (y + iz)^3 - (y - iz)^3 \qquad (1.22)$$

and the triple product in V is determined by the cubes x^3.

Given a real Jordan triple $(V, \{\cdot, \cdot, \cdot\})$, it can be complexified to a complex Jordan triple or a Hermitian Jordan triple. If the complexification $V_c = V \oplus iV$ of the real vector space V is furnished with the triple product

$$\{x \oplus iu, y \oplus iv, x \oplus iu\}_c = (\{x, y, x\} - \{u, y, u\} - 2\{x, v, u\})$$
$$\oplus\, i\, (\{x, v, x\} - \{u, v, u\} + 2\{x, y, u\}), \quad (1.23)$$

then $(V_c, \{\cdot, \cdot, \cdot\}_c)$ is a complex Jordan triple, called the *complexification* of $(V, \{\cdot, \cdot, \cdot\})$. On the other hand, the following triple product turns V_c into a Hermitian Jordan triple:

$$\{x \oplus iu, y \oplus iv, x \oplus iu\}_h = (\{x, y, x\} - \{u, y, u\} + 2\{x, v, u\})$$
$$\oplus\, i\, (-\{x, v, x\} + \{u, v, u\} + 2\{x, y, u\}). \quad (1.24)$$

We shall call $(V_c, \{\cdot, \cdot, \cdot\}_h)$ the *hermitification* of V.

Jordan triple systems are equivalent to Jordan pairs with involution. The concept of a Jordan pair was introduced by Loos [84] in connection with a pair of mutually dual Riemannian symmetric spaces. A pair (V_{-1}, V_1) of real or complex vector spaces is called a *Jordan pair* if there are two trilinear maps,

$$\{\cdot, \cdot, \cdot\}_\alpha : V_\alpha \times V_{-\alpha} \times V_\alpha \longrightarrow V_\alpha \qquad (\alpha = \pm 1),$$

which are symmetric in the outer variables and satisfy

$$\{a, b, \{x, y, z\}_\alpha\}_\alpha = \{\{a, b, x\}_\alpha, y, z\}_\alpha$$
$$- \{x, \{b, a, y\}_{-\alpha}, z\}_\alpha + \{x, y, \{a, b, z\}_\alpha\}_\alpha \quad (1.25)$$

for $a, x, z \in V_\alpha$ and $b, y \in V_{-\alpha}$.

An *involution* θ of a Jordan pair (V_{-1}, V_1) is a linear isomorphism $\theta : V_{-1} \longrightarrow V_1$ satisfying

$$\theta\{x, \theta y, x\}_{-1} = \{\theta x, y, \theta x\}_1 \qquad (x, y \in V_{-1}).$$

If instead, such θ is *conjugate linear* for a *complex* Jordan pair (V_{-1}, V_1), then it is called a *Hermitian involution*.

Evidently, if (V_{-1}, V_1) is a Jordan pair with involution θ, then V_{-1} is a Jordan triple with triple product $\{x, \theta y, z\}_{-1}$ for $x, y, z \in V_{-1}$. Also, V_{-1} is a Hermitian Jordan triple if the involution θ is Hermitian.

Conversely, given a Jordan triple $(V, \{\cdot, \cdot, \cdot\})$, let $V_{-1} = V_1 = V$. Then (V_{-1}, V_1) is a Jordan pair with trilinear maps $\{x, y, z\}_\alpha = \{x, y, z\}$ for $\alpha = \pm 1$. For a Hermitian Jordan triple $(V, \{\cdot, \cdot, \cdot\})$, we let $V_{-1} = V$ and V_1 be the complex vector space obtained from V but replacing its scalar multiplication by

$(\beta, v) \in \mathbb{C} \times V \mapsto \bar{\beta}v \in V$. Then the identity map $\theta : V_{-1} \longrightarrow V_1$ is conjugate linear and is a Hermitian involution of the Jordan pair (V_{-1}, V_1) with the trilinear map $\{x, y, z\}_\alpha = \{x, \theta^{\pm\alpha}y, z\}$.

Example 1.2.6 Let $M_{mn}(\mathbb{C})$ be the vector space of $m \times n$ complex matrices. Then $(M_{mn}(\mathbb{C}), M_{nm}(\mathbb{C}))$ is a Jordan pair with trilinear maps

$$\{A, B, C\}_\alpha = \frac{1}{2}(ABC + CBA) \qquad (A, C \in V_{-\alpha}, B \in V_\alpha),$$

where $V_{-1} = M_{mn}(\mathbb{C})$ and $V_1 = M_{nm}(\mathbb{C})$. The pair has a Hermitian involution

$$\theta : M_{mn}(\mathbb{C}) \longrightarrow M_{nm}(\mathbb{C})$$

given by $\theta(B) = B^*$, where $B^* = (\overline{b_{ji}})$ is the adjoint of the matrix $B = (b_{ij})$.

The complex vector space $M_{mn}(\mathbb{C})$ is a Hermitian Jordan triple with triple product

$$\{A, B, C\} = \{A, \theta(B), C\}_{-1} = \frac{1}{2}(AB^*C + CB^*A)$$

and $M_{mm}(\mathbb{C})$ is a Jordan algebra with involution * and Jordan product $A \circ B = (AB + BA)/2$. We shall write $M_m(\mathbb{C})$ for $M_{mm}(\mathbb{C})$ hereafter.

One can consider an infinite-dimensional extension of this example. Let H and K be complex Hilbert spaces and let $L(H, K)$ be the Banach space of bounded linear operators between H and K. Then it is a Hermitian Jordan triple in the triple product

$$\{R, S, T\} = \frac{1}{2}(RS^*T + TS^*R) \qquad (R, S, T \in L(H, K)),$$

where $S^* : K \longrightarrow H$ is the adjoint of S.

In fact, for any associative algebra $\mathcal{A}$ with special Jordan product $a \circ b = (ab + ba)/2$, the canonical Jordan triple product of the Jordan algebra $(\mathcal{A}, \circ)$ is given by

$$\{a, b, c\} = \frac{1}{2}(abc + cba).$$

If $\mathcal{A}$ is over $\mathbb{C}$ and equipped with an involution *, then the canonical Hermitian Jordan triple product is given by

$$\{a, b, c\} = \frac{1}{2}(ab^*c + cb^*a).$$

Example 1.2.7 Let $M_{12}(\mathcal{O}) = \{(z_1, z_2) : z_1, z_2 \in \mathcal{O}\}$ be the complex vector space of 1×2 matrices over $\mathcal{O}$. Given $y = (y_1, y_2)$ with $y_1 = \sum_{0}^{7} \alpha_{1,k}e_k$ and

$y_2 = \sum_0^7 \alpha_{2,k} e_k$, we define $y_j{}^* = \overline{\alpha}_{j,0} e_0 - \sum_1^7 \overline{\alpha}_{j,k} e_k$ for $j = 1, 2$, and

$$y^* = \begin{pmatrix} y_1{}^* \\ y_2{}^* \end{pmatrix}.$$

The space $M_{1,2}(\mathcal{O})$ is a 16-dimensional Hermitian Jordan triple in the following triple product:

$$\{x, y, z\} = \frac{1}{2}(x(y^*z) + z(y^*x)).$$

Example 1.2.8 Let $H_3(\mathcal{O})$ be the complex vector space of 3×3 matrices over the Cayley algebra $\mathcal{O}$, Hermitian with respect to the standard involution in $\mathcal{O}$; that is, (a_{ij}) belongs to $H_3(\mathcal{O})$ if and only if

$$(a_{ij}) = (\tilde{a}_{ji}),$$

where the usual linear involution $\sim$ on $\mathcal{O}$ is defined by

$$\overset{\sim}{\left(\sum_{k=0}^7 \alpha_k e_k \right)} = \alpha_0 e_0 - \sum_{k=1}^7 \alpha_k e_k$$

for $\sum_{k=0}^7 \alpha_k e_k \in \mathcal{O}$. We always equip $H_3(\mathcal{O})$ with the Jordan product

$$A \circ B = \frac{1}{2}(AB + BA) \qquad (A, B \in H_3(\mathcal{O})),$$

where the multiplication on the right is the usual matrix multiplication. The product $\circ$ makes $H_3(\mathcal{O})$ into a complex exceptional Jordan algebra. There is a natural conjugate linear involution $\natural$ on $\mathcal{O}$ defined by

$$\left(\sum_{k=0}^7 \alpha_k e_k \right)^{\natural} = \sum_{k=0}^7 \overline{\alpha}_k e_k,$$

which induces a conjugate linear involution $*$ on $H_3(\mathcal{O})$:

$$(a_{ij})^* := (a_{ij}^{\natural}).$$

We always equip $H_3(\mathcal{O})$ with the triple product

$$\{A, B, C\} = (A \circ B^*) \circ C + A \circ (B^* \circ C) - B^* \circ (A \circ C).$$

With this triple product, $H_3(\mathcal{O})$ becomes a Hermitian Jordan triple.

The exceptional real Jordan algebra $H_3(\mathbb{O})$ in Example 1.1.5 is the *real form* of $H_3(\mathcal{O})$ with respect to the involution $* : H_3(\mathcal{O}) \longrightarrow H_3(\mathcal{O})$; that is,

$$H_3(\mathbb{O}) = \{(a_{ij}) \in H_3(\mathcal{O}) : (a_{ij})^* = (a_{ij})\}$$

and $H_3(\mathcal{O}) = H_3(\mathbb{O}) + i H_3(\mathbb{O})$ is the complexification of $H_3(\mathbb{O})$.

We can embed $M_{12}(\mathcal{O})$ as a Hermitian subtriple of $H_3(\mathcal{O})$ via the triple monomorphism

$$(z_1, z_2) \in M_{12}(\mathcal{O}) \mapsto \begin{pmatrix} 0 & z_1 & z_2 \\ \tilde{z}_1 & 0 & 0 \\ \tilde{z}_2 & 0 & 0 \end{pmatrix} \in H_3(\mathcal{O}).$$

For each element a in a Jordan triple system V, we define a binary product $\circ_a$ in V by

$$x \circ_a y = \{x, a, y\} \qquad (x, y \in V).$$

This product is clearly commutative. It also satisfies the Jordan identity which follows from the identities (1.9), (1.17) and (1.14):

$$\begin{aligned}
x^2 \circ_a (x \circ_a y) &= \{\{x, a, x\}, a, \{x, a, y\}\} \\
&= \{\{\{x, a, x\}, a, x\}, a, y\} - \{x, \{a, \{x, a, x\}, a\}, y\} \\
&\quad + \{x, a, \{\{x, a, x\}, a, y\}\} \\
&= x \circ_a (x^2 \circ_a y),
\end{aligned}$$

where, by identities (1.17) and (1.14), we have

$$\begin{aligned}
\{x, &\{a, \{x, a, x\}, a\}, y\} \\
&= 2\{\{x, a, \{x, a, x\}\}, a, y\} - \{\{x, a, x\}, \{a, x, a\}, y\} \\
&= 2\{\{x, a, \{x, a, x\}\}, a, y\} - 2\{x, \{\{a, x, a\}, x, a\}, y\} \\
&\quad + \{\{x, \{a, x, a\}, x\}, a, y\} \\
&= 3\{\{x, a, \{x, a, x\}\}, a, y\} - 2\{x, \{a, \{x, a, x\}, a\}, y\} \\
&= \{\{x, a, \{x, a, x\}\}, a, y\}.
\end{aligned}$$

Definition 1.2.9 Let V be a Jordan triple system and $a \in V$. The *a-homotope* $V^{(a)}$ of V is defined to be the Jordan algebra $(V, \circ_a)$.

Using identities (1.13) and (1.17), we see that the canonical triple product $\{\cdot, \cdot, \cdot\}_a$ in the *a*-homotope $V^{(a)} = (V, \circ_a)$ is given by

$$\begin{aligned}
\{x, y, x\}_a &= 2(x \circ_a y) \circ_a x - (x \circ_a x) \circ_a y \\
&= 2\{\{x, a, y\}, a, x\} - \{\{x, a, x\}, a, y\} \\
&= \{x, \{a, y, a\}, x\}.
\end{aligned}$$

In particular, if $\{a, y, a\} = y$ for all $y \in V$, then we have $\{x, y, x\}_a = \{x, y, x\}$ for all $x, y \in V$ and the Jordan triple system $(V^{(a)}, \{\cdot, \cdot, \cdot\}_a)$ is just V itself. Therefore Jordan triple systems generalize Jordan algebras in that unital Jordan algebras are the Jordan triples containing a *unit element*, that is, an element e satisfying $\{e, y, e\} = y$ for all elements y.

One can deduce power associativity in V via homotopes.

Lemma 1.2.10 *Let V be a Jordan triple system and let $a \in V$. For odd natural numbers m, n and p, we have*

$$\{a^m, a^n, a^p\} = a^{m+n+p}.$$

Proof Let $V^{(a)}$ be the a-homotope and $x^{(n)}$ the nth power of x in the Jordan algebra $(V^{(a)}, \circ_a)$. By (1.20) and induction, we have

$$a^{2n-1} = a^{(n)}.$$

By (1.13) and power associativity in Jordan algebras, we have

$$\{a^m, \{a, a, a\}, a^p\} = \{\{a^m, a, a\}, a, a^p\} = a^{(m+3/2)} \circ_a a^{(p+1/2)}$$
$$= a^{(m+3+p+1/2)} = a^{m+3+p}.$$

One concludes the proof by induction, using the identity

$$\{a^m, \{a, a^{2k-1}, a\}, a^p\} = \{\{a, a, a^m\}, a^{2k-1}, a^p\} + \{a^m, a^{2k-1}, \{a, a, a^p\}\}$$
$$- \{a, a, \{a^m, a^{2k-1}, a^p\}\}. \qquad \square$$

Definition 1.2.11 Let V be a Jordan triple system and $x, y \in V$. Extending the definition of a box operator on a Jordan algebra, we define the box operator $x \,\square\, y : V \longrightarrow V$ by

$$(x \,\square\, y)(v) = \{x, y, v\} \qquad (v \in V).$$

The triple identity (1.10) can be written in terms of Lie brackets of box operators:

$$[a \,\square\, b, x \,\square\, y] = \{a, b, x\} \,\square\, y - x \,\square\, \{y, a, b\}. \qquad (1.26)$$

Definition 1.2.12 A Jordan triple system V is called *abelian* or *commutative* if

$$[a \,\square\, b, x \,\square\, y] = 0$$

for all $a, b, x, y \in V$; in other words, if the operators $a \,\square\, b$ and $x \,\square\, y$ commute for all $a, b, x, y \in V$. We call V *flat* if $x \,\square\, y = y \,\square\, x$ for all $x, y \in V$.

Lemma 1.2.13 *Let V be a Jordan triple system. Then V is abelian if, and only, if*

$$\{a, b, \{x, y, z\}\} = \{a, \{b, x, y\}, z\} = \{\{a, b, x\}, y, z\}$$

for all $a, b, x, y, z \in V$.

Proof By the triple identity (1.26), the stated identities are equivalent to

$$[x \,\square\, y, z \,\square\, b](a) = 0 = [x \,\square\, b, a \,\square\, y](z)$$

for all $a, b, x, y, z \in V$. $\square$

Example 1.2.14 Let V be a Jordan triple system and let $a \in V$. It is plain from power associativity in Lemma 1.2.10 that the linear span $V(a)$ of odd powers of a is the subtriple of V generated by a, that is, the smallest subtriple of V containing a. Moreover, power associativity implies that $V(a)$ is abelian. If V is real, then $V(a)$ is flat, a consequence of power associativity again.

Example 1.2.15 Flat Jordan triple systems must be abelian. However, an abelian Jordan triple system need not be flat. For instance, $\mathbb{C}$ with the triple product $\{x, y\, z\} = x\overline{y}z$ is abelian but not flat.

Apart from the box operator, there are two important basic operators on Jordan triple systems, namely, the *quadratic operator* and the *Bergmann operator*. Let V be a Jordan triple system and let $a, b \in V$. The quadratic operator $Q_a : V \longrightarrow V$, induced by a, is defined by

$$Q_a(x) = \{a, x, a\} \qquad (x \in V),$$

which is linear if V is real or complex, but conjugate linear if V is Hermitian.

The Bergmann operator $B(a, b) : V \longrightarrow V$, induced by (a, b), is defined by

$$B(a, b)(x) = x - 2(a \,\square\, b)(x) + Q_a Q_b(x) \qquad (x \in V).$$

In terms of quadratic operators, the identity (1.14) can be formulated as

$$Q_a(b \,\square\, a) = (a \,\square\, b)Q_a \qquad (a, b \in V). \tag{1.27}$$

One also deduces from (1.14) that

$$B(a, b)Q_a = Q_{a - Q_a(b)} \qquad (a, b \in V).$$

The identity (1.16) can be formulated as

$$Q_{Q_a(b)} = Q_a Q_b Q_a \qquad (a, b \in V). \tag{1.28}$$

We need to derive a few more identities for later applications. Given $a, b \in V$, let us define the map $Q(a, b) : V \longrightarrow V$ by

$$Q(a, b)(x) = \{a, x, b\} \qquad (x \in V).$$

We have of course $Q_a = Q(a, a)$, and it is easy to verify that

$$Q(a + tx, a + tx) = Q_a + 2t\, Q(a, x) + t^2 Q_x$$

for all scalars t. Hence we have, from the identity (1.28),

$$Q(Q_{a+tx}(b), Q_{a+tx}(b)) = Q_{a+tx}Q_bQ_{a+tx}$$
$$= (Q_a + 2tQ(a, x) + t^2Q_x)(Q_bQ_a + 2tQ_bQ(a, x) + t^2Q_bQ_x).$$

Comparing the coefficients of t on both sides, we obtain

$$2Q(Q_a(b), \{a, b, x\}) = Q(a, x)Q_bQ_a + Q_aQ_bQ(a, x). \tag{1.29}$$

Also, comparing the coefficients of t^2 on both sides of the previous identity, we obtain

$$2Q(Q_a(b), Q_x(b)) + 4Q(\{a, b, x\}, \{a, b, x\})$$
$$= Q_aQ_bQ_x + Q_xQ_bQ_a + 4Q(a, x)Q_bQ(a, x). \tag{1.30}$$

Replacing b by $b + c$ in (1.29) and expanding, we get

$$Q(Q_a(b), \{a, c, x\}) + Q(Q_a(c), \{a, b, x\})$$
$$= Q_aQ(b, c)Q(a, x) + Q(a, x)Q(b, c)Q_a. \tag{1.31}$$

Lemma 1.2.16 *For a, c and x in a Jordan triple system V, we have*

$$2Q(Q_aQ_c(x), \{a, c, x\}) = Q_aQ_cQ_x(c \square a) + (a \square c)Q_xQ_cQ_a.$$

Proof Substituting $Q_c(x)$ for b in (1.31), we have

$$2Q(Q_aQ_c(x), \{a, c, x\})$$
$$= 2Q_aQ(Q_c(x), c)Q(a, x) + 2Q(a, x)Q(Q_c(x), c)Q_a$$
$$\quad - 2Q(Q_a(c), \{a, Q_c(x), x\})$$
$$= 2Q_aQ_c(x \square c)Q(a, x) + 2Q(a, x)(c \square x)Q_cQ_a$$
$$\quad - 2Q(Q_a(c), \{a, c, Q_x(c)\}),$$

where the second identity follows from (1.13). When (1.18) and (1.19) are applied to the first two terms, the last formula becomes

$$Q_aQ_c(Q(Q_c(x), a) + Q_x(c \square a)) + (Q(Q_c(x), a) + (a \square c)Q_x)Q_cQ_a$$
$$\quad - 2Q(Q_a(c), \{a, c, Q_x(c)\})$$
$$= Q_aQ_cQ_x(c \square a) + (a \square c)Q_xQ_cQ_a$$
$$\quad + Q_aQ_cQ(Q_c(x), a) + Q(Q_c(x), a)Q_cQ_a - 2Q(Q_a(c), \{a, c, Q_x(c)\})$$
$$= Q_aQ_cQ_x(c \square a) + (a \square c)Q_xQ_cQ_a,$$

where the last three terms in the second formula sum to 0, by (1.29). $\square$

Lemma 1.2.17 *For a, c and x in a Jordan triple system V, we have*

$$2Q(x, Q_a Q_b(x)) + 4Q(\{a, b, x\}, \{a, b, x\})$$
$$= Q_a Q_b Q_x + Q_x Q_b Q_a + 4(a \,\square\, b)Q_x(b \,\square\, a).$$

Proof Considering the last term in the equation and applying (1.19), we have

$$2(a \,\square\, b)Q_x(b \,\square\, a) = 4Q(x, a)(b \,\square\, x)(b \,\square\, a) - 2Q(Q_x(b), a)(b \,\square\, a),$$

which, by (1.18), equals

$$2Q(x, a)(Q_b(x) \,\square\, a) + 2Q(x, a)Q_b Q(x, a) - 2Q(Q_x(b), a)(b \,\square\, a),$$

and by (1.19) again, this is equal to

$$Q(x, Q_a Q_b(x)) + (x \,\square\, Q_b(x))Q_a + 2Q(x, a)Q_b Q(x, a)$$
$$- Q(Q_a(b), Q_x(b)) - (Q_x(b) \,\square\, b)Q_a$$
$$= Q(x, Q_a Q_b(x)) + 2Q(x, a)Q_b Q(x, a) - Q(Q_a(b), Q_x(b)),$$

where the last identity follows from (1.13). It follows from (1.30) that

$$4(a \,\square\, b)Q_x(b \,\square\, a)$$
$$= 2Q(x, Q_a Q_b(x)) + 4Q(x, a)Q_b Q(x, a) - 2Q(Q_a(b), Q_x(b))$$
$$= 2Q(x, Q_a Q_b(x)) + 4Q(\{a, b, x\}, \{a, b, x\}) - Q_a Q_b Q_x - Q_x Q_b Q_a,$$

which completes the proof. $\square$

We are now ready to prove an important identity for the Bergmann operator.

Theorem 1.2.18 *Let V be a Jordan triple system and let $x, y, z \in V$. The Bergmann operator $B(x, y)$ satisfies*

$$Q(B(x, y)z, B(x, y)z) = B(x, y)Q_z B(y, x). \tag{1.32}$$

Proof The identity can be proved by comparing the expansions of both sides. Indeed, the left-hand side is equal to

$$Q_z - 4Q(z, \{x, y, z\}) + 2Q(z, Q_x Q_y(z)) + 4Q(\{x, y, z\}, \{x, y, z\})$$
$$- 4Q(Q_x Q_y(z), \{x, y, z\}) + Q(Q_x Q_y(z), Q_x Q_y(z)),$$

whereas the right-hand side equals

$$Q_z - 2(x \,\square\, y)Q_z - 2Q_z(y \,\square\, x) + Q_x Q_y Q_z + Q_z Q_y Q_x + 4(x \,\square\, y)Q_z(y \,\square\, x)$$
$$- 2Q_x Q_y Q_z(y \,\square\, x) - 2(x \,\square\, y)Q_z Q_y Q_x + Q_x Q_y Q_z Q_y Q_x,$$

which is identical to the left-hand side by the triple identity (1.9), Lemma 1.2.4, Lemma 1.2.16 and Lemma 1.2.17. $\square$

Definition 1.2.19 A Jordan triple system V is called *non-degenerate* if

$$Q_a = 0 \implies a = 0$$

for each $a \in V$.

Lemma 1.2.20 *Let V be a non-degenerate Jordan triple system and let $a \in V$. If $x \,\square\, a = 0$ for all $x \in V$, then $a = 0$.*

Proof We have

$$
\begin{aligned}
0 &= \{x, a, \{y, a, y\}\} \\
&= \{\{x, a, y\}, a, y\} - \{y, \{a, x, a\}, y\} + \{y, a, \{x, a, y\}\} \\
&= -\{y, \{a, x, a\}, y\}
\end{aligned}
$$

for all $y \in V$, which implies that $Q_{\{a,y,a\}} = 0$ for all $y \in V$. Hence $\{a, y, a\} = 0$ for all $y \in V$ and $a = 0$. $\qquad\square$

Lemma 1.2.21 *Let V be a non-degenerate Jordan triple system and let $a \in V$. If $\{x, a, x\} = 0$ for all $x \in V$, then $a = 0$.*

Proof We have $\{x + v, a, x + v\} = 0$ for all $x, v \in V$, which implies $x \,\square\, a = 0$ for all $x \in V$. Hence $a = 0$. $\qquad\square$

Let V be a finite-dimensional real or complex Jordan triple. We define a bilinear form $\langle \cdot, \cdot \rangle : V \times V \longrightarrow \mathbb{F}$, where $\mathbb{F}$ is the underlying scalar field, by

$$\langle x, y \rangle = \mathrm{Trace}\,(x \,\square\, y).$$

If V is a finite-dimensional Hermitian Jordan triple, then

$$\langle x, y \rangle = \mathrm{Trace}\,(x \,\square\, y)$$

is a complex sesquilinear form. We call $\langle \cdot, \cdot \rangle$ the *trace form* on V.

If the trace form on a Jordan triple system V is non-degenerate, that is, $\langle x, y \rangle = 0$ for all $y \in V$ implies $x = 0$, then every linear map $T : V \longrightarrow V$ has an adjoint $T^* : V \longrightarrow V$ with respect to the trace form:

$$\langle Tx, y \rangle = \langle x, T^*y \rangle \qquad (x, y \in V).$$

In this case, the quadratic form $q(x) = \mathrm{Trace}\,(x \,\square\, x)$ is also called the trace form, as q determines $\langle \cdot, \cdot \rangle$ completely. We call q, or $\langle \cdot, \cdot \rangle$, *positive definite* if $q(x) > 0$ for all $x \neq 0$.

Lemma 1.2.22 *Let V be a Jordan triple system which admits a non-degenerate trace form. Then we have, for every $a, b \in V$,*

$$(a \,\square\, b)^* = b \,\square\, a,$$

and hence the trace form $\langle \cdot, \cdot \rangle$ *is symmetric if V is real or complex. If V is Hermitian, the trace form* $\langle \cdot, \cdot \rangle$ *is Hermitian.*

Proof By the triple identity (1.26), we have

$$\langle (a \,\square\, b)x, y \rangle - \langle x, (b \,\square\, a)y \rangle = \text{Trace}\,((a \,\square\, b)x \,\square\, y) - \text{Trace}\,(x \,\square\, (b \,\square\, a)y)$$
$$= \text{Trace}\,[a \,\square\, b, x \,\square\, y] = 0.$$

Hence $\langle a, b \rangle = \text{Trace}\,(a \,\square\, b) = \text{Trace}\,(a \,\square\, b)^* = \text{Trace}\,(b \,\square\, a) = \langle b, a \rangle$ if V is real or complex, whereas $\text{Trace}\,(a \,\square\, b) = \overline{\text{Trace}\,(a \,\square\, b)^*} = \overline{\text{Trace}\,(b \,\square\, a)}$ if V is Hermitian. $\square$

Lemma 1.2.23 *Let V be a real or complex Jordan triple. The following conditions are equivalent.*

 (i) *The trace form is non-degenerate.*
(ii) *The bilinear form* $(x, y) \in V^2 \mapsto \text{Trace}\,(x \,\square\, y + y \,\square\, x)$ *is non-degenerate.*

Proof (i) $\Rightarrow$ (ii). This follows from Lemma 1.2.22.

(ii) $\Rightarrow$ (i). The bilinear form $\ll x, y \gg = \text{Trace}\,(x \,\square\, y + y \,\square\, x)$ on V is non-degenerate and symmetric. Using $\text{Trace}\,[x \,\square\, y, u \,\square\, v] = 0$ and the triple identity (1.26) as before, we have

$$\ll (x \,\square\, y)u, v \gg = \text{Trace}\,((x \,\square\, y)u \,\square\, v) + v \,\square\, (x \,\square\, y)u)$$
$$= \text{Trace}\,(u \,\square\, (y \,\square\, x)v + (y \,\square\, x)v \,\square\, u)$$
$$= \ll u, (y \,\square\, x)v \gg .$$

Hence the box operator $y \,\square\, x$ is the adjoint of $x \,\square\, y$ with respect to the bilinear form $\ll \cdot, \cdot \gg$. Therefore $\text{Trace}\,(y \,\square\, x) = \text{Trace}\,(x \,\square\, y)$ and (i) follows. $\square$

Definition 1.2.24 A Jordan triple system V is called *semisimple* if for each $a \in V$, we have

$$a \,\square\, x \text{ is nilpotent for all } x \in V \implies a = 0.$$

We say that V is *weakly semisimple* if $a \,\square\, x = 0$ for all $x \in V$ implies $a = 0$.

Definition 1.2.25 A Jordan triple system V is called *anisotropic* if $\{x, x, x\} = 0$ implies $x = 0$ for each $x \in V$.

Lemma 1.2.26 *Let V be a Jordan triple system which admits a positive definite trace form. Then V is anisotropic.*

Proof Let $x \in V$ and $\{x, x, x\} = 0$. Using the identities (1.17) and (1.15), we deduce that the box operator $x \,\square\, x$ is nilpotent; in fact, $(x \,\square\, x)^3 = 0$. Therefore

Trace $(x \square x) = 0$ which implies that $x = 0$, since the trace form is positive definite. □

For finite-dimensional Jordan triples, semisimplicity is equivalent to non-degeneracy of the trace form.

Lemma 1.2.27 *Let V be a finite-dimensional Jordan triple system. The following conditions are equivalent:*

(i) *V is semisimple.*
(ii) *The trace form on V is non-degenerate.*

These conditions imply that V is non-degenerate.

Proof We show that

$$\{a \in V : a \square x \text{ is nilpotent } \forall x \in V\} = \{a \in V : \text{Trace}(a \square x) = 0 \, \forall x \in V\},$$

from which the equivalence of (i) and (ii) follows immediately.

Indeed, if $a \square x$ is nilpotent, then Trace $(a \square x) = 0$. If the operator $a \square x$ is not nilpotent for some $x \in V$, then the left multiplication $L_a : V^{(x)} \longrightarrow V^{(x)}$ on the x-homotope $V^{(x)}$ is not nilpotent, as the two operators are identical. Therefore a is not nilpotent in the Jordan algebra $V^{(x)}$, by Lemma 1.1.9. From Lemma 1.1.8, there is an idempotent e in the subalgebra $\mathcal{A}(a)$ of $V^{(x)}$ generated by a, and e is of the form $e = \sum_k \alpha_k a^{(n_k)}$. It follows that Trace $(e \square x) = \text{Trace } L_e > 0$ and the triple identity gives

$$
\begin{aligned}
\text{Trace}(e \square x) &= \left\langle \sum_k \alpha_k a^{(n_k)}, x \right\rangle \\
&= \sum_k \alpha_k \langle (a^{(n_k-1)} \square x)(a), x \rangle \\
&= \left\langle a, \sum_k \alpha_k (x \square a^{(n_k-1)})(x) \right\rangle \\
&= \text{Trace}\left(a \square \sum_k \alpha_k (x \square a^{(n_k-1)})(x) \right),
\end{aligned}
$$

and hence a does not belong to the above set on the right. This proves the equality of the two sets.

Assume condition (i) and let $Q_a = 0$ for some $a \in V$. For every $x \in V$, we have

$$2(a \square x)^2 = Q_a(x) \square x + Q_a Q_x = 0$$

by the identity (1.18). Hence $a \, \Box \, x$ is nilpotent for all $x \in V$ and $a = 0$ by semisimplicity. This proves non-degeneracy of V. $\qquad\qquad\qquad\qquad\square$

Example 1.2.28 A non-degenerate Jordan triple system V is necessarily weakly semisimple, since $a \, \Box \, x(a) = Q_a(x)$. However, weak semisimplicity is strictly weaker than non-degeneracy. Let

$$V = \left\{ \begin{pmatrix} \alpha & \beta \\ 0 & \alpha \end{pmatrix} : \alpha, \beta \in \mathbb{R} \right\}$$

be equipped with the triple product

$$\{a, b, c\} = abc + cba \qquad (a, b, c \in V),$$

where the product on the right is the matrix product. Then V is a weakly semisimple Jordan triple system in which $Q_a = 0$ for $a = \begin{pmatrix} 0 & 1 \\ 0 & 0 \end{pmatrix}$.

Definition 1.2.29 A finite-dimensional Jordan triple system V is called *positive* if $x \, \Box \, x : V \longrightarrow V$ has a non-negative spectrum for each $x \in V$.

If V is a complex Jordan triple and positive, then $ix \, \Box \, ix = -x \, \Box \, x$ implies that each $x \, \Box \, x$ has zero spectrum. We will therefore only be interested in positive real or Hermitian Jordan triples.

Example 1.2.30 A positive Jordan triple need not be non-degenerate. Let

$$V = \left\{ \begin{pmatrix} 0 & \alpha \\ 0 & \beta \end{pmatrix} : \alpha, \beta \in \mathbb{R} \right\}$$

be equipped with the triple product

$$\{a, b, c\} = abc + cba \qquad (a, b, c \in V),$$

where the product on the right is the matrix product. Then V is a positive real Jordan triple. Indeed, for $x = \begin{pmatrix} 0 & \alpha \\ 0 & \beta \end{pmatrix}$, the eigenvalues of the box operator $x \, \Box \, x$ are β^2 and $2\beta^2$. However, $Q_a = 0$ for $a = \begin{pmatrix} 0 & 1 \\ 0 & 0 \end{pmatrix}$.

The role of idempotents in a Jordan algebra is played by tripotents in a Jordan triple system.

Definition 1.2.31 An element e in a Jordan triple system V is called a *tripotent* if $e = \{e, e, e\}$. A tripotent e is said to be *orthogonal* to a tripotent f if $e \, \Box \, f = 0$.

For tripotents e and f in a Jordan triple system V, it will be shown that the condition $e \square f = 0$ is equivalent to $f \square e = 0$. Two elements $a, b \in V$ are defined to be *orthogonal to each other* if $a \square b = b \square a = 0$.

Lemma 1.2.32 *Let V be an anisotropic Jordan triple system and let $a, b \in V$. Then a and b are orthogonal to each other if, and only if, $a \square b = 0$.*

Proof Let $a \square b = 0$. By the triple identity (1.26), we have

$$x \square \{b, a, y\} = \{a, b, x\} \square y - [a \square b, x \square y] = 0 \qquad (x, y \in V),$$

which gives

$$\{b \square a(y), b \square a(y), b \square a(y)\} = 0 \qquad (y \in V).$$

Hence $b \square a = 0$ by anisotropy. $\qquad\qquad\square$

Lemma 1.2.33 *Let e be a tripotent in a Jordan triple system V. Then the box operator $e \square e : V \longrightarrow V$ has eigenvalues in $\{0, 1/2, 1\}$.*

Proof Let $V^{(e)}$ be the e-homotope of V. Then e is an idempotent in the Jordan algebra $V^{(e)}$ with the Jordan product

$$a \circ_e b = \{a, e, b\}.$$

The box operator $e \square e : V \longrightarrow V$ is the left multiplication operator $L_e : V^{(e)} \longrightarrow V^{(e)}$, which has eigenvalues $0, 1/2$ or 1 by (1.6). $\qquad\square$

We now show the existence of tripotents.

Theorem 1.2.34 *Let V be a finite-dimensional real or Hermitian Jordan triple. The following conditions are equivalent:*

(i) *V is semisimple and positive.*
(ii) *The trace form $q(x) = \text{Trace}\,(x \square x)$ is positive definite.*
(iii) *Each nonzero $x \in V$ admits a unique decomposition $x = \alpha_1 e_1 + \cdots + \alpha_n e_n$, where $0 < \alpha_1 < \cdots < \alpha_n$ and $e_1, \ldots, e_n$ are mutually orthogonal tripotents in V.*

Proof (i) $\Rightarrow$ (ii). We have $q(x) \geq 0$ by positivity of V. If $\text{Trace}\,(x \square x) = 0$, then all eigenvalues of $x \square x$ are zero and we must have $x \square x = 0$, since $x \square x$ is self-adjoint with respect to the trace form on V, by Lemma 1.2.22. It follows that x is nilpotent in the a-homotope $V^{(a)}$ for each $a \in V$ since in the Jordan algebra $V^{(a)}$, we have

$$x^{(4)} = \{\{x, a, \{x, a, x\}\}, a, x\}$$
$$= \{(\{x, a, \{x, x, a\}\} + \{x, \{a, x, x\}, a\} - \{x, x, \{x, a, a\}\}), a, x\} = 0.$$

Hence the left multiplication $L_x : V^{(a)} \longrightarrow V^{(a)}$ is nilpotent by Lemma 1.1.9. It follows that $\text{Trace}\,(x \,\square\, a) = \text{Trace}\,L_x = 0$ and $x = 0$ by semisimplicity of V.

(ii) $\Rightarrow$ (iii). Let q be positive definite and let $x \in V \backslash \{0\}$. Let $V(x)$ be the subtriple of V generated by x. Then $V(x)$ is a finite-dimensional inner product space with the trace form q. Let $\mathbb{L}$ be the real linear span of the box operators $\{a \,\square\, b|_{V(x)} : a, b \in V(x)\}$. Then $\mathbb{L}$ is a commutative algebra with respect to composition, as $V(x)$ is abelian, as noted in Example 1.2.14. By Lemma 1.2.22, $\mathbb{L}$ consists of self-adjoint operators on $V(x)$ and can therefore be simultaneously diagonalised. In other words, there is a basis $\{v_1, \ldots, v_n\}$ in $V(x)$ such that $Lv_k \in \mathbb{R}v_k$ for all $L \in \mathbb{L}$ and $k = 1, \ldots, n$. In particular, we have

$$(v_k \,\square\, v_k)(v_k) = \lambda_k v_k \qquad (k = 1, \ldots, n),$$

where $\lambda_k \neq 0$ since V is anisotropic by Lemma 1.2.26. Moreover, v_k/λ_k is an idempotent in the v_k-homotope $V^{(v_k)}$ on which the left multiplication L_{v_k/λ_k} has eigenvalues 0, $1/2$ or 1. The latter says the same of the box operator $(v_k \,\square\, v_k)/\lambda_k$. It follows that $\text{Trace}\,(v_k \,\square\, v_k)$ is a positive multiple of λ_k and hence $\lambda_k > 0$.

Let $e_k = v_k/\sqrt{\lambda_k}$ for $k = 1, \ldots, n$. Then $e_1, \ldots, e_n$ are mutually orthogonal tripotents in V, since $(e_i \,\square\, e_j)(e_k) = (e_k \,\square\, e_j)(e_i) \in \mathbb{R}e_k \cap \mathbb{R}e_i = \{0\}$ for $i \neq k$.

After permutation and sign change, we can write

$$x = \alpha_1 e_1 + \cdots + \alpha_n e_n \qquad (0 \leq \alpha_1 \leq \cdots \leq \alpha_n).$$

By orthogonality, we have

$$x^p = \alpha_1^p e_1 + \cdots + \alpha_n^p e_n$$

for odd powers x^p of x. Since $V(x)$ is spanned by these odd powers, we must have

$$0 < \alpha_1 < \cdots < \alpha_n.$$

To see the uniqueness of the decomposition, let

$$x = \mu_1 u_1 + \cdots + \mu_m u_m,$$

where $0 < \mu_1 < \cdots < \mu_m$ and $u_1, \ldots, u_m$ are mutually orthogonal tripotents. Then we have

$$\mu_1^{2k+1} u_1 + \cdots + \mu_m^{2k+1} u_m = x^{(2k+1)} \in V(x)$$

for $k = 1, 2, \ldots$. Hence each u_j is in $V(x)$ and $u_1, \ldots, u_m$ form a basis of $V(x)$ with $m = n$. We have $u_j = \sum_k \beta_k e_k$ and

$$u_j = \{u_j, u_j, u_j\} = \sum_k \beta_k^3 e_k$$

which imply $\beta_k = 0$ or ± 1. It follows that u_j equals $\pm e_k$ for some k; but the inequalities on the coefficients imply $u_j = e_j$ and $\mu_j = \alpha_j$ for each j.

(iii) $\Rightarrow$ (i). Given $x = \alpha_1 e_1 + \cdots + \alpha_n e_n$ with orthogonal tripotents $e_1, \ldots, e_n$, we have

$$x \square x = \sum_{k=1}^{n} \alpha_k^2 (e_k \square e_k),$$

where each $e_k \square e_k$ has spectrum in $\{0, 1/2, 1\}$. Hence $x \square x$ has non-negative spectrum. If Trace $(x \square x) = 0$, then $\alpha_k^2 = 0$ for all k and $x = 0$. This proves the semisimplicity of V. $\qquad\square$

Definition 1.2.35 Let V be a semisimple positive finite-dimensional real or Hermitian Jordan triple and let $x \in V$. The decomposition

$$x = \alpha_1 e_1 + \cdots + \alpha_n e_n$$

in Theorem 1.2.34 (iii) is called the *spectral decomposition* of x. We define the *triple spectrum* of x to be the set

$$s(x) = \{\alpha_1, \ldots, \alpha_n\}.$$

Example 1.2.36 The 4-dimensional real vector space

$$V = \{(x, -x, y) : x, y \in \mathbb{C}\}$$

is a real Jordan triple with triple product

$$\{(x, -x, y), (\alpha, -\alpha, \beta), (x, -x, y)\} = (x\alpha x, -x\alpha x, y\beta y).$$

Given $a = (x, -x, y) \in V$, the operator $a \square a$ is represented by the following matrix with respect to the basis $\{(1, -1, 0), (i, -i, 0), (0, 0, 1), (0, 0, i)\}$:

$$\begin{pmatrix} x^2 & 0 & 0 & 0 \\ 0 & x^2 & 0 & 0 \\ 0 & 0 & y^2 & 0 \\ 0 & 0 & 0 & y^2 \end{pmatrix}.$$

Hence V is not a positive Jordan triple. For instance, the box operator $(i, -i, i) \square (i, -i, i)$ has negative spectrum.

However, the 2-dimensional Jordan triple $V_{\mathbb{R}} = \{(x, -x, y) : x, y \in \mathbb{R}\}$ is a positive Jordan triple.

We now discuss the Peirce decomposition of a Jordan triple system V induced by a tripotent $u \in V$. Let $V^{(u)}$ be the u-homotope of V. As noted before, u is an idempotent in the Jordan algebra $V^{(u)}$ with Jordan product $a \circ_u b = \{a, u, b\}$ and the box operator $u \square u : V \longrightarrow V$ is the left multiplication $L_u : V^{(u)} \longrightarrow V^{(u)}$, with eigenvalues in $\{0, 1/2, 1\}$.

Definition 1.2.37 Let u be a tripotent in a Jordan triple system V. The eigenspaces

$$V_k(u) = \left\{ z \in V : (u \square u)(z) = \frac{k}{2} z \right\} \qquad (k = 0, 1, 2)$$

are called the Peirce k-spaces of u, and the eigenspace decomposition

$$V = V_0(u) \oplus V_1(u) \oplus V_2(u)$$

is called the *Peirce decomposition* of V.

By (1.7) and (1.8), the Peirce k-space is the range of the Peirce k-projection

$$P_k(u) : V \longrightarrow V$$

given by, using (1.17),

$$P_2(u)(z) = 2L_u^2(z) - L_u(z) = 2\{u, u, \{u, u, z\}\} - \{u, u, z\} = Q_u^2(z)$$
$$P_1(u) = 4(L_u - L_u^2) = 2(L_u - (2L_u^2 - L_u)) = 2(u \square u - Q_u^2)$$
$$P_0(u) = I - 2(u \square u - Q_u^2) - Q_u^2 = I - 2u \square u + Q_u^2 = B(u, u),$$

where $I : V \longrightarrow V$ is the identity operator.

Trivially, if $u = 0$, then $P_0(u) = I$ and $V = V_0(u)$. We usually consider nonzero tripotents in Perice decompositions.

By Lemma 1.1.10, the Peirce 2-space $V_2(u)$ of a tripotent $u \in V$ is a Jordan subalgebra of the u-homotope $V^{(u)}$, containing the identity u, with Jordan product

$$a \circ_u b = \{a, u, b\} \qquad (a, b \in V).$$

Lemma 1.2.38 *Given a tripotent u in an abelian Jordan triple system V, we have*

$$V = V_0(u) \oplus V_2(u),$$

where $(V_2(u), \circ_u)$ is an (abelian) associative algebra.

Proof Since V is abelian, the box operator $u \square u : V \longrightarrow V$ is a projection; that is, $(u \square u)^2 = u \square u$, and it follows that $2\{u, u, v\} = v$ implies $v = 0$. Hence

$V_1(u) = \{0\}$. The associativity of $(V_2(u), \circ_u)$ follows directly from the abelian condition on the triple product. $\qquad\square$

Definition 1.2.39 A tripotent u in a Jordan triple system V is called *maximal* or *complete* if $V_0(u) = \{0\}$. It is called *unitary* if $V_2(u) = V$.

Neither maximal nor unitary tripotents can be 0 unless $V = \{0\}$. We see from Lemma 1.2.38 that an abelian Jordan triple system becomes an abelian associative algebra if it admits a maximal tripotent.

Example 1.2.40 Consider the Jordan triple $M_2(\mathbb{C})$ of 2×2 complex matrices. For the tripotent

$$u = \begin{pmatrix} 1 & 0 \\ 0 & 0 \end{pmatrix},$$

the Peirce k-projections are given b

$$P_0(u)\begin{pmatrix} a & b \\ c & d \end{pmatrix} = \begin{pmatrix} 0 & 0 \\ 0 & d \end{pmatrix}$$

$$P_1(u)\begin{pmatrix} a & b \\ c & d \end{pmatrix} = \begin{pmatrix} 0 & b \\ c & 0 \end{pmatrix}$$

$$P_2(u)\begin{pmatrix} a & b \\ c & d \end{pmatrix} = \begin{pmatrix} a & 0 \\ 0 & 0 \end{pmatrix}.$$

Example 1.2.41 In the Hermitian Jordan triple $H_3(\mathcal{O})$, introduced in Example 1.2.8, the Peirce 1-space of the tripotent

$$u = \begin{pmatrix} 1 & 0 & 0 \\ 0 & 0 & 0 \\ 0 & 0 & 0 \end{pmatrix}$$

is none other than

$$M_{12}(\mathcal{O}) = P_1(u)(H_3(\mathcal{O})).$$

More generally, consider the Hermitian Jordan triple $L(H, K)$ in Example 1.2.6. An operator $u \in L(H, K)$ is a tripotent if and only if $u = uu^*u$; that is, u is a partial isometry. The operators $l = uu^*$ and $r = u^*u$ are projections on the Hilbert spaces K and H, respectively. They can be represented, with suitable orthonormal bases, by square block matrices

$$\ell = \begin{pmatrix} \mathbf{1}_K & O \\ O & O \end{pmatrix}, \quad r = \begin{pmatrix} \mathbf{1}_H & O \\ O & O \end{pmatrix},$$

where $\mathbf{1}_H$ and $\mathbf{1}_K$ are identities. In this representation, each operator in the Peirce 2-space

$$L(H, K)_2(u) = P_2(u)L(H, K) = \{\ell \, T r : T \in L(H, K)\}$$

has a rectangular matrix representation

$$P_2(u)T = \ell \, T r = \begin{pmatrix} [\ell \, T r] & O \\ O & O \end{pmatrix}.$$

The other two Peirce projections of u are given by

$$P_0(u)T = (\mathbf{1}_K - \ell)T(\mathbf{1}_H - r) = \begin{pmatrix} O & O \\ O & [(\mathbf{1}_K - \ell)T(\mathbf{1}_H - r)] \end{pmatrix},$$

$$P_1(u)T = \ell \, T(\mathbf{1}_H - r) + (\mathbf{1}_K - \ell)T r = \begin{pmatrix} O & [\ell \, T(\mathbf{1}_H - r)] \\ [(\mathbf{1}_K - \ell)T r] & O \end{pmatrix}.$$

The matrix form of the Peirce decomposition of $T \in L(H, K)$ is given by

$$T = \begin{pmatrix} [\ell \, T r] & [\ell \, T(\mathbf{1}_H - r)] \\ [(\mathbf{1}_K - \ell)T r] & [(\mathbf{1}_K - \ell)T(\mathbf{1}_H - r)] \end{pmatrix}.$$

Definition 1.2.42 Given two tripotents u and v in a Jordan triple system V, we write $u \leq v$ if $v - u$ is a tripotent orthogonal to u.

Proposition 1.2.43 *Let u, v be tripotents in a Jordan triple system V. The following conditions are equivalent:*

(i) $u \leq v$.
(ii) $\{v, u, v\} = u$.

Proof (i) $\Rightarrow$ (ii). We have $0 = \{v - u, u, u + v\} = \{v, u, v\} - \{u, u, u\}$, which gives $\{v, u, v\} = u$.

   (ii) $\Rightarrow$ (i). We use the identities (1.13), (1.14), (1.16) and (1.17) repeatedly in the following computation.

Since $u = P_2(v)u \in V_2(v)$, we have $\{v, v, u\} = u$. Also

$$
\begin{aligned}
\{u, v, u\} &= \{\{v, u, v\}, v, \{v, u, v\}\} \\
&= \{v, \{u, \{v, v, v\}, u\}, v\} \\
&= \{v, \{u, v, u\}v\} \\
&= \{v, u, \{v, u, v\}\} \\
&= \{v, u, u\}.
\end{aligned}
$$

Hence

$$
\{u, \{u, v, u\}, u\} = \{u, \{v, u, u\}, u\} = \{u, v, \{u, u, u\}\} = \{u, v, u\}.
$$

We note that

$$
\{u, v, \{u, v, u\}\} = \{u, \{v, u, v\}, u\} = \{u, u, u\} = u
$$

and also

$$
2\{v, u, \{u, v, u\}\} = 2\{v, u, \{v, u, u\}\} = \{v, \{u, v, u\}, u\} + \{v, \{u, u, u\}, v\} \\
= 2u,
$$

where

$$
\{v, \{u, v, u\}, u\} = \{\{v, u, v\}, u, u\} = u.
$$

It follows that

$$
\begin{aligned}
u = \{u, \{v, u, v\}, u\} &= \{u, \{v, \{u, u, u\}, v\}, u\} \\
&= \{\{u, v, u\}, u, \{u, v, u\}\} \\
&= \{\{v, u, u\}, u, \{u, u, v\}\} \\
&= \frac{1}{2}\{\{v, u, u\}, \{u, u, u\}, v\} + \frac{1}{2}\{\{v, u, u\}, \{u, v, u\}, u\} \\
&= \frac{1}{2}\{\{v, u, u\}, u, v\} + \frac{1}{2}\{u, \{u, v, u\}, \{u, v, u\}\} \\
&= \frac{1}{2}u + \frac{1}{2}(\{u, \{u, v, u\}, u\} - \{u, \{\{u, v, u\}, u, v\}, u\} \\
&\quad + \{u, v, \{u, \{u, v, u\}, u\}\}) \\
&= \frac{1}{2}u + \frac{1}{2}(\{u, v, u\} - \{u, u, u\} + \{u, v, \{u, v, u\}\}),
\end{aligned}
$$

which yields

$$
\{u, v, u\} = u = \{v, u, u\}.
$$

It is now readily verified that

$$
\{v - u, v - u, v - u\} = v - u;
$$

that is, $v - u$ is a tripotent, and

$$\{u, u, v - u\} = 0.$$

$\square$

We note from the preceding proof that $u \leq v$ implies $\{u, v, u\} = u$. One deduces readily from condition (ii) that the relation $\leq$ in Definition 1.2.42 is a partial ordering on the set of tripotents. Indeed, if $u \leq v \leq u$, then $u = \{v, u, v\} = v$. If $u \leq v \leq e$, then $u \leq e$ because, by (1.17), we have

$$\{e, u, e\} = \{e, \{v, u, v\}, e\} = 2\{e, v, \{u, v, e\}\} - \{e, \{v, e, v\}, u\}$$
$$= 2\{v, v, \{u, v, v\}\} - \{e, v, u\} = 2u - \{v, v, u\} = u.$$

Let u be a tripotent in a Jordan triple system V. For any real scalar $t \neq 0$, the Bergmann operator

$$B(u, (1 - t)u) : V \longrightarrow V$$

is invertible. Indeed, a simple calculation gives

$$B(u, (1 - t)u) = P_0(u) + t P_1(u) + t^2 P_2(u) \tag{1.33}$$

and therefore $B(u, (1 - t)u)$ has inverse

$$B(u, (1 - t^{-1})u) = P_0(u) + \frac{1}{t} P_1(u) + \frac{1}{t^2} P_2(u)$$

by mutual orthogonality of the Peirce projections.

Since

$$B(u, (1 - t)u)v = \sum_{k=0}^{2} t^k P_k(u)v \tag{1.34}$$

for each $v \in V$, it follows that, for $t \neq 1$, we have $B(u, (1 - t)u)v = t^k v$ if and only if $v \in V_k(u)$ for $k = 0, 1, 2$. However, for $k \notin \{0, 1, 2\}$, we have $B(u, (1 - t)u)v = t^k v$ if and only if $v = 0$.

We now derive the basic Peirce multiplication rules.

Theorem 1.2.44 *Let u be a tripotent of a Jordan triple system V. Then the Peirce k-spaces $V_k(u)$ satisfy*

$$\{V_0(u), V_2(u), V\} = \{V_2(u), V_0(u), V\} = \{0\}$$

$$\{V_i(u), V_j(u), V_k(u)\} \subset V_{i-j+k}(u), \tag{1.35}$$

where $V_\alpha(u) = \{0\}$ for $\alpha \notin \{0, 1, 2\}$.

Proof Let $x \in V_2(u)$ and $y \in V_0(u)$. We first observe that

$$\{u, y, u\} = Q_u^3 y = Q_u P_2(u) y = Q_u P_2(u) P_0(u) y = 0.$$

Hence, by (1.17), we have

$$\{z, u, y\} = \{z, \{u, u, u\}, y\} = 2\{z, u, \{u, u, y\}\} - \{z, \{u, y, u\}, u\} = 0$$

for any $z \in V$. It follows from (1.19) that

$$\{x, y, z\} = \{\{u, \{u, x, u\}, u\}, y, z\} = 2\{u, \{y, u, \{u, x, u\}\}, z\}$$
$$- \{\{u, y, u\}, \{u, x, u\}, z\} = 0$$

for all $z \in V$. Likewise $\{V_0(u), V_2(u), V\} = \{0\}$.

To show (1.35), we make use of invertibility of $B(u, (1 - t)u)$ for real scalars $t \neq 0$, with inverse $B(u, (1 - t^{-1})u)$, and deduce from Theorem 1.2.18 that

$$B(u, (1 - t)u)\{z, x, z\}$$
$$= \{B(u, (1 - t)u)z, B(u, (1 - t^{-1})u)x, B(u, (1 - t)u)z\}$$

for $x, z \in V$. By polarization in (1.21), we have

$$B(u, (1 - t)u)\{x, y, z\}$$
$$= \{B(u, (1 - t)u)x, B(u, (1 - t^{-1})u)y, B(u, (1 - t)u)z\} \quad (1.36)$$

for $x, y, z \in V$. In particular, for $v_\alpha \in V_\alpha(u)$, the preceding remarks imply

$$B(u, (1 - t)u)\{v_i, v_j, v_k\}$$
$$= \{B(u, (1 - t)u)v_i, B(u, (1 - t^{-1})u)v_j, B(u, (1 - t)u)v_k\}$$
$$= \{t^i v_i, t^{-j} v_j, t^k v_k\}$$
$$= t^{i-j+k}\{v_i, v_j, v_k\}$$

and $\{v_i, v_j, v_k\} \in V_{i-j+k}(u)$. $\square$

The Peirce multiplication rules reveal immediately that the Peirce k-spaces $V_k(u)$ of a tripotent u in a Jordan triple system V are subtriples of V. These rules also entail the following useful results.

Corollary 1.2.45 *Let u, e be tripotents in a Jordan triple system V such that $u \in V_2(e)$. Then we have $V_2(u) \subset V_2(e)$ and $V_0(e) \subset V_0(u)$.*

Proof We have $V_2(u) = P_2(u)(V)$. Each $x \in V$ has a Peirce decomposition $x = x_0 + x_1 + x_2 \in V_0(e) \oplus V_1(e) \oplus V_2(e)$ with respect to e, and the Peirce rules imply $P_2(u)(x) = P_2(u)(x_2) \in V_2(e)$.

The Peirce rules also imply $(u \square u)(V_0(e)) = \{0\}$. $\square$

Corollary 1.2.46 *Let* u, v *be tripotents in a Jordan triple system* V. *The following conditions are equivalent:*

(i) *u and v are orthogonal to each other.*
(ii) $v \square u = 0$.
(iii) $\{u, u, v\} = 0$.
(iv) $\{v, v, u\} = 0$.

Proof This follows easily from Theorem 1.2.44. Indeed, orthogonality implies (iv), which in turn implies that u is in the Peirce 0-space $V_0(v)$ of v and therefore $v \square u = 0$ by Theorem 1.2.44. By the same token, (ii) is equivalent to (iii). $\square$

Given two mutually orthogonal tripotents u and v in a Jordan triple system V, it is evident that $u + v$ is also a tripotent. One can form the Peirce decomposition of V with respect to a family of orthogonal tripotents. We show the construction for two orthogonal tripotents e_1 and e_2. It can be extended naturally to the case of a finite number of orthogonal tripotents. We first observe that, by orthogonality,

$$B(e_j, te_j)e_k = e_k \qquad (j, k \in \{1, 2\}, \, j \neq k \text{ and } t \in \mathbb{R}),$$

and a direct computation using (1.36) yields

$$B(e_1, (1 - t)e_1)B(e_2, (1 - s)e_2) = B(e_2, (1 - s)e_2)B(e_1, (1 - t)e_1). \quad (1.37)$$

Since $B(e_1, (1 - t)e_1) = P_0(e_1) + tP_1(e_1) + t^2 P_2(e_1)$ and $B(e_2, (1 - s)e_2) = P_0(e_2) + sP_1(e_2) + s^2 P_2(e_2)$, comparing coefficients in the equation (1.37), one finds that the Peirce projections $P_j(e_1)$ and $P_k(e_2)$ commute. Therefore, we have the decomposition

$$V = \bigoplus_{0 \leq j \leq k \leq 2} V_{j,k} = V_{0,0} \oplus V_{0,1} \oplus V_{0,2} \oplus V_{1,1} \oplus V_{1,2} \oplus V_{2,2},$$

where $V_{k,k} = V_2(e_k)$ for $k \neq 0$ and $V_{0,0} = V_0(e_1) \cap V_0(e_2)$ is the range of the projection $P_0(e_1)P_0(e_2)$, and

$$V_{0,1} = V_0(e_2) \cap V_1(e_1), \quad V_{0,2} = V_0(e_1) \cap V_1(e_2), \quad V_{1,2} = V_1(e_1) \cap V_1(e_2)$$

are ranges of mutually orthogonal projections $P_j(e_{j'})P_k(e_{k'})$ for suitably chosen indices j, j', k and k'.

A subspace J of a Jordan triple system V is called a *triple ideal*, or simply, an *ideal*, if it satisfies the condition

$$\{J, V, V\} + \{V, J, V\} \subset J.$$

If a subspace $J \subset V$ satisfies only

$$\{J, V, J\} \subset J,$$

then it is called an *inner ideal*. The concept of an inner ideal is important in modern Jordan structure theory. Inner ideals are substitutes for *one-sided ideals*; the latter are absent in Jordan systems. Actually, every left or right ideal, or their intersection, in an associative algebra $\mathcal{A}$ is an inner ideal in the special Jordan algebra $(\mathcal{A}, \circ)$ with the Jordan product $a \circ b = (ab + ba)/2$. So is any subspace of the form $a\mathcal{A}b$. In a Jordan triple system V, the subspace $\{v, V, v\}$ is an inner ideal, called the *principal inner ideal* determined by v.

Given an ideal J of a Jordan triple system V, the quotient space V/J is naturally a Jordan triple system with the triple product

$$\{x + J, y + J, z + J\} = \{x, y, z\} + J.$$

The kernel $\varphi^{-1}(0)$ of a triple homomorphism $\varphi : V \longrightarrow W$ is an ideal of V. On the other hand, an ideal J of a Jordan triple system V is the kernel of the quotient map $q : V \longrightarrow V/J$, which is a triple homomorphism.

Let u be a tripotent in a Jordan triple system V. Applying the Peirce multiplication rules in Theorem 1.2.44 to the Peirce decomposition $V = V_0(u) \oplus V_1(u) \oplus V_2(u)$, one deduces the following fact readily.

Proposition 1.2.47 *Given a tripotent u of a Jordan triple system V, the Peirce spaces $V_0(u)$ and $V_2(u)$ are inner ideals of V.*

1.3 Lie algebras and the Tits–Kantor–Koecher construction

In this section, we show an important connection between Jordan triple systems and Lie algebras via the Tits–Kantor–Koecher (TKK) construction. Lie algebras play an important role in geometry and this connection provides us with a useful link to apply Jordan theory to geometry. We will only be concerned with real or complex Lie algebras which, however, can be infinite-dimensional.

In what follows, a *Lie algebra* is a real or complex vector space $\mathfrak{g}$ of any dimension, with a bilinear multiplication, called the *Lie brackets*,

$$(x, y) \in \mathfrak{g} \times \mathfrak{g} \mapsto [x, y] \in \mathfrak{g},$$

satisfying $[x, x] = 0$ and the Jacobi identity

$$[[x, y], z] + [[y, z], x] + [[z, x], y] = 0$$

for all $x, y, z \in \mathfrak{g}$. We note that the multiplication is not associative but is anticommutative:

$$[x, y] = -[y, x].$$

On any associative algebra $\mathfrak{a}$, one can define the Lie brackets by *commutation*,

$$[x, y] = xy - yx,$$

where the product on the right-hand side is the original product in $\mathfrak{a}$. Then $(\mathfrak{a}, [\cdot, \cdot])$ is a Lie algebra. Unlike Jordan algebras, a theorem of Poincaré, Birkhoff and Witt asserts that *any* Lie algebra can be obtained in this way from an associative algebra. Nevertheless, reducing Lie questions to associative ones via this device seldom seems to be conclusive.

Given subspaces h and k of a Lie algebra $\mathfrak{g}$, we define

$$[h, k] = \{[x_1, y_1] + \cdots + [x_n, y_n] : x_1, \ldots, x_n \in h; y_1, \ldots y_n \in k\},$$

which is a subspace of $\mathfrak{g}$. We note that $[h, k] = [k, h]$. A Lie algebra $\mathfrak{g}$ is called *abelian* if $[\mathfrak{g}, \mathfrak{g}] = 0$. An *ideal* of $\mathfrak{g}$ is a subspace $\mathfrak{h}$ of $\mathfrak{g}$ satisfying $[\mathfrak{g}, \mathfrak{h}] \subset \mathfrak{h}$. For instance, $[\mathfrak{g}, \mathfrak{g}]$ is an ideal of $\mathfrak{g}$. Given an ideal $\mathfrak{h}$ of a Lie algebra $\mathfrak{g}$, the quotient space $\mathfrak{g}/\mathfrak{h}$ is a Lie algebra in the product

$$[x + \mathfrak{h}, y + \mathfrak{h}] = [x, y] + \mathfrak{h} \qquad (x + \mathfrak{h}, y + \mathfrak{h} \in \mathfrak{g}/\mathfrak{h}).$$

A *homomorphism* between two Lie algebras $\mathfrak{g}$ and $\mathfrak{h}$ is a linear map $\theta : \mathfrak{g} \longrightarrow \mathfrak{h}$ satisfying $\theta[x, y] = [\theta x, \theta y]$ for all $x, y \in \mathfrak{g}$. Given an ideal $\mathfrak{h}$ of $\mathfrak{g}$, the quotient map $\theta : x \in \mathfrak{g} \mapsto x + \mathfrak{h} \in \mathfrak{g}/\mathfrak{h}$ is a homomorphism. A bijective homomorphism between Lie algebras is called an *isomorphism*. An isomorphism from $\mathfrak{g}$ onto itself is called an *automorphism* of $\mathfrak{g}$.

Definition 1.3.1 Let $\mathfrak{g}$ be a Lie algebra. The set Aut $\mathfrak{g}$ of all automorphisms $\theta : \mathfrak{g} \longrightarrow \mathfrak{g}$ forms a group with composition as group product, called the *automorphism group* of $\mathfrak{g}$.

A *derivation* of a Lie algebra $\mathfrak{g}$ is a linear map $\delta : \mathfrak{g} \longrightarrow \mathfrak{g}$ satisfying

$$\delta[x, y] = [\delta x, y] + [x, \delta y] \qquad (x, y \in \mathfrak{g}).$$

The vector space aut $\mathfrak{g}$ of all derivations of $\mathfrak{g}$ is a Lie algebra in the Lie brackets

$$[\delta, \gamma] = \delta\gamma - \gamma\delta.$$

For each element $x \in \mathfrak{g}$, the map ad$(x) : \mathfrak{g} \longrightarrow \mathfrak{g}$ defined by

$$\mathrm{ad}(x)(y) = [x, y] \qquad (y \in \mathfrak{g})$$

is a derivation of $\mathfrak{g}$, and the Jacobi identity implies that the map

$$\mathrm{ad} : \mathfrak{g} \longrightarrow \mathrm{aut}\,\mathfrak{g}$$

is a homomorphism, called the *adjoint representation* of $\mathfrak{g}$. The kernel of ad is the *centre* of $\mathfrak{g}$:

$$\mathfrak{z}(\mathfrak{g}) = \{x \in \mathfrak{g} : [x, y] = 0 \;\forall y \in \mathfrak{g}\}.$$

The range of ad, denoted by ad $\mathfrak{g}$, is an ideal of aut $\mathfrak{g}$, since

$$[\delta, \mathrm{ad}(x)] = \mathrm{ad}(\delta x) \qquad (\delta \in \mathrm{aut}\,\mathfrak{g}, x \in \mathfrak{g}).$$

The elements of ad $\mathfrak{g}$ are called the *inner derivations* of $\mathfrak{g}$.

A Lie algebra $\mathfrak{g}$ is called *solvable* if its *derived series*

$$\mathfrak{g} \supset \mathfrak{g}^{(1)} = [\mathfrak{g}, \mathfrak{g}] \supset \mathfrak{g}^{(2)} = [\mathfrak{g}^{(1)}, \mathfrak{g}^{(1)}] \supset \cdots \supset \mathfrak{g}^{(n+1)} = [\mathfrak{g}^{(n)}, \mathfrak{g}^{(n)}] \supset \cdots$$

eventually terminates, that is, $\mathfrak{g}^{(n)} = \{0\}$ for some n. We refer to Kaplansky [68, theorem 33] for a proof of the following criterion of solvability due to Cartan.

Theorem 1.3.2 *Let $\mathfrak{g}$ be a Lie algebra of real or complex matrices. Then $\mathfrak{g}$ is solvable if* Trace $(AB) = 0$ *for all $A, B \in \mathfrak{g}$.*

Given a finite-dimensional Lie algebra $\mathfrak{g}$, the symmetric bilinear form

$$\beta : \mathfrak{g} \times \mathfrak{g} \longrightarrow \mathbb{F} \qquad (\mathbb{F} = \mathbb{R} \text{ or } \mathbb{C})$$

defined by

$$\beta(x, y) = \mathrm{Trace}\,(\mathrm{ad}(x)\mathrm{ad}(y))$$

is called the *Killing form* of $\mathfrak{g}$. The quadratic form $\beta(x, x)$ is known as the *Casimir polynomial* of $\mathfrak{g}$. The Killing form β is *invariant*; that is,

$$\beta([x, y], z) = \beta(x, [y, z]) \qquad (x, y, z \in \mathfrak{g}),$$

which is equivalent to

$$\beta([x, y], z) + \beta(y, [x, z]) = 0 \qquad (x, y, z \in \mathfrak{g}).$$

The latter condition says that $\mathrm{ad}(x)$ is skew-symmetric with respect to β.

A Lie algebra $\mathfrak{g}$ is called *semisimple* if $\mathfrak{g}$ contains no nonzero abelian ideal, which is equivalent to the condition that $\mathfrak{g}$ contains no nonzero solvable ideal. We prove the Cartan–Killing criterion for semisimplicity as follows.

Theorem 1.3.3 *A finite-dimensional Lie algebra $\mathfrak{g}$ is semisimple if, and only if, its Killing form β is non-degenerate.*

Proof Let the Killing form $\beta(x, y) = \mathrm{Trace}\,(\mathrm{ad}(x)\mathrm{ad}(y))$ be non-degenerate. Let $\mathfrak{h}$ be an abelian ideal of $\mathfrak{g}$. We show that $\mathfrak{h} = \{0\}$. Let $x \in \mathfrak{h}$ and let $y \in \mathfrak{g}$.

Since $\mathfrak{h}$ is an ideal, we have $\mathrm{ad}(x)\mathrm{ad}(y)(\mathfrak{g}) \subset \mathfrak{h}$ and hence

$$(\mathrm{ad}(x)\mathrm{ad}(y))^2(\mathfrak{g}) \subset \mathrm{ad}(x)\mathrm{ad}(y)(\mathfrak{h}) \subset [x, \mathfrak{h}] = \{0\},$$

since $\mathfrak{h}$ is abelian. This shows that the linear map $\mathrm{ad}(x)\mathrm{ad}(y) : \mathfrak{g} \longrightarrow \mathfrak{g}$ is nilpotent and therefore $\mathrm{Trace}\,(\mathrm{ad}(x)\mathrm{ad}(y)) = 0$. Non-degeneracy of β gives $x = 0$.

Conversely, let $\mathfrak{g}$ be semisimple. Let

$$\mathfrak{k} = \{x \in \mathfrak{g} : \beta(x, \mathfrak{g}) = \{0\}\}.$$

We need to show $\mathfrak{k} = \{0\}$. Since β is invariant, $\mathfrak{k}$ is an ideal of $\mathfrak{g}$. We have $\mathrm{Trace}\,(AB) = 0$ for $A, B \in \mathrm{ad}(\mathfrak{k})$. It follows from Cartan's solvability criterion that $\mathrm{ad}(\mathfrak{k})$ is a solvable ideal in $\mathrm{aut}\,\mathfrak{g}$. Semisimplicity of $\mathfrak{g}$ implies that the homomorphism ad has zero kernel and therefore $\mathfrak{k}$ is a solvable ideal in $\mathfrak{g}$. Hence $\mathfrak{k} = \{0\}$ by semisimplicity again. $\qquad \square$

The Tits–Kantor–Koecher construction originally relates Jordan algebras to finitely graded Lie algebras. This construction has been extended to Jordan triple systems by Meyberg [90]. We will describe the construction for Jordan triple systems (see also [22]). Let $\mathbb{Z}$ be the ring of integers. By a $\mathbb{Z}$-grading of a Lie algebra $\mathfrak{g}$, we mean a decomposition of $\mathfrak{g}$ into a direct sum of vector subspaces,

$$\mathfrak{g} = \bigoplus_{n \in \mathbb{Z}} g_n,$$

such that $[g_n, g_m] \subset g_{n+m}$. The grading is said to be *finite* if the set $\{n : g_n \neq 0\}$ is finite. It is said to be *nontrivial* if $\oplus_{n \neq 0} g_n \neq 0$. Lie algebras with a nontrivial finite $\mathbb{Z}$-grading have been classified by Zelmanov [122], in which the TKK construction plays an important part. If $A \subset \mathbb{Z}$, a Lie algebra

$$\mathfrak{g} = \bigoplus_{\alpha \in A} \mathfrak{g}_\alpha$$

is said to be graded if $[\mathfrak{g}_\alpha, \mathfrak{g}_\beta] \subset \mathfrak{g}_{\alpha+\beta}$, where $\mathfrak{g}_{\alpha+\beta} = \{0\}$ if $\alpha + \beta \notin A$. There is a one-to-one correspondence between 3-graded Lie algebras $\mathfrak{g}_{-1} \oplus \mathfrak{g}_0 \oplus \mathfrak{g}_1$ and Jordan pairs [87]. We consider only 3-graded Lie algebras with an involution, called the Tits–Kantor–Koecher Lie algebras, because of their connections to symmetric spaces.

By an *involutive Lie algebra* $(\mathfrak{g}, \theta)$, we mean a Lie algebra $\mathfrak{g}$ equipped with an involution θ (i.e., an involutive automorphism $\theta : \mathfrak{g} \longrightarrow \mathfrak{g}$). We will always denote the 1-eigenspace of θ by $\mathfrak{k}$, and by $\mathfrak{p}$ the (-1)-eigenspace of θ such that $\mathfrak{g}$ has the decomposition $\mathfrak{g} = \mathfrak{k} \oplus \mathfrak{p}$, where

$$[\mathfrak{k}, \mathfrak{k}] \subset \mathfrak{k}, \quad [\mathfrak{p}, \mathfrak{p}] \subset \mathfrak{k} \quad \text{and} \quad [\mathfrak{k}, \mathfrak{p}] \subset \mathfrak{p}.$$

If $(\mathfrak{g}, \theta)$ is finite-dimensional, the involution θ is called a *Cartan involution* if the symmetric bilinear form β_θ defined by

$$\beta_\theta(x, y) = -\beta(x, \theta y) \qquad (x, y \in \mathfrak{g})$$

is positive definite.

Example 1.3.4 Let V be a normed vector space and $\mathfrak{gl}(V)$ the normed algebra of continuous linear self-maps on V. Then $\mathfrak{gl}(V)$ is a Lie algebra in the usual Lie brackets,

$$[X, Y] = XY - YX \qquad (X, Y \in \mathfrak{gl}(V)).$$

If $n = \dim V < \infty$, we often denote $\mathfrak{gl}(V)$ as $\mathfrak{gl}(n, \mathbb{F})$ where $\mathbb{F} = \mathbb{R}$ or $\mathbb{C}$.

If V is a Hilbert space, then the subspace $\mathfrak{gl}_{hs}(V)$ of $\mathfrak{gl}(V)$, consisting of all Hilbert–Schmidt operators, is an ideal and is equipped with a natural complete inner product,

$$\langle X, Y \rangle_2 = \mathrm{Trace}\,(XY^*) \qquad (X, Y \in \mathfrak{gl}_{hs}(V)).$$

Of course, $\mathfrak{gl}_{hs}(V) = \mathfrak{gl}(V)$ if $\dim V < \infty$.

We can define an involution $\theta : \mathfrak{gl}(V) \longrightarrow \mathfrak{gl}(V)$ by

$$\theta(X) = -X^* \qquad (X \in \mathfrak{gl}(V)),$$

where $X^* : V \longrightarrow V$ denotes the adjoint operator of X. If $\dim V < \infty$, then θ is a Cartan involution, since

$$-\beta(X, \theta X) = -\mathrm{Trace}\,(\mathrm{ad}(X)\mathrm{ad}(\theta X)) = \mathrm{Trace}\,(\mathrm{ad}(X)\mathrm{ad}(X^*))$$
$$= \mathrm{Trace}\,(\mathrm{ad}(X)\mathrm{ad}(X)^*) \geq 0,$$

where $\mathrm{ad}\,(X)^* : \mathfrak{gl}(V) \longrightarrow \mathfrak{gl}(V)$ is the adjoint operator of $\mathrm{ad}\,(X)$ with respect to the inner product $\langle \cdot, \cdot \rangle_2$.

In fact, the Killing form of $\mathfrak{gl}(V)$ can be computed explicitly, it is given by

$$\beta(X, Y) = 2n\,\mathrm{Trace}\,(XY) - 2\,\mathrm{Trace}\,(X)\mathrm{Trace}\,(Y) \qquad (X, Y \in \mathfrak{gl}(V)).$$

Definition 1.3.5 A graded Lie algebra $\mathfrak{g} = \mathfrak{g}_{-1} \oplus \mathfrak{g}_0 \oplus \mathfrak{g}_1$ is called a *Tits–Kantor–Koecher Lie algebra* (*TKK Lie algebra*) if $\mathfrak{g}$ admits an involution θ satisfying

$$\theta(\mathfrak{g}_\alpha) = \mathfrak{g}_{-\alpha}.$$

We call $\mathfrak{g}$ *canonical* if $[\mathfrak{g}_{-1}, \mathfrak{g}_1] = \mathfrak{g}_0$.

We define the *canonical* part of $\mathfrak{g}$ to be the Lie subalgebra

$$\mathfrak{g}^c = \mathfrak{g}_{-1} \oplus [\mathfrak{g}_{-1}, \mathfrak{g}_1] \oplus \mathfrak{g}_1,$$

which is also a TKK Lie algebra with the restriction of θ as its involution.

The *symmetric part* of $\mathfrak{g}$ is defined to be the following Lie subalgebra:

$$\mathfrak{g}_s = \{a \oplus h \oplus -\theta a : a \in \mathfrak{g}_{-1}, \theta h = h \in \mathfrak{g}_0\},$$

where

$$[a \oplus h \oplus -\theta a, \, b \oplus k \oplus -\theta b]$$
$$= ([a, k] + [h, b]) \oplus [h, k] \oplus ([-\theta a, k] - [h, \theta b]).$$

The restriction of θ to $\mathfrak{g}_s$ is an involution. The Lie subalgebra

$$\mathfrak{g}_s^* = \{a \oplus h \oplus \theta a : a \in \mathfrak{g}_{-1}, \theta h = h \in \mathfrak{g}_0\}$$

is called the *dual symmetric part* of $\mathfrak{g}$, which is the 1-eigenspace of θ.

We define the *dual involution* θ^* on $\mathfrak{g}$ by $\theta^*(a \oplus h \oplus b) = -\theta b \oplus \theta h \oplus -\theta a$, which restricts to an involution on $\mathfrak{g}_s^*$.

Remark 1.3.6 With the dual involution, $(\mathfrak{g}, \theta^*)$ is also a TKK Lie algebra and $\mathfrak{g}_s^*$ now becomes the symmetric part of $(\mathfrak{g}, \theta^*)$.

We will now show the correspondence between Jordan triple systems and TKK Lie algebras. We first consider non-degenerate Jordan triples.

Lemma 1.3.7 *Let V be a non-degenerate Jordan triple system and let $\sum_j a_j \,\square\, b_j = \sum_k u_k \,\square\, v_k$. Then we have $\sum_j b_j \,\square\, a_j = \sum_k v_k \,\square\, u_k$.*

Proof We have

$$\left[\sum_j a_j \,\square\, b_j, \, x \,\square\, y \right] = \left(\sum_j (a_j \,\square\, b_j) x \right) \,\square\, y - x \,\square\, \left(\sum_j b_j \,\square\, a_j \right) y$$
$$= \left[\sum_k u_k \,\square\, v_k, \, x \,\square\, y \right] = \left(\sum_k (u_k \,\square\, v_k) x \right) \,\square\, y - x \,\square\, \left(\sum_k v_k \,\square\, u_k \right) y,$$

which gives $x \,\square\, \left(\sum_j b_j \,\square\, a_j \right) y = x \,\square\, \left(\sum_k v_k \,\square\, u_k \right) y$ for all $x, y \in V$. By Lemma 1.2.20, we conclude that $\sum_j b_j \,\square\, a_j = \sum_k v_k \,\square\, u_k$. $\qquad\square$

Theorem 1.3.8 *Let V be a non-degenerate Jordan triple. Then there is a canonical Tits–Kantor–Koecher Lie algebra $\mathfrak{L}(V)$ with grading*

$$\mathfrak{L}(V) = \mathfrak{L}(V)_{-1} \oplus \mathfrak{L}(V)_0 \oplus \mathfrak{L}(V)_1$$

and an involution θ such that $\mathfrak{L}(V)_{-1} = V = \mathfrak{L}(V)_1$ and

$$\{x, y, z\} = [[x, \theta y], \, z]$$

for $x, y, z \in \mathfrak{L}(V)_{-1}$.

Proof Form the algebraic direct sum

$$\mathfrak{L}(V) = V_{-1} \oplus V_0 \oplus V_1,$$

where $V_{-1} = V_1 = V$ and V_0 is the linear span of $V \,\square\, V$ in the space $L(V)$ of linear self-maps on V. The Jordan triple identity (1.26) implies that V_0 is a Lie algebra in the bracket product

$$[h, k] = hk - kh.$$

By Lemma 1.3.7, the mapping

$$x \,\square\, y \in V \,\square\, V \mapsto y \,\square\, x \in V \,\square\, V$$

is well defined and extends to an involution $^\natural : V_0 \longrightarrow V_0$ satisfying

$$[x \,\square\, y, u \,\square\, v]^\natural = -[y \,\square\, x, v \,\square\, u].$$

This enables us to define an involutive automorphism $\theta : \mathfrak{L}(V) \longrightarrow \mathfrak{L}(V)$ by

$$\theta(x \oplus h \oplus y) = y \oplus -h^\natural \oplus x \qquad (x \oplus h \oplus y \in V_{-1} \oplus V_0 \oplus V_1),$$

where we also write (x, h, y) for $x \oplus h \oplus y$, x for $(x, 0, 0)$, $\overline{y}$ for $(0, 0, y)$, and h for $(0, h, 0)$ if there is no confusion. By identifying V_α naturally as subspaces of $\mathfrak{L}(V)$, we see immediately that $\theta(V_\alpha) = V_{-\alpha}$ for $\alpha = 0, \pm 1$.

We show that $\mathfrak{L}(V)$ is a Lie algebra in the following product:

$$[x \oplus h \oplus y, u \oplus k \oplus v]$$
$$= (h(u) - k(x), [h, k] + x \,\square\, v - u \,\square\, y, k^\natural(y) - h^\natural(v)).$$

Given $x \in V_{-1}$ and $\overline{y} \in V_1$, we have $[x, \overline{y}] = [(x, 0, 0), (0, 0, y)] = (0, x \,\square\, y, 0)$ and hence $[V_{-1}, V_1] = V_0$. We also have $[V_{-1}, V_{-1}] = [V_1, V_1] = 0$.

For $x, y, z \in V_{-1}$, we have

$$\{x, y, z\} = [[x, \theta y], z],$$

and the properties of the Jordan triple product translate into the Jacobi identity. Indeed, the bracket product has zero square and the Jacobian $J(x, y, z) = [[x, y], z] + [[y, z], x] + [[z, x], y]$ alternates, where for $h \in V \,\square\, V$, the preceding remarks imply that $J(h, V_{-1}, V_{-1}) = J(h, V_1, V_1) = J(V_{-1}, V_{-1}, V_{-1}) = J(V_1, V_1, V_1) = J(h, h, V_{-1}) = J(h, h, V_1) = 0$. Hence Jacobi identity holds if and only if

$$J(h, V_{-1}, V_1) = J(V_{-1}, V_{-1}, V_1) = J(V_{-1}, V_1, V_1) = 0,$$

where $J(h, V_{-1}, V_1) = 0$ is the same as

$$[h, u \,\square\, v] = h(u) \,\square\, v + u \,\square\, (\theta h)(v) \qquad ((u, 0, 0) \in V_{-1}, (0, 0, v) \in V_1),$$

which holds since, putting $h = x \,\square\, y$, this identity is just the Jordan triple identity

$$[x \,\square\, y, u \,\square\, v] = \{x, y, u\} \,\square\, v - u \,\square\, \{y, x, v\}.$$

Likewise, the vanishing of $J(V_{-1}, V_{-1}, V_1)$ and $J(V_{-1}, V_1, V_1)$ follows from the symmetry of the Jordan triple product.

Finally, $\mathfrak{L}(V)$ is canonical since $[V_{-1}, V_1] = V_0$.   □

Remark 1.3.9 The involution θ in the TKK Lie algebra $\mathfrak{g} = \mathfrak{L}(V)$ is the unique involution satisfying

$$a \,\square\, b = [a, \theta b] = -\theta(b \,\square\, a) \qquad (a, b \in V_{-1} = V)$$

and is called the *main involution* (cf. [77, p. 793]). Indeed, given an involution σ satisfying $\sigma(\mathfrak{g}_\alpha) = \mathfrak{g}_{-\alpha}$ and $x \,\square\, y = [x, \sigma y] = -\sigma(y \,\square\, x)$, for $x, y \in \mathfrak{g}_{-1} = V$, we have $\sigma \theta y \in \mathfrak{g}_{-1}$ and $x \,\square\, \sigma \theta y = [x, \sigma^2 \theta y] = [x, \theta y] = x \,\square\, y$, which implies $\sigma \theta$ is the identity on $\mathfrak{g}_{-1}$, by Lemma 1.2.20. We also have $\sigma = \theta$ on $\mathfrak{g}_0$, since $\sigma(x \,\square\, y) = -y \,\square\, x = \theta(x \,\square\, y)$. Finally, for $x \in \mathfrak{g}_1$, we have $\theta x \in \mathfrak{g}_{-1}$ and $\sigma \theta x = \theta^2 x = x$ and hence $\sigma = \theta$ on $\mathfrak{g}_1$.

The *dual involution* $\theta^* : \mathfrak{L}(V) \longrightarrow \mathfrak{L}(V)$ is given by

$$\theta^*(x \oplus h \oplus y) = -y \oplus -h^\natural \oplus -x$$

and we have $a \,\square\, b = -[a, \theta^* b]$.

Remark 1.3.10 The previous construction translates the non-degeneracy of a Jordan triple V into the following property of its TKK Lie algebra $\mathfrak{L}(V)$:

$$[[a, \theta y], a] = 0 \quad \text{for all } a, y \in \mathfrak{L}(V)_{-1} \Longrightarrow a = 0,$$

which is equivalent to the condition

$$(\operatorname{ad} a)^2 = 0 \Longrightarrow a = 0 \qquad (a \in \mathfrak{L}(V)_{-1}) \tag{1.38}$$

since $(\operatorname{ad} a)^2(x \oplus h \oplus y) = -Q_a(y)$ for $a \in \mathfrak{L}(V)_{-1}$ and $x \oplus h \oplus y \in \mathfrak{L}(V)$.

A TKK Lie algebra $\mathfrak{g} = \mathfrak{g}_{-1} \oplus \mathfrak{g}_0 \oplus \mathfrak{g}_1$ is called *non-degenerate* if $(\operatorname{ad} a)^2 = 0 \Longrightarrow a = 0$ for $a \in \mathfrak{g}_{-1}$. We now show the one–one correspondence between non-degenerate Jordan triples and non-degenerate TKK Lie algebras.

Two TKK Lie algebras $(\mathfrak{g}, \theta)$ and $(\mathfrak{g}', \theta')$ are said to be *isomorphic* if there is a graded isomorphism $\psi : \mathfrak{g} \longrightarrow \mathfrak{g}'$ which commutes with involutions:

$$\psi \theta = \theta' \psi.$$

Given a TKK Lie algebra $\mathfrak{g} = \mathfrak{g}_{-1} \oplus \mathfrak{g}_0 \oplus \mathfrak{g}_1$ with involution θ, we can identify $\mathfrak{g}_1$ with $\mathfrak{g}_{-1}$ by θ. In fact, every TKK Lie algebra $(\mathfrak{g}, \theta)$ is isomorphic to, and hence identified with, a TKK Lie algebra

$$\mathfrak{g}' = \mathfrak{g}'_{-1} \oplus \mathfrak{g}'_0 \oplus \mathfrak{g}'_1$$

in which $\mathfrak{g}'_{-1} = \mathfrak{g}'_1 = \mathfrak{g}_{-1}$ and $\mathfrak{g}'_0 = \mathfrak{g}_0$, with involution $\theta'(x \oplus h \oplus y) = y \oplus \theta h \oplus x$ and product $[\cdot, \cdot]'$ defined by

$$[x, y]' = [x, \theta y], \quad [h, y]' = (0, 0, [\theta h, y]) \qquad ((x, h, y) \in \mathfrak{g}'_{-1} \times \mathfrak{g}'_0 \times \mathfrak{g}'_1)$$

but otherwise identical to the product $[\cdot, \cdot]$ of $\mathfrak{g}$. The graded isomorphism $\psi : \mathfrak{g} \longrightarrow \mathfrak{g}'$ is given by $\psi(x \oplus h \oplus y) = x \oplus h \oplus \theta y$.

With this identification, let $\mathcal{G}$ be the category of non-degenerate canonical TKK Lie algebras in which the morphisms are graded isomorphisms commuting with involutions. Let $\mathcal{V}$ be the category of non-degenerate Jordan triples in which the morphisms are triple isomorphisms.

Theorem 1.3.11 *For each V in the category $\mathcal{V}$ of non-degenerate Jordan triples, let $\mathfrak{L}(V) \in \mathcal{G}$ be the TKK Lie algebra constructed in Theorem 1.3.8. Then $\mathfrak{L} : \mathcal{V} \longrightarrow \mathcal{G}$ is an equivalence of the two categories $\mathcal{V}$ and $\mathcal{G}$.*

Proof Given a triple isomorphism $\varphi : V \longrightarrow V'$ between two non-degenerate Jordan triples, we have

$$\varphi a \,\square\, \varphi b = \varphi(a \,\square\, b)\varphi^{-1} \qquad (a, b \in V).$$

Hence there is a graded isomorphism $\widetilde{\varphi} : (\mathfrak{L}(V), \theta) \longrightarrow (\mathfrak{L}(V'), \theta')$ defined by

$$\widetilde{\varphi}(a \oplus h \oplus b) = \varphi a \oplus \varphi h \varphi^{-1} \oplus \varphi b$$

which satisfies

$$\widetilde{\varphi}\theta = \theta'\widetilde{\varphi}.$$

Conversely, given a non-degenerate TKK Lie algebra $\mathfrak{g} = \mathfrak{g}_{-1} \oplus \mathfrak{g}_0 \oplus \mathfrak{g}_1$ with involution θ and $\mathfrak{g}_1 = \mathfrak{g}_{-1}$, we let $V = \mathfrak{g}_{-1}$. Then it follows from the Jacobi identity and $[\mathfrak{g}_{-1}, \mathfrak{g}_{-1}] = 0$ that V is a non-degenerate Jordan triple with the Jordan triple product defined by

$$\{x, y, z\} = [[x, \theta y], z]$$

and we have $\mathfrak{L}(V) = \mathfrak{g}$ if $\mathfrak{g}$ is canonical.

If $\psi : (\mathfrak{L}(V), \theta) \longrightarrow (\mathfrak{L}(V'), \theta')$ is a graded isomorphism satisfying $\psi\theta = \theta'\psi$, then the restriction $\psi|_V : V \longrightarrow V'$ defines a triple isomorphism. $\square$

We note that TKK Lie algebras are reduced. We recall that an involutive Lie algebra $(\mathfrak{g}, \theta)$ with eigenspace decomposition $\mathfrak{g} = \mathfrak{k} \oplus \mathfrak{p}$ is *reduced* if $\mathfrak{k}$ does not contain any nonzero ideal of $\mathfrak{g}$, which is equivalent to the condition that the isotropy representation $\mathrm{ad}_{\mathfrak{k}} : X \in \mathfrak{k} \mapsto ad\, X|_{\mathfrak{p}} \in End\,(\mathfrak{p})$ is faithful (cf. [11, p. 21]).

Lemma 1.3.12 *The TKK Lie algebra* $(\mathfrak{L}(V), \theta)$ *of a non-degenerate Jordan triple* V *is reduced.*

Proof Let $\mathfrak{L}(V) = \mathfrak{k} \oplus \mathfrak{p}$ be the decomposition into eigenspaces of θ, where

$$\mathfrak{k} = \{u \oplus h \oplus u : u \in V, \theta h = h\}.$$

Let $X = u \oplus h \oplus u \in \mathfrak{k}$ be such that $[X, Y] = 0$ for all $Y \in \mathfrak{p}$. For each $g \in V_0$ satisfying $\theta g = -g$, and for each $v \in V$, we have

$$[u \oplus h \oplus u, v \oplus g \oplus -v] = 0$$
$$= (hv - gu) \oplus ([h, g] - u \,\square\, v - v \,\square\, u) \oplus (gu - hv),$$

which gives $hv = gu$ and in particular $hv = 0$ for all $v \in V$ if $g = 0$. Hence $h = 0$ and $u \,\square\, v + v \,\square\, u = 0$ for all $v \in V$. Choose $Y = (0, g, 0)$ with $g = v \,\square\, v$; then $(v \,\square\, v)(u) = 0$ and hence $(v \,\square\, u)(v) = -(u \,\square\, v)(v) = -(v \,\square\, v)(u) = 0$ for all $v \in V$. This implies $v \,\square\, u = 0$ for all $v \in V$, since $\{x + v, u, x + v\} = 0$ for all $x, v \in V$. Therefore $u = 0$ by Lemma 1.2.20, which proves $X = 0$. $\square$

Proposition 1.3.13 *Let* $(\mathfrak{L}(V), \theta)$ *be the TKK Lie algebra of a non-degenerate Jordan triple* V, *with eigenspace decomposition* $\mathfrak{L}(V) = \mathfrak{k} \oplus \mathfrak{p}$ *of* θ. *Then the centralizer*

$$\mathfrak{z}(\mathfrak{p}) = \{X \in \mathfrak{L}(V) : [X, \mathfrak{p}] = 0\}$$

is trivial.

Proof We have $\mathfrak{z}(\mathfrak{p}) = \mathfrak{z}(\mathfrak{p}) \cap \mathfrak{k} \oplus \mathfrak{z}(\mathfrak{p}) \cap \mathfrak{p}$. By Ji [63, lemma 4.2], $\mathfrak{z}(\mathfrak{p}) \cap \mathfrak{k}$ is an ideal of $\mathfrak{L}(V)$. By Lemma 1.3.12, $\mathfrak{L}(V)$ is reduced. Hence $\mathfrak{z}(\mathfrak{p}) \cap \mathfrak{k} = \{0\}$ and $\mathfrak{z}(\mathfrak{p}) \subset \mathfrak{p}$.

We show that $\mathfrak{z}(\mathfrak{p}) = \{0\}$. Fix $X = a \oplus h \oplus -a \in \mathfrak{z}(\mathfrak{p})$ where $\theta h = -h$. For $u \oplus p \oplus -u \in \mathfrak{p}$, we have

$$0 = [a \oplus h \oplus -a, u \oplus p \oplus -u]$$
$$= (hu - pa, [h, p] - a \,\square\, u + u \,\square\, a, hu - pa).$$

Choose $p = 0$; then $hu = pa = 0$ for all $u \in V$ and hence $h = 0$ and $a \,\square\, u = u \,\square\, a$. If we choose $p = u \,\square\, u$, then $u \,\square\, u(a) = 0$ implies $u \,\square\, a(u) = $

$a \,\square\, u(u) = 0$ for all $u \in V$. By Lemma 1.2.21, we have $a = 0$. This proves $\mathfrak{z}(\mathfrak{p}) = \{0\}$. $\square$

To remove the non-degeneracy assumption in Theorem 1.3.8, the TKK construction given in the proof reveals that all one needs is a well-defined involution θ for arbitrary Jordan triples V. To achieve this, we replace the previous involution $\natural : x \,\square\, y \mapsto y \,\square\, x$ by the involution $(x \,\square\, y, y \,\square\, x) \mapsto (y \,\square\, x, x \,\square\, y)$, which is well defined for any V. Motivated by this, we define an *inner derivation pair* d_{xy} by

$$d_{xy} = (x \,\square\, y, -y \,\square\, x) \in V_0 \times V_0 \qquad (x, y \in V)$$

for any Jordan triple system V, where $V_0 \times V_0$ is a Lie algebra in the coordinatewise bracket product. Since

$$[d_{ab}, d_{xy}] = d_{\{abx\}y} - d_{x\{bay\}},$$

the linear span V_{00} of

$$\{d_{xy} : x, y \in V\}$$

is a Lie subalgebra of $V_0 \times V_0$. The Lie algebra V_{00} can be written in the form

$$V_{00} = \{(h^+, h^-) : h^+ = \sum_j a_j \,\square\, b_j, \ h^- = -\sum_j b_j \,\square\, a_j, \ a_j, b_j \in V\}.$$

We write $h = (h^+, h^-)$ for an element in V_{00}.

To construct the TKK Lie algebra $\mathfrak{L}(V)$ from any Jordan triple V, we form

$$\mathfrak{L}(V) = V \oplus V_{00} \oplus V \qquad (1.39)$$

and define the Lie product

$$\begin{aligned}
&[x \oplus (h^+, h^-) \oplus y, \ u \oplus (k^+, k^-) \oplus v] \\
&= (h^+(u) - k^+(x), \ [(h^+, h^-), (k^+, k^-)] + d_{xv} - d_{uy}, \ h^-(v) - k^-(y)).
\end{aligned}$$
$$(1.40)$$

We define the main involution $\theta : \mathfrak{L}(V) \longrightarrow \mathfrak{L}(V)$ by

$$\theta(x \oplus (h^+, h^-) \oplus y) = y \oplus (h^-, h^+) \oplus x.$$

Then $(\mathfrak{L}(V), \theta)$ is a canonical TKK Lie algebra with $\mathfrak{L}(V)_{-1} = \mathfrak{L}(V)_1 = V$ and $\mathfrak{L}(V)_0 = V_{00}$ such that

$$\{x, y, z\} = [[x, \theta y], z] \qquad (x, y, z \in V_{-1}).$$

Conversely, given a TKK Lie algebra $\mathfrak{g} = \mathfrak{g}_{-1} \oplus \mathfrak{g}_0 \oplus \mathfrak{g}_1$ with involution θ and $\mathfrak{g}_1 = \mathfrak{g}_{-1}$, $V = \mathfrak{g}_{-1}$ is a Jordan triple with triple product

$$\{x, y, z\} = [[x, \theta y], z] \qquad (x, y, z \in \mathfrak{g}_{-1})$$

and $\mathfrak{L}(V) = \mathfrak{g}$ if $\mathfrak{g}$ is canonical.

We note that the above construction for an arbitrary Jordan triple V is identical to the one in Theorem 1.3.8 if V is non-degenerate since, in this case, V_0 identifies with V_{00} via the map

$$\sum_j a_j \square b_j \in V_0 \mapsto \left(\sum_j a_j \square b_j, -\sum_j b_j \square a_j \right) \in V_{00},$$

which is well defined by Lemma 1.3.7.

1.4 Matrix Lie groups

A remarkable feature in Lie theory is that one can study a curved object such as a Lie group via a flat space, namely, a Lie algebra. The close relationship between Jordan and Lie structures having been seen, it should not be surprising that Jordan theory has played a useful role in geometry. We are going to show that the Lie algebras of many familiar classical Lie groups are TKK Lie algebras and can therefore be constructed from Jordan triple systems. The Lie groups to be discussed are the general linear groups and their subgroups $SL(n, \mathbb{R})$, $SL(n, \mathbb{C})$, $O(n)$, $SO(n)$, $U(n)$, $SU(n)$ and $Sp(n)$, which are matrix Lie groups. Infinite-dimensional Lie groups will be considered later.

We denote by $GL(n, \mathbb{R})$ the general linear group of invertible $n \times n$ real matrices, and by $GL(n, \mathbb{C})$ the general linear group of invertible $n \times n$ complex matrices, with identity I_n. These two groups are the groups of invertible elements, equipped with the relative topology, in the Banach algebras $L(\mathbb{R}^n)$ and $L(\mathbb{C}^n)$ of linear operators on $\mathbb{R}^n$ and $\mathbb{C}^n$, respectively, or equivalently, in the algebras $M_n(\mathbb{R})$ and $M_n(\mathbb{C})$ of $n \times n$ real and complex matrices, respectively. We can consider $GL(n, \mathbb{R})$ as an open subspace of $\mathbb{R}^{n^2}$, with the induced differential structure. Likewise the complex Lie group $GL(n, \mathbb{C})$ can be viewed as a $2n^2$-dimensional real Lie group.

All Lie groups discussed in the following are closed subgroups, and hence Lie subgroups, of the general linear group and are equipped with the relative topology. The *special linear groups* $SL(n, \mathbb{R})$ and $SL(n, \mathbb{C})$ are the subgroups of $GL(n, \mathbb{R})$ and $GL(n, \mathbb{C})$, respectively, consisting of matrices with determinant 1.

The *orthogonal group* $O(n)$ of $n \times n$ orthogonal real matrices is the compact group of linear isometries of $\mathbb{R}^n$. The connected identity component of $O(n)$ is the *special orthogonal group* $SO(n)$ consisting of matrices in $O(n)$ with determinant 1. The groups $SO(2k + 1)$ are simple and the only nontrivial normal subgroup of $SO(2k)$ is the group $\{\pm I_{2k}\}$.

Unitary operators on a complex Hilbert space H are the surjective linear isometries of H. The unitary operators on $\mathbb{C}^n$ form a compact group, namely, the *unitary group* $U(n)$ of $n \times n$ (complex) unitary matrices. The *special unitary group* is the subgroup $SU(n)$ of $U(n)$ consisting of matrices with determinant 1. Although the orthogonal group $O(n)$ has two connected components, the unitary group $U(n)$ is connected.

Let $\mathbb{H}$ be the quaternions with basis $\{\mathbf{1}, \mathbf{i}, \mathbf{j}, \mathbf{k}\}$ and consider the cartesian product $\mathbb{H}^n$, for $n = 1, 2, \ldots$, as a right $\mathbb{H}$-module with coordinatewise operations. There is a natural positive definite $\mathbb{H}$-Hermitian form

$$\langle \cdot, \cdot \rangle : \mathbb{H}^n \times \mathbb{H}^n \longrightarrow \mathbb{H}$$

given by

$$\langle (a_1, \ldots, a_n), (b_1, \ldots, b_n) \rangle = \sum_{j=1}^{n} a_j \bar{b}_j,$$

which is called the *standard* Hermitian form on $\mathbb{H}^n$. The metric defined by the norm

$$\|(a_1, \ldots, a_n)\| = \sum_{j=1}^{n} a_j \bar{a}_j$$

is complete, and by a slight abuse of language, we call $\mathbb{H}^n$ a *quaternion Hilbert space* or a *Hilbert space over* $\mathbb{H}$, equipped with the standard Hermitian form. More generally, on a right $\mathbb{H}$-module V, a positive definite *Hermitian form* or *symplectic inner product* is a positive definite real bilinear form

$$\langle \cdot, \cdot \rangle : V \times V \longrightarrow \mathbb{H}$$

which is $\mathbb{H}$-linear in the first variable, and conjugate $\mathbb{H}$-linear in the second such that $\langle u, v \rangle = \overline{\langle v, u \rangle}$ for $u, v \in V$. A right $\mathbb{H}$-module V is called an *inner product* $\mathbb{H}$-module if it admits a symplectic inner product, and V is called a *Hilbert space over* $\mathbb{H}$ if the inner product norm $\|v\|^2 = \langle v, v \rangle$ induces a complete metric on V, in which case the $\mathbb{H}$-linear continuous operators from V to itself form a real Banach algebra $L(V)$ in the usual operator norm

$$\|T\| = \sup\{\|T(v)\| : v \in V, \|v\| \leq 1\}.$$

An operator $T \in L(V)$ is called *quaternion unitary* if it is a surjective isometry.

We can consider V as a real Hilbert space with inner product Re $\langle \cdot, \cdot \rangle$. Each $T \in L(V)$ is a continuous linear operator on the real Hilbert space V and there is a real continuous linear operator $t^* : V \longrightarrow V$ satisfying

$$\mathrm{Re}\,\langle u, Tv \rangle = \mathrm{Re}\,\langle t^*u, v \rangle \qquad (u, v \in V).$$

Let

$$T^* = \frac{1}{4}(t^*(\cdot) - \mathbf{i}t^*(\mathbf{i}\cdot) - \mathbf{j}t^*(\mathbf{j}\cdot) - \mathbf{k}t^*(\mathbf{k}\cdot)).$$

Then $T^* \in L(V)$ is the adjoint of T; that is, $\langle u, Tv \rangle = \langle T^*u, v \rangle$ for all $u, v \in V$. The quaternion unitary operators in $L(V)$, with identity operator I, are the ones satisfying the identity $T^*T = TT^* = I$.

One can represent $L(\mathbb{H}^n)$ as the algebra of $n \times n$ matrices with entries in $\mathbb{H}$, acting in the usual way on $\mathbb{H}^n$ on the left, with $\mathbb{H}^n$ regarded as the space of column vectors. The adjoint of a matrix $(a_{ij}) \in L(\mathbb{H}^n)$ is then $(a_{ij})^* = (\bar{a}_{ji})$. The group $GL(n, \mathbb{H})$ of invertible matrices in $L(\mathbb{H}^n)$ is a Lie subgroup of the general linear group $GL(4n, \mathbb{R})$ by identifying $\mathbb{H}$ with $\mathbb{R}^4$ as real vector spaces. To see more details, we represent $\mathbb{H}$, as in (1.4), by the complex matrices

$$\mathbb{H} = \left\{ \begin{pmatrix} a & b \\ -\bar{b} & \bar{a} \end{pmatrix} : (a, b) \in \mathbb{C}^2 \right\} \subset M_2(\mathbb{C}).$$

Let $\mathbf{j} = \begin{pmatrix} 0 & 1 \\ -1 & 0 \end{pmatrix}$ so that each quaternion in $\mathbb{H}$ can be written as

$$\begin{pmatrix} a & b \\ -\bar{b} & \bar{a} \end{pmatrix} = \begin{pmatrix} a & 0 \\ 0 & \bar{a} \end{pmatrix} + \mathbf{j} \begin{pmatrix} \bar{b} & 0 \\ 0 & b \end{pmatrix}$$

and denoted by $a + \mathbf{j}\bar{b}$. Considering $\mathbb{H} = \mathbb{C}^2$, each matrix A in $GL(1, \mathbb{H}) = \mathbb{H}\backslash\{0\} \subset GL(2, \mathbb{C})$ is a complex linear map $A : \mathbb{C}^2 \to \mathbb{C}^2$ and an $\mathbb{H}$-linear map $A : \mathbb{H} \to \mathbb{H}$. A matrix B in $M_2(\mathbb{C})$ is a complex linear map $B : \mathbb{H} \to \mathbb{H}$. It is $\mathbb{H}$-linear precisely when $B(v\mathbf{j}) = B(v)\mathbf{j}$ for $v = a + \mathbf{j}b \in \mathbb{H}$. When $v = a + \mathbf{j}b$ is regarded as a column vector $\begin{pmatrix} a \\ b \end{pmatrix}$ in $\mathbb{C}^2$, the multiplication $v \mapsto v\mathbf{j} = -\bar{b} + \mathbf{j}\bar{a}$ in $\mathbb{H}$ corresponds to the matrix multiplication

$$\begin{pmatrix} a \\ b \end{pmatrix} \in \mathbb{C}^2 \mapsto \begin{pmatrix} 0 & -1 \\ 1 & 0 \end{pmatrix} \begin{pmatrix} \bar{a} \\ \bar{b} \end{pmatrix} \in \mathbb{C}^2.$$

Therefore $\mathbb{H}$-linearity of B is equivalent to the matrix multiplication

$$B\left(-\mathbf{j}\left(\overline{\begin{pmatrix} a \\ b \end{pmatrix}}\right)\right) = -\mathbf{j}\left(\overline{B\begin{pmatrix} a \\ b \end{pmatrix}}\right) = -\mathbf{j}\bar{B}\begin{pmatrix} \bar{a} \\ \bar{b} \end{pmatrix}.$$

Hence we have

$$GL(1, \mathbb{H}) = \{B \in GL(2, \mathbb{C}) : B\mathbf{j} = \mathbf{j}\overline{B}\} = \{B \in GL(2, \mathbb{C}) : \mathbf{j}B\mathbf{j} = -\overline{B}\}.$$

If we identify $\mathbb{C}^2$ with $\mathbb{R}^4$ via

$$\begin{pmatrix} s + it \\ x + iy \end{pmatrix} \mapsto \begin{pmatrix} s \\ x \\ t \\ y \end{pmatrix},$$

then multiplication in $\mathbb{C}^2$ by i corresponds to the matrix multiplication

$$\begin{pmatrix} s \\ x \\ t \\ y \end{pmatrix} \mapsto \begin{pmatrix} 0 & I_2 \\ -I_2 & 0 \end{pmatrix} \begin{pmatrix} s \\ x \\ t \\ y \end{pmatrix}.$$

It follows that

$$GL(2, \mathbb{C}) = \{A \in GL(4, \mathbb{R}) : A\mathbf{j}_4 = \mathbf{j}_4 A\},$$

where

$$\mathbf{j}_{2n} = \begin{pmatrix} 0 & I_n \\ -I_n & 0 \end{pmatrix} \qquad (n = 1, 2, \ldots).$$

Further, multiplying $\begin{pmatrix} s + it \\ x + iy \end{pmatrix}$ by $\mathbf{j} \in \mathbb{H}$ corresponds to the matrix multiplication

$$-\begin{pmatrix} \mathbf{j}_2 & 0 \\ 0 & -\mathbf{j}_2 \end{pmatrix} \begin{pmatrix} s \\ x \\ t \\ y \end{pmatrix}.$$

Letting

$$\mathbf{J}_{4n} = \begin{pmatrix} \mathbf{j}_{2n} & 0 \\ 0 & -\mathbf{j}_{2n} \end{pmatrix} \qquad (n = 1, 2, \ldots),$$

we have

$$GL(1, \mathbb{H}) = \{A \in GL(4, \mathbb{R}) : A\mathbf{j}_4 = \mathbf{j}_4 A, \ A\mathbf{J}_4 = \mathbf{J}_4 A\}$$
$$= \{A \in GL(4, \mathbb{R}) : A = -\mathbf{j}_4 A\mathbf{j}_4 = -\mathbf{J}_4 A\mathbf{J}_4\}$$

and a quaternion can be represented as a 4×4 real matrix

$$
\begin{pmatrix}
s & t & x & y \\
-t & s & y & -x \\
-x & -y & s & t \\
-y & x & -t & s
\end{pmatrix}.
$$

Extending the preceding identifications $(a, b) \in \mathbb{C}^2 \mapsto a + \mathbf{j}b \in \mathbb{H}$ and $(s, x, t, y) \in \mathbb{R}^4 \mapsto (s + it, x + iy) \in \mathbb{C}^2$ to

$$
(a_1, a_2, \ldots, a_{2n}) \in \mathbb{C}^{2n} \mapsto (a_1 + \mathbf{j}a_{n+1}, \ldots, a_n + \mathbf{j}a_{2n}) \in \mathbb{H}^n = \mathbb{C}^n + \mathbf{j}\mathbb{C}^n
$$

and

$$
(a_1, a_2, \ldots, a_{2n}) \in \mathbb{R}^{2n} \mapsto (a_1 + ia_{n+1}, \ldots, a_n + ia_{2n}) \in \mathbb{C}^n = \mathbb{R}^n + i\mathbb{R}^n,
$$

we arrive at

$$
\begin{aligned}
GL(n, \mathbb{H}) &= \{A \in GL(2n, \mathbb{C}) : A\mathbf{j}_{2n} = \mathbf{j}_{2n}\overline{A}\} \\
&= \{A \in GL(4n, \mathbb{R}) : A = -\mathbf{j}_{4n}A\,\mathbf{j}_{4n} = -\mathbf{J}_{4n}A\mathbf{J}_{4n}\}.
\end{aligned}
$$

In this identification, the standard Hermitian form of $\mathbb{H}^n$ can be expressed as

$$
\begin{aligned}
&\langle(a_1 + \mathbf{j}a_{n+1}, \ldots, a_n + \mathbf{j}a_{2n}), (b_1 + \mathbf{j}b_{n+1}, \ldots, b_n + \mathbf{j}b_{2n})\rangle \\
&= \sum_{\iota=1}^{2n} a_\iota \overline{b}_\iota - \mathbf{j} \sum_{\iota=1}^{n} (a_\iota b_{n+\iota} - a_{n+\iota}b_\iota).
\end{aligned}
$$

Therefore a quaternion matrix in $L(\mathbb{H}^n)$, when regarded as a $2n \times 2n$ complex matrix, preserves the standard Hermitian form if, and only if, it preserves the standard inner product in $\mathbb{C}^{2n}$ as well as the standard skew-symmetric form

$$
((a_1, \ldots, a_{2n}), (b_1, \ldots, b_{2n})) \in \mathbb{C}^{2n} \times \mathbb{C}^{2n} \mapsto \sum_{\iota=1}^{n} (a_\iota b_{n+\iota} - a_{n+\iota}b_\iota).
$$

The compact subgroup $Sp(n)$ of $GL(n, \mathbb{H})$ consisting of quaternion unitary matrices, the ones preserving the standard Hermitian form, is called the *symplectic group*. The symplectic group $Sp(1)$ consists of quaternions $\alpha\mathbf{1} + x\mathbf{i} + y\mathbf{j} + z\mathbf{k} \in \mathbb{H}$ of unit norm and identifies with the unit sphere $S^3 = \{(\alpha, x, y, z) \in \mathbb{R}^4 : \alpha^2 + x^2 + y^2 + z^2 = 1\}$ in the Euclidean space $\mathbb{R}^4$. The unit quaternions are the complex matrices $\begin{pmatrix} a & b \\ -\bar{b} & \bar{a} \end{pmatrix}$ with determinant 1. Therefore we have the identification

$$
S^3 = Sp(1) = SU(2).
$$

The group of complex matrices in $GL(2n, \mathbb{C})$ preserving the standard skew-symmetric form is denoted by $Sp(n, \mathbb{C})$. We have $Sp(n) = Sp(n, \mathbb{C}) \cap SU(2n)$.

The Lie algebras of the classical Lie groups can be described in the following way. Given a real matrix Lie group G as a smooth manifold, let $T_e G$ be the tangent space at the identity $e \in G$, consisting of tangent vectors $X = \gamma'(0)$ at e, where $\gamma : (-c, c) \to G$ is a smooth curve with $c > 0$ and $\gamma(0) = e$. One can refer to Section 2.1 for the basics of differentiable manifolds.

The linear automorphisms of $T_e G$ form an open subspace $GL(T_e G)$ of the space $L(T_e G)$ of linear self-maps on $T_e G$.

For each $g \in G$, let $\ell_g : x \in G \mapsto gx \in G$ and $r_g : x \in G \mapsto xg \in G$ be left and right multiplication by g, respectively. The *adjoint representation* of G,

$$\mathrm{Ad} : G \longrightarrow GL(T_e G),$$

is defined by the differential $d(\ell_g r_{g^{-1}})_e$ of the conjugation map $\ell_g r_{g^{-1}}$ at e,

$$\mathrm{Ad}(g) = d(\ell_g r_{g^{-1}})_e : T_e G \longrightarrow T_e G \qquad (g \in G),$$

which is a linear isomorphism. Taking the differential at e once more, we let

$$\mathrm{ad} = d(\mathrm{Ad})_e : T_e G \longrightarrow L(T_e G).$$

For $X, Y \in T_e G$, we define

$$[X, Y] = \mathrm{ad}(X)(Y).$$

Then $\mathfrak{g} = (T_e G, [\cdot, \cdot])$ is the Lie algebra of G.

Let us now apply this construction to some concrete examples. For the general linear group $GL(n, \mathbb{R})$, each matrix $X \in M_n(\mathbb{R})$ defines a smooth curve $\gamma : (-c, c) \to GL(n, \mathbb{R})$ by

$$\gamma(t) = \exp(tX) = \sum_{n=0}^{\infty} \frac{(tX)^n}{n!}$$

with $\gamma'(0) = X$. Therefore $M_n(\mathbb{R})$ is the tangent space $T_e GL(n, \mathbb{R})$ at the identity $e = I_n$. For each $A \in GL(n, \mathbb{R})$, we have

$$\mathrm{Ad}(A)(X) = \frac{d}{dt}(A\gamma(t)A^{-1})|_{t=0} = A\gamma'(0)A^{-1} = AXA^{-1}.$$

Hence $\mathrm{Ad}(AB) = \mathrm{Ad}(A)\mathrm{Ad}(B)$. Further, for $Y \in M_n(\mathbb{R})$, we have

$$
\begin{aligned}
[X, Y] = \mathrm{ad}(X)(Y) &= \frac{d}{dt}(\mathrm{Ad}(\gamma(t))(Y))|_{t=0} \\
&= \frac{d}{dt}(\gamma(t)Y\gamma(t)^{-1})|_{t=0} \\
&= (\gamma'(t)Y\gamma(t)^{-1} + \gamma(t)Y(-\gamma(t)^{-1}\gamma'(t)\gamma(t)^{-1}))|_{t=0} \\
&= \gamma'(0)Y\gamma(0)^{-1} + \gamma(0)Y(-\gamma(0)^{-1}\gamma'(0)\gamma(0)^{-1}) \\
&= XY - YX
\end{aligned}
$$

and $(M_n(\mathbb{R}), [\cdot, \cdot])$ is indeed a Lie algebra, which is denoted by $\mathfrak{gl}(n, \mathbb{R})$.

For each $A \in GL(n, \mathbb{R})$ and $X, Y \in \mathfrak{gl}(n, \mathbb{R})$, we see plainly that

$$
\mathrm{Ad}(A)[X, Y] = A[X, Y]A^{-1} = [\mathrm{Ad}(A)(X), \mathrm{Ad}(A)(Y)].
$$

Hence $\mathrm{Ad}(A) : \mathfrak{gl}(n, \mathbb{R}) \to \mathfrak{gl}(n, \mathbb{R})$ is a Lie algebra automorphism. Therefore the adjoint representation of $GL(n, \mathbb{R})$ is a group homomorphism,

$$
\mathrm{Ad} : GL(n, \mathbb{R}) \longrightarrow \mathrm{Aut}\,\mathfrak{gl}(n, \mathbb{R}).
$$

We note that the exponential map $\exp : \mathfrak{gl}(n, \mathbb{R}) \longrightarrow GL(n, \mathbb{R})$ is homeomorphic on a neighbourhood of 0. Indeed, it has a local inverse defined on the neighbourhood $\{A \in GL(n, \mathbb{R}) : \|A - I_n\| < 1\}$ by

$$
\log A = \sum_{k=1}^{\infty} (-1)^{k+1} \frac{(A - I_n)^k}{k}.
$$

The Lie algebras of the other classical Lie groups are Lie subalgebras of $\mathfrak{gl}(n, \mathbb{R})$. For each matrix $X \in M_n(\mathbb{C})$, we have

$$
\det(\exp X) = \exp\mathrm{Trace}\,(X) = 1 \quad \text{if and only if} \quad \mathrm{Trace}\,(X) = 0.
$$

Hence $\gamma(t) = \exp(tX) \in SL(n, \mathbb{R})$ for $t \neq 0$ if and only if $\mathrm{Trace}\,(X) = 0$. It follows that the tangent space $T_eSL(n, \mathbb{R})$ at the identity of $SL(n, \mathbb{R})$ is the space

$$
\{X \in M_n(\mathbb{R}) : \mathrm{Trace}\,(X) = 0\},
$$

which is the Lie algebra of $SL(n, \mathbb{R})$, with brackets $[X, Y] = XY - YX$, denoted by $\mathfrak{sl}(n, \mathbb{R})$.

Now consider the tangent space T_eO_n, where $e = I_n$. Let $X \in T_eO_n$ with $X = \gamma'(0)$ for some smooth curve $\gamma : (-c, c) \to O_n$ such that $\gamma(0) = I_n$. Since $\gamma(t)\gamma(t)^T = I_n$, differentiating gives

$$
\gamma'(t)\gamma(t) + \gamma(t)\gamma'(t)^T = 0
$$

and hence $X + X^T = \gamma'(0) + \gamma'(0)^T = 0$; that is, X is skew-symmetric. On the other hand, given a skew-symmetric matrix $X \in M_n$, we have $XX^T = X^T X$ and hence

$$\exp(tX)\exp(tX)^T = \exp(tX)\exp(tX^T) = \exp(t(X + X^T)) = \exp 0 = I_n.$$

We can therefore define a smooth curve $\gamma : (-\varepsilon, \varepsilon) \to O_n$ by $\gamma(t) = \exp(tX)$, which satisfies $\gamma(0) = I_n$ and $\gamma'(0) = X$. It follows that the Lie algebra of O_n is

$$\mathfrak{so}(n, \mathbb{R}) = \{A \in \mathfrak{gl}(n, \mathbb{R}) : A + A^T = 0\},$$

which is also the Lie algebra of $SO(n)$, since in the above construction, the curve $\gamma(t) = \exp(tX)$ actually lies in $SO(n)$ if X is skew-symmetric, by continuity of the function $t \in (-\varepsilon, \varepsilon) \mapsto \det(\exp(tX)) \in \{\pm 1\}$. The real Lie algebra $\mathfrak{so}(n, \mathbb{R})$ has dimension $n(n-1)/2$.

Analogous to $GL(n, \mathbb{R})$, the Lie algebra of $GL(n, \mathbb{C})$ is $M_n(\mathbb{C})$ with the Lie product $[X, Y] = XY - YX$, and we write $\mathfrak{gl}(n, \mathbb{C})$ for the complex Lie algebra $(M_n(\mathbb{C}), [\cdot, \cdot])$. The Lie algebra of $SL(n, \mathbb{C})$ is the Lie subalgebra $\mathfrak{sl}(n, \mathbb{C})$ of $\mathfrak{gl}(n, \mathbb{C})$, consisting of complex matrices with trace 0. The Lie algebra $\mathfrak{gl}(n, \mathbb{H})$ of $GL(n, \mathbb{H})$ is $L(\mathbb{H}^n)$ with the same Lie brackets.

As with the orthogonal group O_n, the tangent vectors of the unitary group $U(n)$ at the identity are the complex skew-Hermitian matrices and the Lie algebra

$$\mathfrak{u}(n) = \{X \in \mathfrak{gl}(n, \mathbb{C}) : X + X^* = 0\}$$

of $U(n)$ is a real Lie algebra of dimension n^2, since each $X \in \mathfrak{u}(n)$ has $n(n-1)/2$ complex entries above the diagonal and n pure imaginary entries on the diagonal. However, in contrast to the case of $SO(n)$, the determinant $\det(\exp tX)$ of a skew-Hermitian matrix X is a complex number of unit modulus and $\det(\exp(tX)) = 1$ if and only if Trace $(X) = 0$. Therefore the Lie algebra of $SU(n)$ is

$$\mathfrak{su}(n) = \{X \in \mathfrak{gl}(n, \mathbb{C}) : \text{Trace }(X) = 0, X + X^* = 0\} = \mathfrak{sl}(n, \mathbb{C}) \cap \mathfrak{u}(n).$$

We have $\dim \mathfrak{su}(n) = n^2 - 1$, since each $X \in \mathfrak{su}(n)$ has trace 0.

We note that the Lie algebras of the two non-isomorphic Lie groups $SU(2)$ and $SO(3)$ can be identified by the isomorphism

$$\begin{pmatrix} ai & b+ci \\ -b+ci & -ai \end{pmatrix} \in \mathfrak{su}(2) \mapsto \begin{pmatrix} 0 & -2a & -2b \\ 2a & 0 & -2c \\ 2b & 2c & 0 \end{pmatrix} \in \mathfrak{so}(3) \quad (a, b, c \in \mathbb{R}).$$

In fact, they both identify with the cross product Lie algebra $\mathbb{R}^3$, with the vector cross product as the Lie algebra product, via

$$\begin{pmatrix} ai & b+ci \\ -b+ci & -ai \end{pmatrix} \in \mathfrak{su}(2) \mapsto (2a, 2b, 2c) \in \mathbb{R}^3.$$

One sees readily that the exponential map $\exp : \mathfrak{su}(2) \longrightarrow SU(2)$ is surjective. Indeed, given

$$A = \begin{pmatrix} a+bi & c+di \\ -c+di & a-bi \end{pmatrix} \in SU(2)\backslash\{I_2\},$$

where $a^2 + b^2 + c^2 + d^2 = 1$, let $\sin\theta = \sqrt{b^2+c^2+d^2}$ and

$$X = \frac{1}{\sqrt{b^2+c^2+d^2}} \begin{pmatrix} bi & c+di \\ -c+di & -bi \end{pmatrix} \in \mathfrak{su}(2).$$

We have $X^2 = -I_2$ and therefore

$$\exp(\theta X) = \left(1 - \frac{\theta^2}{2!} + \cdots\right) I_2 + \left(\theta - \frac{\theta^3}{3!} + \cdots\right) X$$

$$= (\cos\theta)I_2 + (\sin\theta)X = aI_2 + \sqrt{b^2+c^2+d^2}\, X = A.$$

For the symplectic group $Sp(n) = \{A \in GL(n, \mathbb{H}) : AA^* = A^*A = I_n\}$, its Lie algebra is

$$\mathfrak{sp}(n) = \{X \in \mathfrak{gl}(n, \mathbb{H}) : X + X^* = 0\}.$$

Since each $X \in \mathfrak{sp}(n)$ has $n(n-1)/2$ quaternion entries above the diagonal and n pure imaginary quaternion entries on the diagonal, the dimension of the real Lie algebra $\mathfrak{sp}(n)$ is given by

$$\dim \mathfrak{sp}(n) = 2n(n-1) + 3n = 2n^2 + n.$$

When a quaternion matrix $X \in L(\mathbb{H}^n)$ is represented in $M_{2n}(\mathbb{C})$, its adjoint X^* is represented as the usual conjugate transpose in $M_{2n}(\mathbb{C})$. In $M_{4n}(\mathbb{R})$, the adjoint X^* is represented by the transpose. As a Lie algebra of real or complex matrices, we have

$$\mathfrak{sp}(n) = \{X \in \mathfrak{gl}(2n, \mathbb{C}) : X = -X^* = \mathbf{j}_{2n} X^T \mathbf{j}_{2n}\}$$
$$= \{X \in \mathfrak{gl}(4n, \mathbb{R}) : X = -X^T = -\mathbf{j}_{4n} X \mathbf{j}_{4n} = -\mathbf{J}_{4n} X \mathbf{J}_{4n}\}.$$

Most Lie groups and Lie algebras discussed in the above are real. One can complexify a real Lie algebra $\mathfrak{g}$ to form a complex Lie algebra $\mathfrak{g} + i\mathfrak{g}$ in the usual way. For instance, we have

$$\mathfrak{gl}(n, \mathbb{C}) = \mathfrak{u}(n) + i\mathfrak{u}(n),$$

where each complex matrix X can be written as

$$X = \frac{1}{2}(X - X^*) + i\frac{1}{2i}(X + X^*).$$

Likewise, we have $\mathfrak{sl}(n, \mathbb{C}) = \mathfrak{su}(n) + i\mathfrak{su}(n)$. However, by comparing dimensions, there is no similar relationship between $\mathfrak{sp}(n)$ and the Lie algebra $\mathfrak{gl}(n, \mathbb{H})$ of quaternion matrices. The following fact explains why there are fewer complex Lie groups.

Proposition 1.4.1 *Let G be a connected complex Lie group. If G is compact, then it must be abelian.*

Proof We note that the adjoint representation

$$\mathrm{Ad} : G \longrightarrow \mathrm{Aut}\,\mathfrak{g}$$

is holomorphic and therefore must be constant by compactness of G. It follows that

$$d(\ell_g r_{g^{-1}})_e = \mathrm{Ad}(g) = \mathrm{Ad}(e)$$

is the identity map for each $g \in G$. Hence $\ell_g r_{g^{-1}}(\exp X) = \exp(d(\ell_g r_{g^{-1}})_e(X)) = \exp X$ for all $X \in \mathfrak{g} = T_e G$, which implies that G is abelian. $\qquad\square$

Example 1.4.2 The Lie algebra $\mathfrak{gl}(2, \mathbb{R})$ is a TKK Lie algebra. We have the grading

$$\mathfrak{gl}(2, \mathbb{R}) = \left\{ \begin{pmatrix} 0 & x \\ 0 & 0 \end{pmatrix} : x \in \mathbb{R} \right\} \oplus \left\{ \begin{pmatrix} a & 0 \\ 0 & b \end{pmatrix} : a, b \in \mathbb{R} \right\} \oplus \left\{ \begin{pmatrix} 0 & 0 \\ y & 0 \end{pmatrix} : y \in \mathbb{R} \right\}$$

with involution $\theta : \mathfrak{gl}(2, \mathbb{R}) \longrightarrow \mathfrak{gl}(2, \mathbb{R})$ defined by

$$\theta \begin{pmatrix} a & x \\ y & b \end{pmatrix} = \begin{pmatrix} b & y \\ x & a \end{pmatrix}.$$

The canonical part of $\mathfrak{gl}(2, \mathbb{R})$ is

$$\mathfrak{sl}(2, \mathbb{R}) = \left\{ \begin{pmatrix} 0 & x \\ 0 & 0 \end{pmatrix} : x \in \mathbb{R} \right\} \oplus \left\{ \begin{pmatrix} a & 0 \\ 0 & -a \end{pmatrix} : a \in \mathbb{R} \right\} \oplus \left\{ \begin{pmatrix} 0 & 0 \\ y & 0 \end{pmatrix} : y \in \mathbb{R} \right\}.$$

Identifying $V = \left\{ \begin{pmatrix} 0 & x \\ 0 & 0 \end{pmatrix} : x \in \mathbb{R} \right\}$ with $\theta(V)$, we see that $\mathfrak{sl}(2, \mathbb{R})$ is the TKK Lie algebra $\mathfrak{L}(V)$ of V with Jordan triple product given by

$$\left[\left[\begin{pmatrix} 0 & x \\ 0 & 0 \end{pmatrix}, \theta \begin{pmatrix} 0 & y \\ 0 & 0 \end{pmatrix} \right], \begin{pmatrix} 0 & z \\ 0 & 0 \end{pmatrix} \right] = \begin{pmatrix} 0 & 2xyz \\ 0 & 0 \end{pmatrix}.$$

Since $\theta \begin{pmatrix} a & 0 \\ 0 & -a \end{pmatrix} = \begin{pmatrix} a & 0 \\ 0 & -a \end{pmatrix}$ if and only if $a = 0$, the symmetric part of $\mathfrak{sl}(2, \mathbb{R})$ is just

$$\mathfrak{so}(2, \mathbb{R}) = \left\{ \begin{pmatrix} 0 & x \\ -x & 0 \end{pmatrix} : x \in \mathbb{R} \right\}.$$

Example 1.4.3 The Lie algebras $\mathfrak{su}(2)$ and $\mathfrak{so}(3, \mathbb{R})$ are isomorphic to the cross product Lie algebra $\mathbb{R}^3$, which is a canonical TKK Lie algebra with grading

$$\mathbb{R}^3 = \{(a, 0, 0) : a \in \mathbb{R}\} \oplus \{(0, b, 0) : b \in \mathbb{R}\} \oplus \{(0, 0, c) : c \in \mathbb{R}\}$$

and involution $\theta(a, b, c) = (c, b, a)$. The Jordan triple product in $V = \{(a, 0, 0) : a \in \mathbb{R}\}$ is given by

$$\{(x, 0, 0), (y, 0, 0), (z, 0, 0)\} = [[(x, 0, 0), \theta(y, 0, 0)], (z, 0, 0)] = xyz$$

and hence V is just $\mathbb{R}$ with the usual triple product. We note that the two Jordan triple products $\{x, y, z\} = xyz$ and $\{x, y, z\}' = 2xyz$ on $\mathbb{R}$ are non-isomorphic.

Example 1.4.4 We can also view $\mathfrak{gl}(n, \mathbb{R})$ as a TKK Lie algebra with grading $\mathfrak{gl}(n, \mathbb{R}) = \mathfrak{gl}(n, \mathbb{R})_{-1} \oplus \mathfrak{gl}(n, \mathbb{R})_0 \oplus \mathfrak{gl}(n, \mathbb{R})_1$ and matrix transpose as involution, where

$$\mathfrak{gl}(n, \mathbb{R})_{-1} = \left\{ \begin{pmatrix} \mathbf{0} & B \\ \mathbf{0} & \mathbf{0} \end{pmatrix} : B \in M_{n-1, 1}(\mathbb{R}) \right\}$$

$$\mathfrak{gl}(n, \mathbb{R})_0 = \left\{ \begin{pmatrix} A & \mathbf{0} \\ \mathbf{0} & a \end{pmatrix} : a \in \mathbb{R}, A \in M_{n-1, n-1}(\mathbb{R}) \right\}$$

$$\mathfrak{gl}(n, \mathbb{R})_1 = \left\{ \begin{pmatrix} \mathbf{0} & \mathbf{0} \\ C & \mathbf{0} \end{pmatrix} : C \in M_{1, n-1}(\mathbb{R}) \right\}$$

and $M_{m,n}(\mathbb{R})$ denotes the space of $m \times n$ real matrices. Since

$$\left[\begin{pmatrix} \mathbf{0} & B \\ \mathbf{0} & \mathbf{0} \end{pmatrix}, \begin{pmatrix} \mathbf{0} & \mathbf{0} \\ C & \mathbf{0} \end{pmatrix} \right] = \begin{pmatrix} BC & \mathbf{0} \\ & \\ \mathbf{0} & -\text{Trace}\,(BC) \end{pmatrix},$$

the canonical part of $\mathfrak{gl}(n, \mathbb{R})$ is the Lie algebra $\mathfrak{sl}(n, \mathbb{R})$.

If we define an involution θ on $\mathfrak{gl}(n, \mathbb{R})$ by $\theta(X) = -X^T$, then the dual symmetric part of $\mathfrak{gl}(n, \mathbb{R})$ with respect to θ is the Lie algebra $\mathfrak{so}(n, \mathbb{R})$.

As in the above example, one can view $\mathfrak{gl}(n, \mathbb{C})$ and $\mathfrak{gl}(n, \mathbb{H})$ as TKK Lie algebras with matrix transpose as involution. The Lie algebra $\mathfrak{sl}(n, \mathbb{C})$ is the canonical part of $\mathfrak{gl}(n, \mathbb{C})$. With the involution $\theta(X) = -X^*$, the dual symmetric parts of $\mathfrak{gl}(n, \mathbb{C})$ and $\mathfrak{gl}(n, \mathbb{H})$ are $\mathfrak{u}(n)$ and $\mathfrak{sp}(n)$, respectively.

Notes

The basic results of Jordan algebras presented in this chapter are classical. Further details can be found in the books by Braun and Koecher [13], Jacobson [62], Schafer [104] and Springer [105], as well as the recent one by McCrimmon [88], who has also given in another work [87] an interesting account of the development of Jordan algebras up to the late 1970s. The book by Zhevlakov *et al.* [123] also discusses Jordan algebras, although it is devoted primarily to the theory of alternative algebras.

The construction of Lie algebras from Jordan algebras was discovered independently by Tits [108], Kantor [66, 67] and Koecher [77, 78]. Meyberg introduced the concept of a Jordan triple system in his work [90] and extended the construction of Koecher to the wider class of Jordan triple systems. The version of Tits–Kantor–Koecher construction in this chapter is taken from Chu [22]. Loos showed in [82] the correspondence between a class of finite-dimensional Jordan triple systems and the class of symmetric R-spaces, which is explained further in Example 2.4.33, and introduced the concept of a Jordan pair [83, 84, 85]. Various identities for Jordan triple systems proved in this chapter are based on Loos [84, 85]. We note that the name *Bergmann* referred to in this book is sometimes written as *Bergman* in the literature because Stefan Bergmann and Stefan Bergman are the names of the same author.

2

Jordan structures in geometry

2.1 Banach manifolds and Lie groups

We introduce in this section manifolds of any dimension, including infinite-dimensional Lie groups, and their basic properties. We begin by reviewing differential calculus in Banach spaces which are *real* or *complex* throughout.

Let V and W be (real or complex) Banach spaces, and let $\mathcal{U}$ be an open subset of V. A function $f : \mathcal{U} \longrightarrow W$ is said to be *differentiable* at a point $a \in \mathcal{U}$ if there is a continuous linear map $f'(a) : V \longrightarrow W$ satisfying

$$\lim_{h \to 0} \frac{\| f(a + h) - f(a) - f'(a)(h) \|}{\|h\|} = 0.$$

The map $f'(a)$ is called the *(Fréchet) derivative of f at a*. The function f is called *differentiable* in $\mathcal{U}$ if it is differentiable at every point in $\mathcal{U}$, in which case the *derivative f'* is a mapping

$$f' : \mathcal{U} \longrightarrow L(V, W),$$

where $L(V, W)$ is the Banach space of continuous linear operators from V to W. We say that f is *continuously differentiable* in $\mathcal{U}$, or of *class C^1*, if the derivative f' is continuous on $\mathcal{U}$. One defines k-times continuously differentiable functions, or C^k-functions, for $k \in \mathbb{N}$, by iteration. A *smooth function* on $\mathcal{U}$ is one that is infinitely differentiable, that is, it is in the class C^k for all $k \in \mathbb{N}$.

If $V = \mathbb{R}$ or $\mathbb{C}$, then we identify the derivative $f'(a) : V \longrightarrow W$ with the vector $f'(a)(1) \in W$, and often write $f'(a) \in W$ by abuse of notation.

A basic rule in differentiation is the chain rule, which states that the composite $f \circ g$ of two differentiable functions f and g, whenever this is well defined, is differentiable and the derivative is given by

$$f'(g(a)) \circ g'(a)$$

for a in the domain of g. One very useful theorem on differentiation is undoubtedly the following mean value theorem (cf. [33, 8.5.4]).

Theorem 2.1.1 *Let* $f : \mathcal{U} \longrightarrow W$ *be a differentiable function on an open set* $\mathcal{U}$ *which contains the segment* $\{a + sh : a, h \in \mathcal{U}, 0 \le s \le 1\}$. *Then we have*

$$\|f(a + h) - f(h)\| \le \|h\| \sup_{0 \le s \le 1} \|f'(a + sh)\|.$$

If V and W are complex Banach spaces, a differentiable function $f : \mathcal{U} \longrightarrow W$ is smooth and is usually called *holomorphic*. In addition, it has a local power series expansion, which is made precise below.

First, for $n \in \mathbb{N}$, the vector space $L^n(V, W)$ of all continuous n-linear maps

$$F : \underbrace{V \times \cdots \times V}_{n\text{-times}} \longrightarrow W$$

is a Banach space in the norm

$$\|F\| = \sup_{v_k \ne 0} \frac{\|F(v_1, \ldots, v_n)\|}{\|v_1\| \cdots \|v_n\|}.$$

We define $L^0(V, W) = W$. If W is the underlying scalar field of the Banach space V, then $L^1(V, W)$ is just the dual space V^* of V.

Definition 2.1.2 A continuous map $p : V \longrightarrow W$ between Banach spaces is called a *homogeneous polynomial of degree* n if there exists $P \in L^n(V, W)$ such that

$$p(v) = P(v, \ldots, v) \qquad (v \in V).$$

If P is chosen to be *symmetric*, that is, invariant under permutation of variables, then P is uniquely determined by p and is called the *polar form* of p (cf. [54]).

We denote by $\mathcal{P}^n(V, W)$ the vector space of all homogeneous polynomials of degree n from V to W, and equip it with the norm

$$\|p\| = \sup_{v \ne 0} \frac{\|p(v)\|}{\|v\|} \qquad (p \in \mathcal{P}^n(V, W)).$$

We note that $\mathcal{P}^n(V, W)$ is a closed subspace of the space $C(V, W)$ of continuous maps from V to W, in the pointwise topology.

Definition 2.1.3 Let V and W be Banach spaces. A *power series* from V to W is a formal sum

$$\sum_{n=0}^{\infty} p_n$$

where $p_n \in \mathcal{P}^n(V, W)$. Its *radius of convergence* is defined to be the largest non-negative number $R \leq \infty$ such that the series

$$\sum_{n=0}^{\infty} p_n(v) \qquad (v \in V)$$

converges uniformly in W for $\|v\| \leq r$ and $r < R$.

As in the scalar case, the radius of convergence R can be obtained by

$$R = \frac{1}{\limsup_n \|p_n\|^{1/n}}.$$

Likewise, one can define the *radius of convergence* R for the series $\displaystyle\sum_{n=0}^{\infty} F_n$, where $F_n \in L^n(V, W)$, for which we also have

$$R = \frac{1}{\limsup_n \|F_n\|^{1/n}},$$

so that $\displaystyle\sum_{n=0}^{\infty} F_n(v_1, \ldots, v_n)$ converges uniformly whenever $\max(\|v_1\|, \ldots, \|v_n\|) \leq r$ and $r < R$.

A function $f : \mathcal{U} \longrightarrow W$ from an open set $\mathcal{U}$ in a Banach space V to another Banach space W is said to be *analytic* at a point $a \in \mathcal{U}$ if it can be expressed as a convergent power series about a, which means that there is a power series $\sum_n p_n$ with positive radius of convergence such that

$$f(v) = \sum_{n=0}^{\infty} p_n(v - a)$$

for each v in some neighbourhood of a. An *analytic* function $f : \mathcal{U} \longrightarrow W$ is one that is analytic at every point in $\mathcal{U}$. If an analytic function $f : \mathcal{U} \longrightarrow W$ is bijective and the inverse f^{-1} is analytic, then f is called *bianalytic*.

Plainly, homogeneous polynomials are analytic. Analytic functions are smooth, whereas, for instance, the function

$$f(x) = \begin{cases} \exp x^{-1} & (x > 0) \\ 0 & (x \leq 0) \end{cases}$$

is smooth on $\mathbb{R}$ but not analytic. For complex Banach spaces, however, holomorphic functions are analytic and the term "*biholomorphic*" is a synonym of "bianalytic".

If an analytic function f has a power series representation $\sum_n p_n$ about $a \in \mathcal{U}$ with $p_n(v) = P_n(v, \ldots, v)$, then its nth derivative at a is given by

$$f^{(n)}(a) = n! P_n \in L^n(V, W).$$

Given a real Banach space V, one can equip its complexification $V_c = V \oplus iV$ with a norm $\| \cdot \|_c$ so that $(V_c, \| \cdot \|_c)$ is a complex Banach space and the isometric embedding $v \in V \mapsto (v, 0) \in V \oplus iV$ identifies V as a real closed subspace of V_c. Although there are many choices of the norm $\| \cdot \|_c$, they are all equivalent if we require

$$\max(\|u\|, \|v\|) \leq \|(u, v)\|_c \leq \|u\| + \|v\|$$

for all $u, v \in V$. We will always assume V_c is equipped with such a norm and by a slight abuse of language, call $(V_c, \| \cdot \|_c)$ *the* complexification of V.

Let V and W be *real* Banach spaces with their respective complexifications $(V_c, \| \cdot \|_c)$ and $(W_c, \| \cdot \|_c)$. Then every polynomial $p \in \mathcal{P}^n(V, W)$ has a unique extension to $p^c \in \mathcal{P}^n(V_c, W_c)$ satisfying $p^c|_V = p$ and $\|p\| \leq \|p^c\|$. For each n-linear map $P \in L^n(V, W)$ with complex extension $P^c \in L^n(V_c, W_c)$ we have

$$\|P^c(u + iv, \ldots, u + iv)\| = \left\| \sum_{k=0}^{n} \binom{n}{k} i^k P(\overbrace{u, \ldots, u}^{n-k}, \overbrace{v \ldots v}^{k}) \right\|$$

$$\leq \sum_{k=0}^{n} \binom{n}{k} \|P\| \|u\|^{n-k} \|v\|^k$$

$$= \|P\|(\|u\| + \|v\|)^n \leq 2^n \|P\| \|u + iv\|_c.$$

Hence, if $p(v) = P(v, \ldots, v)$ for $P \in L^n(V, W)$, then $\|p^c\| \leq 2^n \|P\|$. It follows that, given a series $\sum_n p_n$ with radius of convergence R_p, where $p_n(v) = P_n(v, \ldots, v)$ and $P_n \in L^n(V, W)$, the radius of convergence R_c for the complexified series $\sum_n p_n^c$ satisfies

$$R_P/2 \leq R_c \leq R_p,$$

where R_P is the radius of convergence of $\sum_n P_n$. If each P_n is the polar form of p_n, then $\|P_n\| \leq \frac{n^n}{n!} \|p_n\|$ (cf. [111, p. 6]), and therefore

$$\limsup_n \|P_n\|^{1/n} \leq \limsup_n \frac{n}{\sqrt[n]{n!}} \|p_n\|^{1/n}$$

gives $R_P \geq R_p/e$. It follows that an analytic function $f : \mathcal{U} \longrightarrow W$ on an open set $\mathcal{U}$ in V has a complex extension f^c which is holomorphic on an open set

$\mathcal{U}_c$ in the complexification V_c such that $\mathcal{U}_c \supset \mathcal{U}$ and $f^c|_{\mathcal{U}} = f$. Hence, many important properties of holomorphic functions can be passed on to real analytic functions. We mention, however, one exception.

Remark 2.1.4 Given a bounded holomorphic function $f : \mathcal{U} \longrightarrow W$ on a bounded open set $\mathcal{U}$ with $a \in \mathcal{U}$, we have the Cauchy inequality

$$\|f'(a)\| \leq \frac{1}{R} \sup\{\|f(z)\| : z \in \mathcal{U}\}, \tag{2.1}$$

where R is the distance between a and the topological boundary of $\mathcal{U}$. However, real analytic functions may not satisfy this inequality. A simple example is the function $f(x) = \sin x$ on $(-1, 1)$, where $|f'(0)| = 1 > \sup\{|\sin x| : x \in (-1, 1)\}$.

Example 2.1.5 For any Banach spaces V, W and Z over $\mathbb{F} = \mathbb{R}, \mathbb{C}$, a bilinear map $f : V \times W \longrightarrow Z$ is analytic. By complexification, it suffices to show that f is holomorphic for $\mathbb{F} = \mathbb{C}$. Indeed, the derivative $f'(a, b) : V \times W \longrightarrow Z$ at any point $(a, b) \in V \times W$ is given by

$$f'(a, b)(x, y) = f(a, y) + f(x, b).$$

The following two fundamental properties of analytic functions are frequently used.

Theorem 2.1.6 (Principle of analytic continuation). *Let* $f : \mathcal{U} \longrightarrow W$ *be an analytic function on an open connected set* $\mathcal{U}$ *such that* $f(x) = 0$ *for all* x *in some non-empty open set* $\mathcal{S} \subset \mathcal{U}$. *Then* f *is identically* 0.

Theorem 2.1.7 (Inverse function theorem). *Let* $f : \mathcal{U} \longrightarrow W$ *be an analytic function on an open set* $\mathcal{U}$ *such that the derivative* $f'(a) : V \longrightarrow W$ *has a continuous inverse for some* $a \in \mathcal{U}$. *Then* f *is bianalytic from a neighbourhood of* a *onto a neighbourhood of* $f(a)$.

We are now ready to introduce Banach manifolds.

Definition 2.1.8 Let $\mathbb{F} = \mathbb{R}$ or $\mathbb{C}$. A *Banach manifold M over* $\mathbb{F}$ is a Hausdorff topological space with a family $\mathcal{A} = \{(\mathcal{U}_\varphi, \varphi, V_\varphi)\}$ of *local charts* $(\mathcal{U}_\varphi, \varphi, V_\varphi)$ satisfying the following conditions:

(i) $\mathcal{U}_\alpha$ is an open subset of M and $M = \bigcup_\varphi \mathcal{U}_\varphi$;
(ii) $\varphi : \mathcal{U}_\varphi \longrightarrow V_\varphi$ is a homeomorphism onto an open subset of a Banach space V_φ over $\mathbb{F}$;

(iii) the local charts are *compatible*, that is, the change of charts

$$\psi\varphi^{-1} : \varphi(\mathcal{U}_\varphi \cap \mathcal{U}_\psi) \longrightarrow \psi(\mathcal{U}_\varphi \cap \mathcal{U}_\psi)$$

is bianalytic;

(iv) the family $\mathcal{A}$ is maximal relative to conditions (i), (ii) and (iii), that is, if $(\mathcal{U}, \varphi, V)$ is a local chart compatible with all the local charts in $\mathcal{A}$, then $(\mathcal{U}, \varphi, V) \in \mathcal{A}$.

A family $\{(\mathcal{U}_\varphi, \varphi, V_\varphi)\}$ satisfying conditions (i), (ii) and (iii) is called an *atlas* of M or an *analytic structure* on M. Since an atlas can always be extended to a maximal one satisfying condition (iv), it is often sufficient to exhibit an atlas for a topological space to be a Banach manifold. A *local chart* (or *system of local coordinates*) at a point $p \in M$ is a chart $(\mathcal{U}_\varphi, \varphi, V_\varphi)$ with $p \in \mathcal{U}_\varphi$. Note that the Banach spaces V_φ in an atlas need not be the same space, but V_φ and V_ψ are isomorphic if $\mathcal{U}_\varphi \cap \mathcal{U}_\psi$ contains a point p, in which case the derivative

$$(\psi\varphi^{-1})'(\varphi(p)) : V_\varphi \longrightarrow V_\psi$$

is an isomorphism. If all Banach spaces V_φ in the atlas are isomorphic, we can always find an equivalent atlas in which they are all equal to some Banach space V, in which case, we say that the manifold M is *modelled on the Banach space V* and that V is a *model space* for M. We define the *dimension* dim M of M to be that of V if dim $V < \infty$. If V is infinite-dimensional, we define dim $M = \infty$. We call M a *Hilbert manifold* if the Banach spaces V_φ in its atlas are Hilbert spaces.

For a local chart $(\mathcal{U}, \varphi, V)$ of a Banach manifold M, the set

$$\{p \in M : \exists \text{ a local chart } (\mathcal{U}_\psi, \psi, V_\psi) \text{ at } p \text{ and } V_\psi \simeq V\}$$

is open and closed in M. Consequently, on a connected component of M, we can choose an atlas in which all V_φ are the same space. In particular, if M is connected, then we can find a model space for M.

We call a manifold *real* or *complex* according to the underlying scalar field of the spaces V_φ. As usual, a complex Banach space can be viewed as a real Banach space when the scalar multiplication is restricted to the real field. A complex Hilbert space V with inner product $\langle \cdot, \cdot \rangle$ can also be viewed as a real Hilbert space with real inner product Re $\langle \cdot, \cdot \rangle$. We call this real Hilbert space the *real restriction of V*. Hence a complex Banach or Hilbert manifold can always be viewed as a real manifold with the underlying real analytic structure.

If, in Definition 2.1.8, the Banach spaces V_φ are real and the coordinate transformations

$$\psi\varphi^{-1} : \varphi(\mathcal{U}_\varphi \cap \mathcal{U}_\psi) \longrightarrow \psi(\mathcal{U}_\varphi \cap \mathcal{U}_\psi)$$

are only smooth, then we call M a (real) *smooth Banach manifold* and its atlas a *differentiable structure*. Assuming analytic structures in our definition of Banach manifolds has the advantage of unifying both real and complex cases. Of course, we can (and will) always regard a Banach manifold (according to our definition) as a (real) smooth Banach manifold. Sometimes we speak of *analytic* Banach manifolds to highlight the underlying analytic structure. For finite-dimensional real Lie groups defined as smooth manifolds in Section 1.4, there is always a unique compatible analytic Lie group structure [112, p. 42].

Example 2.1.9 Every Banach space over $\mathbb{F}$ is itself a Banach manifold over $\mathbb{F}$, with the analytic structure given by the identity map. Also, an open subset U of a Banach manifold M is a Banach manifold, endowed with the atlas $\{(U \cap \mathcal{U}_\varphi, \varphi|_{U \cap \mathcal{U}_\varphi}, V_\varphi)\}$ derived from the atlas $\{(\mathcal{U}_\varphi, \varphi, V_\varphi)\}$ of M.

Example 2.1.10 Given two Banach manifolds M and N over the same field with analytic structures $\{(\mathcal{U}_\varphi, \varphi, V_\varphi)\}$ and $\{(\mathcal{V}_\psi, \psi, W_\psi)\}$, respectively, the Cartesian product $M \times N$ is a Banach manifold with the natural analytic structure $\{(\mathcal{U}_\varphi \times \mathcal{V}_\psi, \varphi \times \psi, V_\varphi \times W_\psi)\}$.

Example 2.1.11 Let V be a real Banach space and let

$$S(V) = \{(\lambda, v) \in \mathbb{R} \times V : \lambda^2 + \|v\|^2 = 1\}$$

be the unit sphere in the product manifold $\mathbb{R} \times V$. Let $p = (1, 0)$ and $q = (-1, 0)$. On $S(V)$, define two local charts $(S(V)\backslash\{p\}, \varphi, V)$ and $(S(V)\backslash\{q\}, \psi, V)$ by

$$\varphi(\lambda, v) = \frac{v}{1 - \lambda}; \quad \psi(\lambda, v) = \frac{v}{1 + \lambda}.$$

We have $\varphi(-1, 0) = 0 = \psi(1, 0)$. For $v \in \varphi(S(V)\backslash\{p\})\backslash\{0\}$, we have $\varphi^{-1}(v) = \left(\frac{\|v\|^2 - 1}{\|v\|^2 + 1}, \frac{2v}{\|v\|^2 + 1} \right)$. Therefore

$$\psi \circ \varphi^{-1}(v) = \frac{v}{\|v\|^2} \qquad (v \in \varphi(S(V)\backslash\{p\} \cap S(V)\backslash\{q\})).$$

Hence these charts define an analytic structure on $S(V)$. If $V = \mathbb{R}^n$, we adopt the usual notation S^n for the unit sphere $S(\mathbb{R}^n)$ in $\mathbb{R}^{n+1}$.

A mapping $f : M \longrightarrow N$ between Banach manifolds (over the same field $\mathbb{F}$) is called *analytic* if, for each $x \in M$, there are charts $(\mathcal{U}, \varphi, V)$ of M and $(\mathcal{V}, \psi, W)$ of N such that $x \in \mathcal{U}$, $f(\mathcal{U}) \subset \mathcal{V}$ and the composed map

$$\psi \circ f \circ \varphi^{-1} : \varphi(\mathcal{U}) \longrightarrow \psi(\mathcal{V}) \subset W$$

is analytic. A bijection $f : M \longrightarrow N$ is called *bianalytic* if both f and the inverse f^{-1} are analytic, in which case M is said to be *bianalytic to* N. The coordinate map $\varphi : \mathcal{U} \longrightarrow \varphi(\mathcal{U})$ of a chart $(\mathcal{U}, \varphi, V)$ is bianalytic.

Holomorphic and *biholomorphic* maps on complex Banach manifolds are defined likewise. Holomorphic maps on complex Banach manifolds are analytic and in the sequel, the terms *holomorphic* (respectively, *biholomorphic*) and *analytic* (respectively, *bianalytic*) are interchangeable for complex manifolds. One can also define smooth maps between real smooth Banach manifolds M and N in the same manner, and a smooth map $f : M \longrightarrow N$ is called a *diffeomorphism* if it is bijective and the inverse f^{-1} is also smooth, in which case we say that M is *diffeomorphic to* N.

A bianalytic map $f : M \longrightarrow M$ is called an *automorphism* of M. The automorphisms of M form a group, with composition product, called the *automorphism group* of M and is denoted by Aut M.

We define tangent vectors on a manifold as directional derivatives along smooth curves. A *smooth curve* in a Banach manifold M over $\mathbb{F}$ is a smooth map $\gamma : (-c, c) \longrightarrow M$ on some open interval in $\mathbb{R}$, where $c > 0$.

Let $p \in M$ and let $\gamma : (-c, c) \longrightarrow M$ be a smooth curve such that $\gamma(0) = p$. Let $(\mathcal{U}, \varphi, V)$ be a local chart at p. Then $\gamma(t) \in \mathcal{U}$ for t near 0 and the derivative $(\varphi \circ \gamma)'(0) : \mathbb{R} \longrightarrow V$ is (identified with) a vector in V. If $\gamma_1 : (-c_1, c_1) \longrightarrow M$ is another smooth curve such that $\gamma_1(0) = p$ and $(\varphi \circ \gamma)'(0) = (\varphi \circ \gamma_1)'(0)$, then for any chart $(\mathcal{U}_\psi, \psi, V_\psi)$ at p, we have $(\psi \circ \gamma)'(0) = (\psi \circ \varphi^{-1})'(\varphi(p))(\varphi \circ \gamma)'(0) = (\psi \circ \gamma_1)'(0)$. Hence we can define an equivalence relation $\sim$ on smooth curves γ in M with $\gamma(0) = p$ by

$$\gamma \sim \gamma_1 \text{ if } (\varphi \circ \gamma)'(0) = (\varphi \circ \gamma_1)'(0)$$

for a local chart $(\mathcal{U}, \varphi, V)$ at p. An equivalence class $[\gamma]_p$ is called a *tangent vector to* M *at* p. We define the *tangent space of* M *at* p to be the set $T_p M$ of all tangent vectors $[\gamma]_p$ at p.

For the local chart $(\mathcal{U}, \varphi, V)$ at p, the map $[\gamma]_p \in T_p M \mapsto (\varphi \circ \gamma)'(0) \in V$ is a bijection. Indeed, given a vector $v \in V$, there is a smooth curve $\gamma_v(t) = \varphi^{-1}(\varphi(p) + tv)$ in M from some interval in $\mathbb{R}$ such that $(\varphi \circ \gamma_v)'(0) = v$. Hence $T_p M$ can be identified with V via the bijection and is equipped with a Banach space structure. In particular, if M is an open subset of a Banach space W and $p \in M$, then $T_p M = W$ via the identification $[\gamma]_p \in T_p M \mapsto \gamma'(0) \in W$.

In the above chart, if $\gamma(t) \in \mathcal{U}$ is defined, then $(\varphi \circ \gamma)'(t) \in V$ is a tangent vector at $\gamma(t)$. Indeed, $(\varphi \circ \gamma)'(t) = (\varphi \circ \alpha)'(0)$ with $[\alpha]_{\gamma(t)} \in T_{\gamma(t)} M$, where $\alpha(s)$ is the curve

$$\alpha(s) = \varphi^{-1}(\varphi(\gamma(t)) + s(\varphi \circ \gamma)'(t))$$

defined on some interval in $\mathbb{R}$.

Remark 2.1.12 In view of the above identification of tangent vectors, we often denote by $\gamma'(t)$ the tangent vector to a curve γ at $\gamma(t)$.

Let $f : M \longrightarrow N$ be an analytic map between Banach manifolds and let $p \in M$. The map

$$df_p : [\gamma]_p \in T_p M \mapsto [f \circ \gamma]_{f(p)} \in T_{f(p)} N$$

is a continuous linear map and is called the *differential of f* at p. The differential of a smooth map is defined in the same manner.

Example 2.1.13 Let M and N be open subsets of Banach spaces V and W, respectively. Let $f : M \longrightarrow N$ be an analytic map and let $p \in M$. Identifying the tangent spaces $T_p M$ and $T_{f(p)} N$ with V and W as before, the differential $df_p : T_p M \longrightarrow T_{f(p)} N$ is the derivative $f'(p) : V \longrightarrow W$. Indeed, given $v \in V$ with $v = \gamma'(0)$ for some $[\gamma]_p \in T_p M$, we have

$$df_p([\gamma]_p) = [f \circ \gamma]_{f(p)} = (f \circ \gamma)'(0) = f'(\gamma(0))\gamma'(0) = f'(p)v.$$

Example 2.1.14 Let $(\mathcal{U}, \varphi, V)$ be a local chart of a Banach manifold M. For $p \in \mathcal{U}$, the differential $d\varphi_p : T_p \mathcal{U} \longrightarrow T_{\varphi(p)} V$ is given by $d\varphi_p([\gamma]_p) = [\varphi \circ \gamma]_{\varphi(p)}$. Under the identification $[\varphi \circ \gamma]_{\varphi(p)} \in T_{\varphi(p)} V \mapsto (\varphi \circ \gamma)'(0) \in V \mapsto [\gamma]_p \in T_p M$, we have $d\varphi_p([\gamma]_p) = [\gamma]_p$.

The following result is an important consequence of the inverse function theorem for Banach spaces.

Theorem 2.1.15 *Let $f : M \longrightarrow N$ be an analytic (a smooth) map between (smooth) Banach manifolds such that the differential $df_p : T_p M \longrightarrow T_{f(p)} N$ is bijective at some point $p \in M$. Then f is locally bianalytic (diffeomorphic) at p; that is, f is bianalytic (diffeomorphic) from a neighbourhood of p onto a neighbourhood of $f(p)$.*

A closed subspace E of a Banach space V is called a *complemented subspace* if there is a closed subspace E^c of V such that $V = E \oplus E^c$ is the direct sum of E and E^c, or equivalently, if there is a continuous linear projection from V onto E. If $\dim V < \infty$, then every subspace of V is complemented.

Definition 2.1.16 Let $f : M \longrightarrow N$ be an analytic (a smooth) map between (smooth) Banach manifolds M and N. Let $p \in M$ and let $df_p : T_p M \longrightarrow T_{f(p)} N$ be the differential of f at p. The map f is called an *immersion* at p if df_p is injective and its image is a complemented subspace of $T_{f(p)} N$. We call f a *submersion* at p if df_p is surjective and its kernel is a complemented subspace of $T_p M$.

In contrast to the case of a bijective differential df_p, where the inverse function theorem implies that f is locally invertible, the weaker condition of injectivity or surjectivity of df_p alone entails the existence of a one-sided inverse of f locally. More precisely, we have the following characterizations of immersion and submersion.

Lemma 2.1.17 *An analytic (a smooth) map $f : M \longrightarrow N$ between (smooth) Banach manifolds is a submersion at $p \in M$ if, and only if, there are an open neighbourhood $\mathcal{V}$ of $f(p) \in N$ and an analytic (a smooth) map $g : \mathcal{V} \longrightarrow M$ such that $g(f(p)) = p$ and $f \circ g$ is the identity map on $\mathcal{V}$.*

Proof We consider the analytic case. The proof for the smooth case is verbatim. Let f be a submersion at $p \in M$. Then the differential $df_p : T_p M \longrightarrow T_{f(p)} N$ is surjective with complemented kernel K. Therefore $T_p M = K \oplus K^c$ and df_p is bijective on K^c.

Let $(\mathcal{U}, \varphi, T_p M)$ be a chart at p with $\varphi(p) = 0$ and $(\mathcal{V}_\psi, \psi, T_{f(p)} N)$ a chart at $f(p)$, with $\psi(f(p)) = 0$ and $f(\mathcal{U}) \subset \mathcal{V}_\psi$, such that $\psi \circ f \circ \varphi^{-1}$ is analytic on $\varphi(\mathcal{U})$.

On an open neighbourhood B of $(0, 0) \in K \times K^c \approx T_p M$, we can define an analytic map $F : B \longrightarrow K \times T_{f(p)} N$ by

$$F(u, v) = (u, \; \psi f \varphi^{-1} v)) \qquad ((u, v) \in B),$$

where $v \in K^c \cap \varphi(\mathcal{U})$. The derivative

$$F'(0, 0) : K \times K^c \longrightarrow K \times T_{f(p)} N$$

is bijective since

$$F'(0, 0)(u, v) = (u, \; (\psi f \varphi^{-1})'(0)(v)),$$

where $(\psi f \varphi^{-1})'(0) = df_p : K^c \longrightarrow T_{f(p)} N$ is an isomorphism. By the inverse function theorem, F is bianalytic from a neighbourhood of $(0, 0) \in K \times K^c$ onto a neighbourhood $S \times O$ of $(0, \psi f(p)) \in K^c \times T_{f(p)} N$, where S and O are open. The inverse F^{-1} induces a well-defined analytic map $h : \psi f \varphi^{-1}(v) \in O \mapsto v \in K^c \cap \varphi(\mathcal{U})$. Let

$$\mathcal{V} = \psi^{-1}(O \cap \psi(\mathcal{V}_\psi))$$

and define $g : \mathcal{V} \longrightarrow \mathcal{U} \subset M$ by

$$g(y) = \varphi^{-1} \circ h \circ \psi(y) \qquad (y \in \mathcal{V})$$

which satisfies $g(f(p)) = p$. For each $y \in \mathcal{V}$, we have $\psi(y) = \psi f \varphi^{-1}(v)$ for some $v \in K^c \cap \varphi(\mathcal{U})$ and

$$f(g(y)) = f(\varphi^{-1}(h(\psi(y)))) = f(\varphi^{-1}(v)) = y \qquad (y \in \mathcal{V}).$$

Conversely, let f have local right inverse $g : \mathcal{V} \longrightarrow M$. Then the differential $df_p \circ dg_{f(p)}$ is the identity map on $T_{f(p)}N$. Hence the differential df_p is surjective and $dg_{f(p)}$ is injective. It follows that

$$dg_{f(p)} \circ df_p : T_p M \longrightarrow T_p M$$

is a continuous projection with kernel $(df_p)^{-1}(0)$ which is complemented in $T_p M$.   □

Remark 2.1.18 Lemma 2.1.17 implies that a submersion $f : M \longrightarrow N$ is an open map.

If $f : M \longrightarrow N$ is an immersion at $p \in M$, then we have $T_{f(p)}N = V \oplus V^c$, where $V = df_p(T_p M)$ is the image of the differential df_p. With local charts $(\mathcal{U}, \varphi, T_p M)$ at p and $(\mathcal{V}_\psi, \psi, T_{f(p)}M)$ at $f(p)$, as in the proof of Lemma 2.1.17, we can define an analytic map $F : T_p M \times V^c \longrightarrow V \times V^c \approx T_{f(p)}N$ *locally* by

$$F(u, v) = ((\psi f \varphi^{-1})(u), v)$$

for (u, v) in a neighbourhood of $(0, 0) \in T_p M \times V^c$. Again, with arguments as before, the derivative $F'(0, 0)$ is bijective and the local inverse F^{-1} induces an analytic map $g : \mathcal{V} \longrightarrow M$ from a neighbourhood $\mathcal{V}$ of $f(p)$ such that $g \circ f$ is the identity map on a neighbourhood of p. This gives the following characterisation of an immersion.

Lemma 2.1.19 *An analytic (a smooth) map $f : M \longrightarrow N$ between (smooth) Banach manifolds is an immersion at $p \in M$ if, and only if, there are open neighbourhoods $\mathcal{U}$ and $\mathcal{V}$ of p and $f(p) \in N$, respectively, and an analytic (a smooth) map $g : \mathcal{V} \longrightarrow M$ such that $f(\mathcal{U}) \subset \mathcal{V}$ and $g \circ f$ is the identity map on $\mathcal{U}$.*

A topological subspace N of a Banach manifold M is called a *submanifold of M* if N is itself a Banach manifold and the inclusion map $\iota : N \hookrightarrow M$ is an immersion. In this case, the tangent space $T_p N$ of N at a point $p \in N$ can be identified as a complemented subspace of the tangent space $T_p M$ of M at p via the differential $d\iota_p : T_p N \longrightarrow T_p M$. An equivalent way of saying that N is a submanifold of M is that at each point $p \in N$, there is a chart $(\mathcal{U}_\varphi, \varphi, V_\varphi)$

of M and a complemented subspace $W_\varphi \subset V_\varphi$ such that

$$\varphi(N \cap \mathcal{U}_\varphi) = W_\varphi \cap \varphi(\mathcal{U}_\varphi),$$

in which case $\{(N \cap \mathcal{U}_\varphi), \varphi|_{(N \cap \mathcal{U}_\varphi)}, W_\varphi\}$ forms an atlas of N.

Every open topological subspace of a Banach manifold M is a submanifold of M.

Let M be a Banach manifold over $\mathbb{F}$ and let

$$T(M) = \bigcup_{p \in M} T_p M$$

be the union of tangent spaces of M. Then $T(M)$ is a Banach manifold over $\mathbb{F}$ with an analytic structure induced from that of M. Indeed, let $\{\mathcal{U}_\varphi, \varphi, V_\varphi)\}$ be an atlas of M and write $T\mathcal{U}_\varphi = \bigcup_{p \in \mathcal{U}_\varphi} T_p M$. Define a combined map

$$T(\varphi) : T\mathcal{U}_\varphi \longrightarrow V_\varphi \times V_\varphi$$

by

$$T(\varphi)[\gamma]_p = (\varphi(p), (\varphi \circ \gamma)'(0)) \qquad ([\gamma]_p \in T_p M).$$

The image of $T\mathcal{U}_\varphi$ is $\varphi(\mathcal{U}_\varphi) \times V_\varphi$. Define a topology on $T(M)$ in which the open sets $\mathcal{O}$ are those such that $T(\varphi)(\mathcal{O} \cap T\mathcal{U}_\varphi)$ is open in $V_\varphi \times V_\varphi$. Then $\{(T\mathcal{U}_\varphi, T(\varphi), V_\varphi \times V_\varphi)\}$ is an analytic structure on $T(M)$.

The Banach manifold $T(M)$ is a vector bundle on M, with bundle projection

$$\pi : [\gamma]_p \in TM \mapsto p \in M,$$

and is called the *tangent bundle of M*.

An *analytic vector field* on M is an analytic map $X : M \longrightarrow TM$ such that $X(p) \in T_p M$; that is, X is a section of the tangent bundle $T(M)$. We often write X_p for $X(p)$. The analytic vector fields on M form a vector space $\mathfrak{X}M$ under pointwise addition and scalar multiplication.

Given an analytic map $f : M \longrightarrow N$ between Banach manifolds M and N, we can define the *tangent map* $df : T(M) \longrightarrow T(N)$ as the combined map

$$df(X_p) = df_p(X_p),$$

which in turn induces the map $df \circ X : M \longrightarrow T(N)$ for each $X \in \mathfrak{X}M$.

Analytic vector fields can be regarded as differential operators in the following way. Given an analytic map $f : \mathcal{U} \longrightarrow W$ from an open subset $\mathcal{U}$ of M to a Banach space W (over the same field of M), and given $X \in \mathfrak{X}M$, one can define an analytic map $Xf : \mathcal{U} \longrightarrow W$ by

$$Xf(p) = df_p(X_p) \qquad (p \in M).$$

This mapping is written as $Xf = df \circ X$. We can define the nth iterate X^n of X inductively by

$$X^0 f = f, \quad X^n f = X(X^{n-1} f) \qquad (n = 1, 2, \ldots).$$

Given two analytic vector fields $X, Y \in \mathfrak{X}M$, there is a unique analytic vector field $[X, Y]$ on M, called the *commutator* (or *brackets*) of X and Y, such that for each open set $\mathcal{U} \subset M$, we have

$$[X, Y]f = X(Yf) - Y(Xf)$$

for all analytic maps f from $\mathcal{U}$ to a Banach space W (cf. [111, p. 60]). Using the above formula, one can verify easily that the brackets $[\cdot, \cdot] : \mathfrak{X}M \times \mathfrak{X}M \longrightarrow \mathfrak{X}M$ is antisymmetric, bilinear and satisfies the Jacobi identity

$$[X, [Y, Z]] + [Y, [Z, X]] + [Z, [X, Y]] = 0 \qquad (X, Y, Z \in \mathfrak{X}M).$$

It follows that $\mathfrak{X}M$ is a Lie algebra in the above brackets.

Let $(\mathcal{U}_\varphi, \varphi, V_\varphi)$ be a local chart at $p \in M$. A tangent vector $[\gamma]_q$ at a point $q \in \mathcal{U}_\varphi$ can be identified with the vector $(\varphi \circ \gamma)'(0) \in V_\varphi$. On this chart, an analytic vector field X can be viewed as an analytic function $X : \mathcal{U}_\varphi \longrightarrow V_\varphi$ and $X\varphi = d\varphi \circ X = X$ by Example 2.1.14. It follows that

$$[X, Y]\varphi = X(Y\varphi) - Y(X\varphi) = dY\varphi \circ X - dX\varphi \circ Y = dY \circ X - dX \circ Y$$

and

$$[X, Y](p) = dY_p(X_p) - dX_p(Y_p) \qquad (p \in \mathcal{U}).$$

A bianalytic map $f : M \longrightarrow N$ between Banach manifolds induces a map

$$f_* : \mathfrak{X}M \longrightarrow \mathfrak{X}N \tag{2.2}$$

defined by

$$(f_* X)_{f(p)} = df_p(X_p) \qquad (X \in \mathfrak{X}M, p \in M).$$

If g is another bianalytic map on N, then we have the tangent map $d(g \circ f) = dg \circ df$ and it follows that $(g \circ f)_* = g_* \circ f_*$. In particular, taking $g = f^{-1}$ implies that f_* is bijective. Further, it is straightforward to verify that f_* is in fact a Lie algebra isomorphism (cf. [111, 4.5]).

Given an analytic vector field X on a Banach manifold M, using the existence theorem for differential equations in Banach spaces, one can find an open interval I_p in $\mathbb{R}$ containing 0 and an analytic curve

$$\gamma_p : I_p \longrightarrow \mathcal{U}_\varphi$$

satisfying the differential equation

$$\frac{d\gamma_p(t)}{dt} = X(\gamma_p(t)) \in V_\varphi$$

and the initial condition $\gamma_p(0) = p$ (cf. Remark 2.1.12). Since the solution also depends analytically on the initial point, the theory of differential equations gives the following further result (cf. [94, §I.7] and [81, chapter IV]).

Theorem 2.1.20 *Given an analytic vector field X on a Banach manifold M and $p \in M$, there is an open neighbourhood $\mathcal{U}$ of p and an open interval I in $\mathbb{R}$ containing 0, such that for all $q \in \mathcal{U}$, the curve γ_q satisfying $\gamma_q'(t) = X(\gamma_q(t))$ and $\gamma_q(0) = q$ is defined on I. The map $\alpha : (t, q) \in I \times \mathcal{U} \mapsto \gamma_q(t) \in M$ is analytic and satisfies $\alpha(s + t, q) = \alpha(s, \alpha(t, q))$ for $s, t, s + t \in I$ and $\alpha(t, q) \in \mathcal{U}$.*

For $t \in I$, let $\alpha_t : \mathcal{U} \longrightarrow M$ be the analytic map

$$\alpha_t(q) = \alpha(t, q) = \gamma_q(t) \qquad (q \in \mathcal{U}).$$

Since $\alpha_0(p) = p \in \mathcal{U}$, we have $\alpha_t(p) \in \mathcal{U}$ for t near 0 and hence the differential $(d\alpha_{-t})_{\alpha_t(p)}$ sends a tangent vector $Y_{\alpha_t(p)}$ to a tangent vector in $T_p M$. Noting that $X(p) = [\gamma_p] \in T_p M$ and that for an analytic map f from $\mathcal{U}$ to a Banach space, we have $Xf(p) = [f \circ \gamma_p] = (f \circ \gamma_p)'(0) = \lim_{t \to 0} \frac{1}{t}(f(\alpha_t(p)) - f(p))$, a direct computation gives

$$[X, Y](p) = \lim_{t \to 0} \frac{1}{t}(d\alpha_{-t}(Y_{\alpha_t(p)}) - Y_p), \tag{2.3}$$

which expresses in some sense the rate of change of Y in the direction of X, known as the *Lie derivative* of Y in the direction of X (cf. [81, p. 121]).

Definition 2.1.21 The map $\alpha : (t, q) \in I \times \mathcal{U} \mapsto \gamma_q(t) \in M$ in Theorem 2.1.20 is called a *local flow* or *an integral curve* of the vector field X. If the flow α is defined on $\mathbb{R} \times M$, then X is called a *complete analytic vector field*.

By the uniqueness theorem in differential equations (cf. [81, p. 88]), there is only one flow $\alpha : \mathbb{R} \times M \longrightarrow M$ of a complete analytic vector field X satisfying $\alpha(0, p) = p$. For each $t \in \mathbb{R}$, we define an analytic map

$$\exp tX : M \longrightarrow M$$

by $\exp tX(p) = \alpha(t, p)$ for each $p \in M$. The notation is suggested by the property

$$\exp(s + t)X = \exp sX \circ \exp tX \qquad (s, t \in \mathbb{R}),$$

which also implies that each map $\exp t X$ is bianalytic, since $\exp t X \circ \exp -t X = \exp 0 X$ is the identity map on M. We call the homomorphism

$$t \in \mathbb{R} \mapsto \exp t X$$

the *one-parameter group* of X. In this notation, we have

$$X(p) = \frac{d}{dt}\bigg|_{t=0} \exp t X(p) \qquad (p \in M).$$

Denote by $\operatorname{aut} M$ the set of all complete analytic vector fields on a Banach manifold M. The map

$$X \in \operatorname{aut} M \mapsto \exp 1 X \in \operatorname{Aut} M$$

is denoted by $\exp$.

Let $g : M \longrightarrow N$ be a bianalytic map between Banach manifolds. Then the induced Lie algebra isomorphism $g_* : \mathfrak{X} M \longrightarrow \mathfrak{X} N$, as defined in (2.2), maps $\operatorname{aut} M$ to $\operatorname{aut} N$ and for $X \in \operatorname{aut} M$, we have the commutative diagram

$$
\begin{array}{ccc}
(t, p) \in \mathbb{R} \times M & \longrightarrow & \exp t X(p) \in M \\
\Big\downarrow \iota \times g & & \Big\downarrow g \\
(t, g(p)) \in \mathbb{R} \times N & \longrightarrow & \exp t g_* X(g(p)) \in N
\end{array}
$$

where $g(\exp t X(p)) = \exp t g_* X(g(p))$ gives

$$g \circ \exp t X \circ g^{-1} = \exp t g_* X. \tag{2.4}$$

Remark 2.1.22 For a smooth Banach manifold M, the tangent bundle $T(M)$ is a smooth Banach manifold, in which case one considers smooth vector fields $X : M \longrightarrow T(M)$, smooth flows and so on. A parallel theory for smooth Banach manifolds can be developed along with that for analytic manifolds. We suppress the repetition but will make use of the results without more ado.

An important class of Banach manifolds is the Lie groups. A Banach manifold G is called a *Banach Lie group* if G is a group and the group operations are analytic, that is, the multiplication $(x, y) \in G \times G \mapsto xy \in G$ and the inverse map $x \in G \mapsto x^{-1} \in G$ are analytic. In practice, to verify analyticity of the group operations, it suffices to verify the following three conditions (cf. [12, III.1.1]):

(i) the left translation $\ell_a : g \in G \mapsto ag \in G$ is analytic, for all $a \in G$;
(ii) the mapping $(x, y) \in G \times G \mapsto xy^{-1} \in G$ is analytic in a neighbourhood of (e, e);

(iii) the conjugation $g \in G \mapsto aga^{-1} \in G$ is analytic in a neighbourhood of e, for all $a \in G$.

Let G be a Banach Lie group with identity e. For each $a \in G$, the right translation $r_a : g \in G \mapsto ga \in G$ is an analytic map. An analytic vector field $X : G \longrightarrow T(G)$ is called *left invariant* (respectively, *right invariant*) if $(d\ell_a)_g(X_g) = X_{ag}$ (respectively, $(dr_a)_g(X_g) = X_{ga}$) for all $a, g \in G$. This can be written as $d\ell_a \circ X = X \circ \ell_a$ and $dr_a \circ X = X \circ r_a$ in terms of the tangent maps $d\ell_a, dr_a : T(G) \longrightarrow T(G)$. An invariant vector field X is determined entirely by its value at the identity e, since $X_a = (d\ell_a)_e(X_e)$ for all $a \in G$ if X is left invariant.

Given two left invariant analytic vector fields X and Y on G, it can be verified that the commutator $[X, Y]$ is also left invariant. Therefore the vector space $\mathfrak{L}$ of all left invariant analytic vector fields on G forms a Lie algebra in the product $[X, Y]$. Given a tangent vector $[\gamma]_e$ in the tangent space $T_e G$ at the identity e, the vector field X defined by $X(g) = (d\ell_g)_e([\gamma]_e)$ is left invariant and $X(e) = [\gamma]_e$. It follows that the map $X \in \mathfrak{L} \mapsto X_e \in T_e G$ is a linear isomorphism, since two left invariant vector fields are identical if they have the same value at e. Therefore the tangent space $T_e G$ is a Lie algebra in the brackets

$$[X_e, Y_e] := [X, Y]_e \qquad (X, Y \in \mathfrak{L}).$$

We call $\mathfrak{g} = T_e G = \{X_e : X \in \mathfrak{L}\}$ the *Lie algebra of G*.

Lemma 2.1.23 *A left invariant analytic vector field X on a Banach Lie group G is complete.*

Proof By left invariance of X, given $X(e) = [\gamma]_e \in T_e G$, we have $X(a) = [a\gamma]_a \in T_a G$. If $\gamma_e : I \longrightarrow G$ satisfies

$$\frac{d\gamma_e(t)}{dt} = X(\gamma_e(t)) \quad \text{and} \quad \gamma_e(0) = e,$$

then the curve $\gamma_a = a\gamma_e$ solves

$$\frac{d\gamma_a(t)}{dt} = X(\gamma_a(t)) \quad \text{and} \quad \gamma_a(0) = a.$$

Let $\alpha_e : I \times \mathcal{U} \longrightarrow G$ be a local flow of X satisfying $\alpha(0, p) = p$ for p in a neighbourhood $\mathcal{U}$ of e. Then the map $\alpha_a : (t, ap) \in I \times a\mathcal{U} \mapsto a\alpha_e(t, p) \in G$ is a local flow of X satisfying $\alpha_a(0, ap) = ap$ for all $ap \in a\mathcal{U}$. Taking the union

$\bigcup_{a \in G} I \times a\mathcal{U}$ and applying the uniqueness theorem, we arrive at a local flow $\alpha : I \times G \longrightarrow G$ of X satisfying $\alpha(0, a) = a$ for $a \in G$. Since

$$\alpha(s + t, \cdot) = \alpha(s, \alpha(t, \cdot))$$

for $s, t, s + t$ in I, one can extend α to a map $\alpha_X : \mathbb{R} \times G \longrightarrow G$ by defining

$$\alpha_X(t, \cdot) = \alpha \overbrace{\left(\frac{t}{n}, \alpha \left(\frac{t}{n}, \left(\cdots \alpha \left(\frac{t}{n}, \cdot \right) \right) \right) \right)}^{n\text{-times}}$$

for sufficiently large n, which can be seen to be well defined. Since $\alpha_X(s + t, a) = \alpha_X(s, \alpha_X(t, a))$ and α_X is analytic on $I \times G$, it follows that α_X is analytic on $\mathbb{R} \times G$ and is the flow of X. $\qquad \square$

A *one-parameter subgroup* of a Banach Lie group G is an analytic homomorphism $\theta : (\mathbb{R}, +) \longrightarrow G$ such that $\theta(0) = e$. Given such a homomorphism θ, we have $d\theta_0(1) \in T_e G$. Conversely, for each left invariant vector field X on G, the homomorphism

$$t \in \mathbb{R} \mapsto \exp t X(e) \in G$$

is a one-parameter subgroup of G satisfying $d(\exp t X(e))_0(1) = X_e$. The *exponential map*

$$\exp : T_e G \longrightarrow G$$

is defined by

$$\exp(X_e) = \exp X(e) \qquad (X \in \mathfrak{L}).$$

The exponential map need not be surjective, but if it is, then G is called *exponential*.

Definition 2.1.24 Let G be a Banach Lie group with Lie algebra $\mathfrak{g} = T_e G$. The *adjoint representation of G* is the homomorphism

$$Ad : G \longrightarrow \operatorname{Aut} \mathfrak{g}$$

defined by the differential of the inner automorphism:

$$Ad(g) = d(r_{g^{-1}} \ell_g)_e : \mathfrak{g} \longrightarrow \mathfrak{g}.$$

We note that Ad is a homomorphism since $r_{(gh)^{-1}} \ell_{gh} = r_{g^{-1}} \ell_g \circ r_{h^{-1}} \ell_h$. In the notation of (2.2), we have $Ad(g) = (r_{g^{-1}} \ell_g)_*$ and by (2.4),

$$\exp Ad(g) t X_e = \exp t Ad(g) X_e = g(\exp t X_e) g^{-1} \qquad (X \in \mathfrak{g}). \qquad (2.5)$$

The automorphism group Aut $\mathfrak{g}$ is contained in the Banach space $L(T_eG)$ of continuous linear operators on the tangent space $\mathfrak{g} = T_eG$. Considering Ad as the map $Ad : G \longrightarrow L(T_eG)$, we can take its differential, which defines the *adjoint representation* $ad : \mathfrak{g} \longrightarrow L(T_eG)$ of $\mathfrak{g}$:

$$ad(X_e) = d(Ad)_e(X_e).$$

Lemma 2.1.25 *Given two left invariant analytic vector fields X and Y on a Banach Lie group G, we have $[X, Y]_e = ad(X_e)(Y_e)$.*

Proof We have $Ad(g)(Y_e) = (dr_{g^{-1}})_g(d\ell_g)_e(Y_e) = (dr_{g^{-1}})_g(Y_g)$ by left invariance. Let $\alpha_t(\cdot) = \alpha_X(t, \cdot)$ be the flow of X in Lemma 2.1.23. Since X is left invariant, we have $\ell_a \circ \alpha_t = \alpha_t \circ \ell_a$ for $a \in G$. Hence we have

$$\alpha_{-t}(a) = \alpha_{-t}(\ell_a(e)) = \ell_a(\alpha_{-t}(e)) = r_{\alpha_{-t}(e)}(a).$$

It follows from (2.3) that

$$[X, Y](e) = \lim_{t \to 0} \frac{1}{t} ((d\alpha_{-t})_{\alpha_t(e)} Y_{\alpha_t(e)} - Y_e)$$
$$= \lim_{t \to 0} \frac{1}{t} ((dr_{\alpha_{-t}(e)})_{\alpha_t(e)} Y_{\alpha_t(e)} - Y_e)$$
$$= \lim_{t \to 0} \frac{1}{t} (Ad(\alpha_t(e)) Y_e - Y_e)$$
$$= ad(X_e)(Y_e). \qquad \square$$

A *Banach Lie algebra* is a Banach space $\mathfrak{g}$ which is also a Lie algebra and is equipped with a continuous Lie product. Continuity of the Lie product is equivalent to the existence of a constant $C > 0$ such that

$$\| [X, Y] \| \leq C \|X\| \|Y\| \qquad (X, Y \in \mathfrak{g}).$$

Our remaining task in this section is to show that the Lie algebra of a Banach Lie group is indeed a Banach Lie algebra. We show that, in local coordinates, the Lie product can be expressed by continuous bilinear maps and is therefore continuous.

Theorem 2.1.26 *The Lie algebra $\mathfrak{g}$ of a Banach Lie group G is a Banach Lie algebra.*

Proof We already know that $\mathfrak{g} = T_eG$ is a Lie algebra and also, it identifies with the Banach space V in a local chart $(\mathcal{U}, \varphi, V)$ at e. We show that the Lie product $(X_e, Y_e) \in T_eG \times T_eG \mapsto [X_e, Y_e] \in T_eG = V$ is continuous.

Let $X_e = [\alpha] \in T_e G$ and $Y_e = [\beta] \in T_e G$. By Lemma 2.1.25, we have

$$[X_e, Y_e] = ad([\alpha])(Y_e)$$

$$= \lim_{t \to 0} \frac{1}{t} (Ad(\alpha(t))Y_e - Ad(\alpha(0))Y_e)$$

$$= \frac{d}{dt}\Big|_{t=0} (Ad(\alpha(t))Y_e)$$

$$= \frac{d}{dt}\Big|_{t=0} \left(d(r_{\alpha(t)^{-1}}\ell_{\alpha(t)})_e Y_e\right)$$

$$= \frac{d}{dt}\Big|_{t=0} \left(\frac{d}{ds}\Big|_{s=0} (\varphi(\alpha(t)\beta(s)\alpha(t)^{-1}))\right),$$

where a tangent vector $[\gamma] \in T_e G$ identifies with $(\varphi \circ \gamma)'(0) \in V$.

Considering $\varphi(\alpha(t)\beta(s)\alpha(t)^{-1}) = F(t, s)$ as a function from some open neighbourhood of $(0, 0) \in \mathbb{R}^2$ to V, we have the Taylor expansion

$$F(t, s) = F(0, 0) + F'(0, 0)(t, s) + \frac{1}{2!} F''(0, 0)((t, s), (t, s)) + \cdots$$

$$= D_1 F(0, 0)t + D_2 F(0, 0)s + \frac{1}{2} (D_1 D_1 F(0, 0)(t, t)$$

$$+ 2D_1 D_2 F(0, 0)(t, s) + D_2 D_2 F(0, 0)(s, s)) + \cdots$$

$$= (\varphi \circ \beta)'(0)s + \frac{1}{2} (D_1 D_1 F(0, 0)(t, t)$$

$$+ 2D_1 D_2 F(0, 0)(t, s) + D_2 D_2 F(0, 0)(s, s)) + \cdots.$$

It follows that

$$\frac{d}{dt}\Big|_{t=0} \left(\frac{d}{ds}\Big|_{s=0} F(t, s)\right) = D_1 D_2 F(0, 0)(1, 1) = f((\varphi \circ \alpha)'(0), (\varphi \circ \beta)'(0)),$$

where $f : V \times V \longrightarrow V$ is a continuous bilinear map. It follows that $[X_e, Y_e] = f(X_e, Y_e)$ and the Lie product is continuous. $\qquad\Box$

Remark 2.1.27 In contrast to the finite-dimensional case, a Banach Lie algebra need not arise as the Lie algebra of a Banach Lie group [115]. However, if the centre

$$\mathfrak{z}(\mathfrak{g}) = \{X \in \mathfrak{g} : [X, \mathfrak{g}] = 0\}$$

of a Banach Lie algebra $(\mathfrak{g}, [\cdot, \cdot])$ is trivial, then $\mathfrak{g}$ is the Lie algebra of a connected Banach Lie group. We refer to Robart [97] for a proof and will make use of this fact in establishing Theorem 2.4.31 later.

Example 2.1.28 A Banach algebra $\mathfrak{A}$ with identity $\mathbf{1}$ is a Banach Lie algebra in the commutator product

$$[a, b] = ab - ba \qquad (a, b \in \mathfrak{A}).$$

The set G of invertible elements is open in $\mathfrak{A}$ and forms a group in the associative multiplication of $\mathfrak{A}$, and G is a Banach Lie group with Lie algebra $T_1 G = \mathfrak{A}$.

Example 2.1.29 The infinite-dimensional analogue of the general linear groups is the group $GL(H)$ of invertible elements in the Banach algebra $L(H)$ of bounded linear operators on a Hilbert space H. The group $GL(H)$ is norm open in $L(H)$ and is a Banach Lie group modelled on $L(H)$, which is a Banach Lie algebra in the commutator product. The exponential map $\exp : L(H) \longrightarrow G(H)$ is the usual one:

$$\exp A = \mathbf{1} + A + \frac{A^2}{2!} + \cdots \qquad (A \in L(H))$$

where $\mathbf{1}$ is the identity operator in $L(H)$. The two-sided ideal $K(H)$ of compact operators in $L(H)$ is a Lie ideal and is the Lie algebra of the Banach Lie group

$$GL_c(H) = \{A \in GL(H) : \mathbf{1} - A \in K(H)\}.$$

In the operator norm topology, the group

$$GL_2(H) = \{A \in GL(H) : \mathbf{1} - A \text{ is Hilbert–Schmidt}\}$$

is a Hilbert Lie group and its Lie algebra is the Hilbert space $L_2(H)$ of Hilbert–Schmidt operators, equipped with the commutator product and the Hilbert–Schmidt norm

$$\|A\|_2 = \left(\sum_\alpha \|A(e_\alpha)\|^2 \right)^{1/2} \qquad (A \in L_2(H)),$$

where $\{e_\alpha\}$ is an orthonormal basis of H. If H is complex, the unitary group

$$U(H) = \{u \in GL(H) : uu^* = u^*u = \mathbf{1}\}$$

is a real Banach Lie group in the norm topology and its Lie algebra is the real Banach space

$$\mathfrak{u}(H) = \{A \in L(H) : A + A^* = 0\}$$

of skew-Hermitian operators, equipped with the commutator product. For a real Hilbert space, the orthogonal group

$$O(H) = \{t \in GL(H) : tt^* = t^*t = \mathbf{1}\}$$

is a Banach Lie group with Lie algebra

$$\mathfrak{o}(H) = \{A \in L(H) : A + A^* = 0\},$$

consisting of skew-symmetric operators.

Let G be a Banach Lie group with identity e and Lie algebra $\mathfrak{g}$. A subgroup K of G is called a *Banach Lie subgroup* if it is a submanifold of G, in which case K is closed and a Banach Lie group in the induced topology of G [111, p. 128] and it can be shown that the subalgebra

$$\mathfrak{k} = \{X \in \mathfrak{g} : \exp tX \in K, \forall t \in \mathbb{R}\} \qquad (2.6)$$

of $\mathfrak{g}$ identifies with the Lie algebra of K. Further, the left coset space

$$G/K = \{gK : g \in G\}$$

carries the structure of a Banach manifold and the quotient map $\pi : G \longrightarrow G/K$ is a submersion [111, theorem 8.19]. Manifolds of the form G/K are called *homogeneous spaces*. Let $p = \pi(e) = K$. Then the differential $d\pi_e : \mathfrak{g} \longrightarrow T_p(G/K)$ has kernel $\ker d\pi_e = \mathfrak{k}$ and gives the canonical isomorphism $\mathfrak{g}/\mathfrak{k} \approx T_p(G/K)$.

2.2 Riemannian manifolds

A Riemannian manifold M is a manifold equipped with a Riemannian metric. A Riemannian metric g on M is a *smooth* choice of inner product $g(p)$ on the tangent space T_pM at each point $p \in M$, where *smoothness* refers to g as a function of p. Let us make this precise first.

Recall that $L^2(V, \mathbb{F})$ denotes the Banach space of continuous bilinear forms on a Banach space V over $\mathbb{F} = \mathbb{R}, \mathbb{C}$. If V is finite-dimensional, we have the identification $L^2(V, \mathbb{F}) = V^* \otimes V^*$, where the latter is equipped with the injective tensor norm. Hence bilinear forms on V are the so-called $(0, 2)$-*type tensors* or *covariant tensors of order* 2. If V is infinite-dimensional, we have the identification $L^2(V, \mathbb{F}) = (V \hat{\otimes} V)^*$ which only contains the algebraic tensor $V^* \otimes V^*$ as a subspace, where $\hat{\otimes}$ denotes the projective tensor product. Let $L_0^2(V, \mathbb{F})$ be the closed subspace of $L^2(V, \mathbb{F})$, consisting of all symmetric bilinear forms on V. One can view $L_0^2(\,\cdot\,, \mathbb{F})$ as a functor on the category of Banach spaces over $\mathbb{F}$.

Let M be a smooth manifold modelled on a real Banach space V. A symmetric bilinear form $g \in L_0^2(V, \mathbb{R})$ is called *positive definite* if

$$g(x, x) > 0 \quad \text{for all} \quad x \in V \backslash \{0\},$$

in which case g is often called an *inner product*. We call g *completely positive definite* if g is a complete inner product on V. By the open mapping theorem for Banach spaces, g is completely positive definite on V if, and only if, there is a constant $c_g > 0$ such that

$$g(x, x) \geq c_g \|x\|^2 \qquad (x \in V).$$

Of course, if $\dim V < \infty$, then complete positive definiteness is the same as positive definiteness.

If V is a real Hilbert space with inner product $\langle \cdot, \cdot \rangle$, then by the Riesz representation theorem, $L_0^2(V, \mathbb{R})$ is linearly isometric to the closed subspace $L(V)_s$ of $L(V)$, consisting of all symmetric operators in $L(V)$. The linear isometry $g \in L_0^2(V, \mathbb{R}) \mapsto L_g \in L(V)_s$ is implemented by

$$g(x, y) = \langle L_g x, y \rangle \qquad (x, y \in V),$$

where $g(x, x) \geq 0$ for all $x \in V$ if, and only if, $L_g \geq 0$. In this case, positive definiteness of g is equivalent to the symmetric operator L_g being positive and injective. Complete positive definiteness of g is equivalent to the existence of a constant $c_g > 0$ such that

$$g(x, x) \geq c_g \langle x, x \rangle \qquad (x \in V),$$

which is also equivalent to L_g being positive and invertible.

We can apply the functor $L_0^2(\cdot, \mathbb{R})$ to the tangent bundle

$$\pi : \bigcup_{p \in M} T_p M \longrightarrow M$$

and form a vector bundle

$$L_0^2(\pi) : \bigcup_{p \in M} L_0^2(T_p M, \mathbb{R}) \longrightarrow M.$$

A *Riemannian metric* on M is a smooth section

$$g : M \longrightarrow \bigcup_{p \in M} L_0^2(T_p M, \mathbb{R})$$

of this bundle such that each $g(p) : T_p M \times T_p M \longrightarrow \mathbb{R}$ is completely positive definite. The differential structure on $\bigcup_{p \in M} L_0^2(T_p M, \mathbb{R})$ can be described as follows (cf. [81, p. 170]).

Let $\{(\mathcal{U}_\varphi, \varphi, V)\}$ be an atlas of M where each tangent space $T_p M$ identifies with the model space $V = \{(\varphi \circ \gamma)'(0) : [\gamma]_p \in T_p M\}$, and $g(p) \in L_0^2(T_p M, \mathbb{R})$ identifies with $g_\varphi(p) \in L_0^2(V, \mathbb{R})$, defined by

$$g_\varphi(p)\big((\varphi \circ \alpha)'(0), (\varphi \circ \beta)'(0)\big) = g(p)([\alpha]_p, [\beta]_p).$$

Write $L_0^2(M) = \bigcup_{p \in M} L_0^2(T_p M, \mathbb{R})$ and $L_0^2 \mathcal{U}_\varphi = \bigcup_{p \in \mathcal{U}_\varphi} L_0^2(T_p M, \mathbb{R})$. Define an injective map

$$L(\varphi) : L_0^2 \mathcal{U}_\varphi \longrightarrow V \times L_0^2(V, \mathbb{R})$$

by

$$L(\varphi)(g(p)) = (\varphi(p), g_\varphi(p)) \qquad (g(p) \in L_0^2(T_p M, \mathbb{R})).$$

The image of $L(\varphi)$ is $\varphi(\mathcal{U}_\varphi) \times L_0^2(V, \mathbb{R})$. Then

$$\{(L_0^2 \mathcal{U}_\varphi, L(\varphi), V \times L_0^2(V, \mathbb{R}))\}$$

is an atlas on $L_0^2(M)$ whose open sets are the sets $\mathcal{O}$ for which $L(\varphi)(\mathcal{O} \cap L_0^2 \mathcal{U}_\varphi)$ is open in $V \times L_0^2(V, \mathbb{R})$.

Smoothness of the Riemannian metric $g : M \longrightarrow L_0^2(M)$ means that in local charts $(\mathcal{U}_\varphi, \varphi, V)$, the map

$$g_\varphi : \mathcal{U}_\varphi \longrightarrow L_0^2(V, \mathbb{R}) = L(V)_s$$

is smooth.

If M admits a Riemannian metric g, we call (M, g), or M (if g is understood), a *Riemannian manifold*. If M is an analytic manifold and the Riemannian metric g is analytic, then we call M an *analytic Riemannian manifold*.

In what follows, we usually denote the Riemannian metric $g(p)$ on $T_p M$ by $\langle \cdot, \cdot \rangle_p$, and if confusion is unlikely, the symbol g is also often used to denote a diffeomorphism between manifolds.

Remark 2.2.1 If a Banach manifold admits a Riemannian metric, then its model space is isomorphic to a Hilbert space and the manifold itself is actually a Hilbert manifold.

It is well known that every finite-dimensional paracompact smooth manifold admits a Riemannian metric. In particular, every finite-dimensional Lie group carries the structure of a Riemannian manifold. We recall that a topological space M is *paracompact* if every open covering of M has a locally finite refinement. Metric spaces are paracompact. We refer to Lang [81, pp. 36, 171] for a proof of the following existence theorem.

Theorem 2.2.2 *Let M be a paracompact smooth manifold modelled on a separable Hilbert space. Then M admits a Riemannian metric.*

Let (M, g) be a Riemannian manifold and let $f : N \longrightarrow M$ be an immersion on a smooth manifold N. Then we can define a Riemannian metric $\widetilde{g}$ on N by

$$\widetilde{g}_p(u, v) = g_{f(p)}(df_p(u), df_p(v)) \qquad (p \in N, u, v \in T_p N).$$

Given a smooth curve $\gamma : [r, s] \longrightarrow M$, we define its length by

$$\ell(\gamma) = \left(\int_r^s g_{\gamma(t)}(\gamma'(t), \gamma'(t)) dt \right)^{1/2}$$

and say that γ is a curve *from* $\gamma(r)$ *to* $\gamma(s)$ in M. The length of a (continuous and) piecewise smooth curve is the sum of the lengths of the smooth pieces. With respect to the metric g, the *Riemannian distance* $d(p, q)$ between two points $p, q \in M$ is defined to be the distance

$$d(p, q) = \inf\{\ell(\gamma) : \gamma \text{ is a piecewise smooth curve from } p \text{ to } q \}. \quad (2.7)$$

To introduce the concept of distance minimizing curves, namely, *geodesics*, as curves of zero *acceleration* we need to consider "derivatives" of vector fields which have values in *different* tangent spaces. This problem leads to the abstract concept of a *connection*.

Given a scalar smooth function f on a smooth manifold M and a vector field $X \in \mathfrak{X}M$, we denote by $f X$ the vector field

$$f X(p) = f(p)X(p) \qquad (p \in M).$$

A *connection* on a smooth manifold M is a mapping

$$\nabla : \mathfrak{X}M \times \mathfrak{X}M \longrightarrow \mathfrak{X}M$$

which satisfies

(i) $\nabla(f X + h Y, Z) = f\nabla(X, Z) + h\nabla(Y, Z)$;
(ii) $\nabla(X, Y + Z) = \nabla(X, Y) + \nabla(X, Z)$;
(iii) $\nabla(X, f Y) = f\nabla(X, Y) + (Xf)Y$

for $X, Y, Z \in \mathfrak{X}M$ and smooth functions f, h on M.

It is customary to write $\nabla_X Y$ for $\nabla(X, Y)$. The first property of a connection ∇ implies that, if $Y, Z \in \mathfrak{X}M$ are such that $Y(p) = Z(p)$ for $p \in M$, then

$$\nabla_Y X(p) = \nabla_Z X(p).$$

This enables us to introduce the definition of the tangent vector $\nabla_v X \in T_p M$ for a given $v \in T_p M$ by defining

$$\nabla_v X = \nabla_Y X(p), \qquad (2.8)$$

where Y is a smooth vector field on M such that $Y(p) = v$.

Let $\gamma : (-c, c) \longrightarrow M$ be a smooth curve. By a smooth vector field Z *along the curve* γ, we mean a smooth map

$$Z : (-c, c) \longrightarrow T M$$

such that $Z(t) \in T_{\gamma(t)}M$ for $t \in (-c, c)$. According to Remark 2.1.12, $\gamma'(t)$ denotes the tangent vector to γ at $\gamma(t)$. In particular, we can consider

$$\gamma' : (-c, c) \longrightarrow TM$$

as a vector field along γ. Given a connection ∇ on M and $X \in \mathfrak{X}M$,

$$\nabla_{\gamma'}X : t \in (-c, c) \mapsto \nabla_{\gamma'(t)}X \in TM$$

is a vector field along γ.

A connection ∇ on M is said to be *torsion-free* if

$$\nabla_X Y - \nabla_Y X - [X, Y] = 0$$

for all $X, Y \in \mathfrak{X}M$. It is called *Riemannian* if

$$Xg(Y, Z) = g(\nabla_X Y, Z) + g(Y, \nabla_X Z) \qquad (X, Y, Z \in \mathfrak{X}M),$$

where $g(Y, Z)$ is the scalar function

$$g(Y, Z)(p) = g(p)(X(p), Z(p)) \qquad (p \in M).$$

A fundamental theorem in the theory of Riemannian manifolds is the existence of a unique connection, called the *Levi–Civita connection*, which is torsion-free and Riemannian. We refer to Klingenberg [75, theorem 1.8.11] for a proof.

Let ∇ be the Levi–Civita connection on a Riemannian manifold M. We can "restrict" ∇ to vector fields along smooth curves in the following sense. Let Z be a smooth vector field along a smooth curve $\gamma : (-c, c) \longrightarrow M$ and let $X \in \mathfrak{X}M$. Then there exists a vector field along γ, also denoted by $\nabla_X Z$, satisfying

(i) $\nabla_X(fZ) = f'Z + f\nabla_X Z$ for a smooth function $f : (-c, c) \to \mathbb{R}$;
(ii) $\nabla_X Z = \nabla_{\gamma'}Y$ for any $Y \in \mathfrak{X}M$ satisfying $Y(\gamma(t)) = Z(t)$ for all $t \in (-c, c)$.

The vector field $\nabla_X Z$ is constructed in Klingenberg [75, proposition 1.5.5]. For $t \in (-c, c)$, the tangent vector $\nabla_{\gamma'(t)}Z \in T_{\gamma(t)}M$ is well defined by (2.8).

Definition 2.2.3 Let M be a Riemannian manifold with the Levi–Civita connection ∇. A smooth curve $\gamma : (-c, c) \longrightarrow M$ is called a *geodesic* (with respect to ∇) if the vector field γ' along γ is *parallel* which means $\nabla_{\gamma'(t)}\gamma' = 0$ for all $t \in (-c, c)$.

The following existence of geodesics is proved in [75, lemma 1.6.7].

Lemma 2.2.4 *Let M be a Riemannian manifold with the Levi–Civita connection. Let $p \in M$. Then there is a neighbourhood $\mathcal{U}$ of p and $\eta > 0$ such that for each $q \in \mathcal{U}$ and $X \in \mathfrak{X}M$ with $0 < \|X_q\| < \eta$, there is a geodesic $\gamma_{q,X} : (-2, 2) \longrightarrow M$ satisfying $\gamma_{q,X}(0) = q$ and $\gamma'_{q,X}(0) = X_q$.*

For each $q \in \mathcal{U}$ and $\|X_q\| < \eta$ in the above lemma, we define $\exp_q(X_q) := \gamma_{q,X}(1) \in M$, where we put $\gamma_{q,X}(\cdot) = q$ if $X = 0$. By Klingenberg [75, theorem 1.8.15], there exists $\varepsilon \in (0, \eta)$ such that the map

$$\exp_q : B(0, \varepsilon) \subset T_q M \longrightarrow M$$

is a diffeomorphism onto an open subset of M, where $B(0, \varepsilon) = \{v \in T_q M : \|v\| < \varepsilon\}$. By a *normal neighbourhood* of q, we mean a neighbourhood of the form $\exp_q(N)$, where N is a neighbourhood of 0 in $T_q M$, on which $\exp_q$ is diffeomorphic. Each point in a normal neighbourhood of q is of the form $\gamma_{q,X}(1)$ which can be joined to q by the unique geodesic $\gamma_{q,X}$. Moreover, $\gamma_{q,X}$ is *locally minimizing*; that is, if the length ℓ of $\gamma_{q,X}$ from q to a point p is less than η, then $\ell \leq \ell(\gamma_1)$ for any smooth curve γ_1 from q to p [75, theorem 1.9.3].

Using geodesics and normal neighbourhoods, it can be be shown readily that the Riemannian distance d on a Riemannian manifold M, defined in (2.7), is a metric on M and that the topology induced by d coincides with the original topology of M [75, theorem 1.9.5].

Definition 2.2.5 Let M and N be Riemannian manifolds. A diffeomorphism $g : M \longrightarrow N$ is called an *isometry* if it satisfies the condition

$$\langle X_p, Y_p \rangle_p = \langle dg_p(X_p), dg_p(Y_p) \rangle_{g(p)} \qquad (p \in M; \ X_p, Y_p \in T_p M).$$

The isometries of a Riemannian manifold M form a group $G(M)$ under composition. We call $G(M)$ the *isometry group* of M. We say that $G(M)$ *acts transitively on M* if given two points $p, q \in M$, there exists $g \in G(M)$ such that $g(p) = q$.

Remark 2.2.6 Isometries between manifolds should not be confused with *isometries between normed vector spaces*, which will be discussed later. However, the distinction should be clear from the context, and we will mostly consider *linear isometries* on normed vector spaces. A map $\varphi : V \longrightarrow W$ between two normed vector spaces V and W is also called an *isometry* if it preserves the norm, that is, $\|\varphi(v)\| = \|v\|$ for all $v \in V$. The celebrated Mazur–Ulam theorem states that a *surjective* isometry between two real normed vector spaces is necessarily linear if it fixes the origin 0.

Example 2.2.7 A Riemannian manifold could have few isometries. In fact, an isometry group could be finite (cf. [80, p. 55]). On the other hand, isometry groups could be large. Consider a real Hilbert space V as an analytic Riemannian manifold with its own inner product as the Riemannian metric. Given an isometry $g : V \longrightarrow V$, the map $g - g(0)$ is an isometry on V vanishing at 0 and is therefore linear. Hence each isometry of V is a translation $g_0 + v : x \in V \mapsto g_0(x) + v \in V$ of a linear isometry g_0 on V, where $v \in V$. It follows that the isometry group $G(V)$ is homeomorphic to the product $O(V) \times V$ of the orthogonal group

$$O(V) = \{g \in GL(V) : gg^* = g^*g = 1\}$$

and the additive Lie group V. As a group, $G(V)$ is isomorphic to the semidirect product $O(V) \ltimes V$, which is the set $O(V) \times V$ with the group product

$$(g, v)(h, w) = (gh, g(w) + v) \qquad (g, h \in O(V), v, w \in V).$$

Thus the isometry group $G(V)$ is a real Banach Lie group with the product manifold structure of $O(V)$ and V (cf. [12, p. 216]).

Example 2.2.8 Let V be a real Hilbert space and let $S(V) = \{(\lambda, v) \in \mathbb{R} \times V : \lambda^2 + \|v\|^2 = 1\}$ be the unit sphere. Then $S(V)$ admits a Riemannian metric induced by the inclusion immersion $\iota : S(V) \hookrightarrow \mathbb{R} \times V$, where $\mathbb{R} \times V$ is a Hilbert space in the usual inner product. The isometry group for this metric of $S(V)$ is the orthogonal group $O(\mathbb{R} \times V)$ which is a real Banach Lie group. In particular, the isometry group of the n-dimensional sphere S^n is $O(n + 1)$.

Given a Riemannian manifold M with $\dim M < \infty$, it is a well-known result of Myers and Steenrod [91] that the isometry group $G(M)$ is a Lie group in the compact-open topology. Our main objective in this section is to derive an analogous result for groups of bianalytic isometries on infinite-dimensional complex manifolds.

The following proposition is a real analogue of Cartan's uniqueness theorem which states that two holomorphic maps f and g, one of which is biholomorphic, from a bounded domain $\mathcal{U}$ to another domain in a complex Banach space must coincide if $f(p) = g(p)$ and $f'(p) = g'(p)$ for some $p \in \mathcal{U}$. H. Cartan originally proved this result in the 2-dimensional case [17], but it extends readily to arbitrary Banach spaces.

Proposition 2.2.9 *Let $g, h : M \longrightarrow M$ be isometries of a connected Riemannian manifold M such that $g(p) = h(p)$ and $dg_p = dh_p$ for some $p \in M$. Then $g = h$.*

Proof Considering the isometry $g \circ h^{-1}$, we only need to show that if g is an isometry such that $g(p) = p$ and $dg_p : T_pM \longrightarrow T_pM$ is the identity map, then g itself is the identity map on M.

Let $\mathcal{U}$ be a normal neighbourhood of p such that each point $q \in \mathcal{U}$ is joined to p by a unique geodesic $\gamma_q(t)$ with $\gamma_q(0) = p$. Since g is an isometry, $g \circ \gamma_q$ is a geodesic satisfying $g \circ \gamma_q(0) = p$ and

$$(g \circ \gamma_q)'(0) = dg_{\gamma_q(0)}(\gamma_q'(0)) = dg_p(\gamma_q'(0)) = \gamma_q'(0),$$

we must have $g \circ \gamma_q = \gamma_q$ by uniqueness of γ_q. It follows that $g(q) = q$ for all $q \in \mathcal{U}$.

Since M is path-connected, every point in M is connected to p by a continuous path, covered by a finite number of overlapping normal neighbourhoods. This shows that g is the identity map on M. $\qquad\square$

Let X be a complete smooth vector field on a Riemannian manifold M, with the one-parameter group

$$\exp tX : M \longrightarrow M \qquad (t \in \mathbb{R}).$$

We call X a *Killing field* if each $\exp tX$ is an isometry. The Killing fields on M form a Lie algebra with the brackets. We refer to Klingenberg [75, theorem 1.10.11] for a proof of this important result.

Theorem 2.2.10 *The Killing fields on a Riemannian manifold M form a real subalgebra $\mathfrak{g}(M)$ of the Lie algebra $\mathfrak{X}M$ of smooth vector fields on M.*

Let M be a connected manifold modelled on a complex Hilbert space V. If M admits a Riemannian metric when regarded as a real smooth manifold, then we show that the group $G_a(M) = \operatorname{Aut} M \cap G(M)$ of biholomorphic isometries of M form a real Banach Lie group. To achieve this, we need to topologize $G_a(M)$ appropriately. If $\dim M < \infty$, one can take the usual compact-open topology and further, the identity components of $G(M)$ and $G_a(M)$ coincide if they are semisimple Lie groups [51, p. 374]. In infinite dimensions, we need a different topology which, nevertheless, coincides with the compact-open topology in finite dimensions.

Example 2.2.11 A complex Hilbert space $(V, \langle \cdot, \cdot \rangle)$ is a complex analytic manifold which is also a Riemannian manifold in the metric $\operatorname{Re} \langle \cdot, \cdot \rangle$, with isometry group $G(V) = O(V, \operatorname{Re} \langle \cdot, \cdot \rangle) \ltimes V$, where $(V, \operatorname{Re} \langle \cdot, \cdot \rangle)$ is the real restriction of V. Let $f : (V, \operatorname{Re} \langle \cdot, \cdot \rangle) \longrightarrow (V, \operatorname{Re} \langle \cdot, \cdot \rangle)$ be a smooth map with derivative $f'(a) : (V, \operatorname{Re} \langle \cdot, \cdot \rangle) \longrightarrow (V, \operatorname{Re} \langle \cdot, \cdot \rangle)$ at a point $a \in V$. Then f is holomorphic at a if, and only if, the derivative $f'(a) : V \longrightarrow V$ is complex linear. Hence a real linear isometry $g \in O(V, \mathrm{Re}\,\langle \cdot, \cdot \rangle)$ is biholomorphic on V if, and only if, g is complex linear and

$$G_a(V) = U(V) \ltimes V,$$

where $U(V)$ is the unitary group of V. For $V = \mathbb{C}$, the real restriction $(V, \mathrm{Re}\,\langle \cdot, \cdot \rangle)$ is the Euclidean space $\mathbb{R}^2$. Each isometry in $O(2)$ is of the form

$$f = \begin{pmatrix} \cos\theta & -\sin\theta \\ \sin\theta & \cos\theta \end{pmatrix} \quad \text{or} \quad g = \begin{pmatrix} \cos\theta & \sin\theta \\ \sin\theta & -\cos\theta \end{pmatrix},$$

where f is a rotation by θ and is complex linear on $\mathbb{C}$, while g is a reflection in the line $y = x\tan(\theta/2)$, or the line $x = 0$ for $\theta = \pi$, and is conjugate linear on $\mathbb{C}$. In this case, $U(2)$ coincides with the connected component $SO(2)$ of the identity in $O(2)$.

In the remainder of this section, we fix a connected manifold M modelled on a complex Hilbert space V, equipped with a Riemannian metric for its underlying real structure. Let $\{(\mathcal{U}_\varphi, \varphi, V)\}$ be the complex analytic structure on M and let d be the Riemannian distance on M, defined by the Riemannian metric. Given two non-empty subsets A and B of M, their distance is denoted by

$$d(A, B) = \inf\{d(a, b) : a \in A, b \in B\}.$$

The topological boundary of a set A is denoted by ∂A. For each $p \in M$ and $r > 0$, we denote by

$$D(p, r) = \{x \in M : d(x, p) < r\}$$

the open ball of radius r centred at p.

Let $p \in M$ and let $(\mathcal{U}, \varphi, V)$ be a chart at p such that $\varphi(\mathcal{U})$ is bounded in V. An open neighbourhood B of p is called *admissible* if there exists $r > 0$ such that

$$D(p, 4r) \subset \mathcal{U} \quad \text{and} \quad d(B, \partial D(p, r)) > 0.$$

The latter condition implies that $B \subset D(p, r)$, and B is said to be *strictly contained* in $D(p, r)$. Indeed, if there exists $q \in B \cap (M \backslash D(p, r))$, then there is a smooth curve $\gamma : [0, 1] \longrightarrow B$ with $\gamma(0) = p$ and $\gamma(1) = q$. Since $d(p, q) \geq r$, the Intermediate Value Theorem implies that $d(p, \gamma(t)) = r$ for some $t \in [0, 1]$ which gives $\gamma(t) \in B \cap \partial D(p, r)$ and $d(B, \partial D(p, r)) = 0$, a contradiction.

Admissible neighbourhoods B are of finite diameter $d(B) = \sup\{d(x, y) : x, y \in B\}$. An example of an admissible neighbourhood of p is the open ball

$D(p, r/2)$, given $D(p, 4r) \subset \mathcal{U}$. Since V is locally connected, we may assume that $\mathcal{U}$ and $D(p, r)$ are connected when admissible neighbourhoods are chosen. Moreover, choosing normal coordinates as in the finite-dimensional case [51, p. 54], we may further assume that in the chart $(\mathcal{U}, \varphi, V)$, both φ and φ^{-1} are uniformly continuous on $\mathcal{U}$ and $\varphi(\mathcal{U})$, respectively [75, 1.9.8]; that is, given $\varepsilon > 0$, there exists $\delta > 0$ such that

$$d(x, y) < \delta \quad \text{for } x, y \in \mathcal{U} \text{ implies} \quad \|\varphi(x) - \varphi(y)\| < \varepsilon$$

$$\text{(and a similar condition applies to } \varphi^{-1}\text{).} \tag{2.9}$$

In other words, the metric d on $\mathcal{U}$ is *uniformly* equivalent to the metric induced by the norm $\| \cdot \|$.

We use an admissible neighbourhood to define a metric on the biholomorphic isometry group $G_a(M)$ of M. We will denote by ι the identity map in $G_a(M)$. For $p \in M$ and an admissible neighbourhood B_p at p, define a metric ρ_{B_p} on $G_a(M)$ by

$$\rho_{B_p}(f, g) := \sup\{d(f(x), g(x)) : x \in B_p\} \quad (f, g \in G_a(M)). \tag{2.10}$$

We first note that this is a well-defined metric since

$$d(f(x), g(x)) \leq d(f(x), f(p)) + d(f(p), g(p)) + d(g(p), g(x))$$
$$= d(x, p) + d(f(p), g(p)) + d(p, x) \leq 2d(B) + d(f(p), g(p))$$

and $f = g$ as soon as they coincide on B_p, by the principle of analytic continuation. Next, we show that this metric is independent of the choice of admissible neighbourhoods.

Lemma 2.2.12 *Let M be a connected Riemannian manifold modelled on a complex Hilbert space $(V, \| \cdot \|)$ and let $p, q \in M$ with admissible neighbourhoods B_p and B_q, respectively. Then ρ_{B_p} is uniformly equivalent to ρ_{B_q}.*

Proof By translation invariance of the two metrics, it suffices to show that for each $\varepsilon > 0$, there exists $\delta > 0$ such that

$$\rho_{B_q}(g, \iota) < \delta \quad \text{implies} \quad \rho_{B_p}(g, \iota) < \varepsilon \quad (g \in G_a(M)).$$

Let $B_p \subset \mathcal{U}$ and $B_q \subset V$ be chosen from the charts $(\mathcal{U}, \varphi, V)$ and (V, ψ, V), with $d(B_p, \partial D(p, r)) > 0$. First, suppose $B_p \cap B_q$ contains a point z. Then for each $g \in G_a(M)$ satisfying $\rho_{B_q}(g, \iota) < r$, we have $g(D(p, r)) \subset \mathcal{U}$, since $x \in D(p, r)$ implies

$$d(g(x), p) \leq d(g(x), g(z)) + d(g(z), z) + d(z, p) < d(x, z) + r + d(z, p) \leq 4r.$$

It follows that the analytic maps

$$\{\varphi g \varphi^{-1} : \rho_{B_q}(g, \iota) < r\}$$

are uniformly bounded on the domain $\varphi(D(p, r)) \subset V$.

By (2.9), there exists $\tau > 0$ such that

$$\|\varphi(x) - \varphi(y)\| < \tau \quad \text{implies} \quad d(x, y) < \varepsilon \qquad (x, y \in \mathcal{U}).$$

By an application of Hadamard's three-circles theorem in Vigué [114, proposition 1.1.3] to the domains $\varphi(B_p)$ and $\varphi(B_p \cap B_q)$ which are strictly contained in $\varphi(D(p, r))$, there exists $\eta > 0$ such that

$$\sup\{\|\varphi g \varphi^{-1}(v) - \varphi \iota \varphi^{-1}(v)\| : v \in \varphi(B_p)\} < \tau \qquad (2.11)$$

whenever

$$\sup\{\|\varphi g \varphi^{-1}(v) - \varphi \iota \varphi^{-1}(v)\| : v \in \varphi(B_p \cap B_q)\} < \eta$$

and $\rho_{B_q}(g, \iota) < r$. Now pick $\delta \in (0, r)$ such that

$$d(x, y) < \delta \quad \text{implies} \quad \|\varphi(x) - \varphi(y)\| < \eta \qquad (x, y \in \mathcal{U}).$$

Let $\rho_{B_q}(g, \iota) < \delta$. For each $y \in B_p \cap B_q$, we have $g(y) \in \mathcal{U}$ and $d(g(y), y) < \delta$. Hence $\|\varphi(g(y)) - \varphi(y)\| < \eta$. It follows from (2.11) that $\|\varphi(g(x)) - \varphi(x)\| < \tau$ for all $x \in B_p$ and therefore $d(g(x), x) < \varepsilon$ for all $x \in B_p$. This proves that $\rho_{B_p}(g, \iota) \leq \varepsilon$.

Now, without the assumption of $B_p \cap B_q \neq \emptyset$, let

$$S = \{x \in M : d_B \text{ is equivalent to } d_{B_q}, \forall \text{ admissible neighbourhoods } B \text{ of } x\}.$$

Then $q \in S$ and S is closed, for if (x_n) is a sequence in S converging to some x, then any admissible neighbourhood B of x contains some x_n, and hence d_B is equivalent to d_{B_q}.

Also, S is open. Indeed, given $s \in S$, let B_s be an admissible neighbourhood of s. Then we have $B_s \subset S$, for if $y \in B_s$ and B is an admissible neighbourhood of y, then $B \cap B_s \neq \emptyset$ and the preceding arguments imply that d_B is equivalent to d_{B_s} which is in turn equivalent to d_{B_q}. Hence $y \in S$.

Finally, connectedness of M gives $M = S$. $\qquad\qquad\square$

Our next task is to show that the biholomorphic isometry group $G_a(M)$, equipped with the topology induced by the metric ρ_{B_p}, has the structure of a real Banach Lie group. We define

$$\mathfrak{g}_a(M) := \mathfrak{g}(M) \cap \text{aut } M$$

and call the vector fields in $\mathfrak{g}_a(M)$ *analytic Killing fields*. By Theorem 2.2.10, $\mathfrak{g}_a(M)$ forms a Lie algebra and is a candidate for the model space of a manifold structure on $G_a(M)$. For this, we first need to introduce a norm on $\mathfrak{g}_a(M)$ to make it into a Banach Lie algebra.

Fix $p \in M$ and a chart $(\mathcal{U}, \varphi, V)$ at p satisfying (2.9), with $\varphi(p) = 0$ and some $\alpha > 0$ such that

$$d(x, y) < \alpha \quad \text{implies} \quad \|\varphi(x) - \varphi(y)\| < 1 \qquad (x, y \in \mathcal{U}). \qquad (2.12)$$

Choose an admissible neighbourhood $B_p \subset D(p, r)$ with $4r < \alpha$. Equip $G_a(M)$ with the metric ρ_{B_p} defined in (2.10). Let

$$G_p(\iota, r/2) = \{g \in G_a(M) : \rho_{B_p}(g, \iota) < r/2\},$$

which is an open neighbourhood of the identity $\iota \in G_a(M)$. For each $g \in G_p(\iota, r/2)$, we have $g(B_p) \subset D(p, r)$.

Given $g \in G_a(M)$, denote by

$$\widetilde{g} = \varphi g \varphi^{-1} : \varphi(\mathcal{U}) \longrightarrow V$$

the local representation of g on $\varphi(\mathcal{U}) \subset V$. Then for all $x \in B_p$ and $g \in G_p(\iota, r/2)$, we have $d(g(x), p) < r$ and hence

$$\|\widetilde{g}\|_{\varphi(B(p,r))} := \sup_{x \in B(p,r)} \|\widetilde{g}(\varphi(x))\| = \sup_{x \in B(p,r)} \|\varphi(g(x)) - \varphi(p)\| \le 1.$$

In other words, the functions

$$\{\widetilde{g} : g \in G_p(\iota, r/2)\}$$

are bounded on the domain $\varphi(D(p, r)) \subset V$ in the supremum norm $\| \cdot \|_{\varphi(D(p,r))}$.

Fix the admissible neighbourhood B_p at p and fix the metric $\rho = \rho_{B_p}$ on $G_a(M)$. Then $G_a(M)$ is a topological group in the metric ρ.

Owing to Lemma 2.2.12, we have shown, in the terminology of [111, 10.3], that the action $(g, x) \in G_a(M) \times M \mapsto g(x) \in M$ is *locally uniform*, which means that

(i) each point $q \in M$ admits a chart $(\mathcal{V}, \psi, V)$ and an open ball $B(q)$ centred at q such that, for some ρ-neighbourhood $G_q(\iota)$ of $\iota \in G_a(M)$, the functions

$$\{\psi g \psi^{-1} : g \in G_q(\iota)\}$$

are bounded on $\psi(B(q))$ in the supremum norm $\| \cdot \|_{\psi(B(q))}$;

(ii) the map $g \in G(\iota) \mapsto \widetilde{g} = \psi g \psi^{-1}$ is continuous, in the sense that if $h \in G_q(\iota)$ is close to $g \in G_q(\iota)$ in the metric ρ, then $\widetilde{h}$ is close to $\widetilde{g}$ in the norm $\| \cdot \|_{\psi(B(q))}$.

We now introduce a norm on the Lie algebra $\mathfrak{g}_a(M)$ of analytic Killing fields on M. Each Killing field $X \in \mathfrak{g}_a(M)$ gives rise to a one-parameter group $\exp tX : M \longrightarrow M$ of biholomorphic isometries, and by the above remark, the group action

$$(t, x) \in \mathbb{R} \times M \mapsto \exp tX(x) \in M$$

is locally uniform since for each $q \in M$, there exists $c > 0$ and an admissible neighbourhood $B \subset B(q)$ of q such that $(t, x) \in (-c, c) \times B$ implies $\exp tX(x) \in B(q)$, where the metric $\rho_{_B}$ induced by the admissible neighbourhood B is equivalent to ρ. It follows from Upmeier [111, theorem 10.8] that, in the chart $(\mathcal{U}, \varphi, V)$ at $p \in M$, the local representation

$$\widetilde{X} : \varphi(B_p) \longrightarrow V$$

of X is bounded, where $\widetilde{X}$ is given by the tangent map $T(\varphi)$:

$$T(\varphi)(X_a) = (\varphi(a), \widetilde{X}(\varphi(a))) \qquad (a \in B_p).$$

For sufficiently small $t > 0$, we have $\exp tX(a) \in B_p$ and

$$\widetilde{X}(\varphi(a)) = \frac{d}{dt}\bigg|_{t=0} \widetilde{\exp tX}(\varphi(a)) = \lim_{t \to 0} \frac{\varphi(\exp tX)(a) - \varphi(a)}{t}. \tag{2.13}$$

We define the norm

$$\|X\| = \|\widetilde{X}\|_{\varphi(B_p)} := \sup\{\|\widetilde{X}(\varphi(a))\| : a \in B_p\}. \tag{2.14}$$

We continue to use the notation of the chart $(\mathcal{U}, \varphi, V)$ and the admissible neighbourhood B_p. Choose $s > 0$ such that

$$C_p := D(p, s) \subset D(p, 2s) \subset B_p.$$

By (2.9), there exists $\beta_0 > 0$ such that

$$\|\varphi(x) - \varphi(y)\| < \beta_0 \quad \text{implies} \quad d(x, y) < s.$$

For each $0 < \beta < \beta_0$, let

$$G(\iota, \beta) = \{g \in G_a(M) : \|\widetilde{g} - \mathbf{1}\|_{\varphi(C_p)} < \beta\},$$

where $\mathbf{1}$ is the identity map on V and $\| \cdot \|_{\varphi(C_p)}$ denotes the supremum norm for V-valued bounded functions on $\varphi(C_p)$, as in (2.14). The metric ρ_{C_p} induced by the admissible neighbourhood C_p is equivalent to ρ, and by (2.9), the open sets

$$\{G(\iota, \beta) : 0 < \beta < \beta_0\}$$

form a neighbourhood base at the identity $\iota \in G_a(M)$.

For each $g \in G(\iota, \beta)$, we have $g(C_p) \subset B_p$, since $x \in C_p$ implies

$$d(g(x), p) \leq d(g(x), x) + d(x, p) < s + s = 2s.$$

Given $X \in \mathfrak{g}_a(M)$, continuity of $t \in \mathbb{R} \mapsto \exp t X(p) \in M$ implies $\exp t X(p) \in C_p$ for t in some interval $(-c, c)$ with $c > 0$, and hence

$$d(\exp t X(x), p) \leq d(\exp t X(x), \exp t X(p)) + d(\exp t X(p), p)$$
$$= d(x, p) + d(\exp t X(p), p) < 2s \qquad (x \in C_p)$$

gives $\exp t X(C_p) \subset B_p$. Applying (2.13) and the Mean Value Theorem to the differentiable map

$$t \in (-c, c) \mapsto \varphi(\exp t X(a)) \in V$$

for each $a \in C_p$, we obtain

$$\|\widetilde{\exp t X}(\varphi(a)) - \varphi(a)\| \leq |t| \sup_{0 \leq s \leq 1} \|\widetilde{X}(\varphi(\exp st X)(a))\|$$
$$\leq \|\widetilde{X}\|_{\varphi(B_p)} |t| = \|X\| |t| \qquad (|t| < c). \quad (2.15)$$

Lemma 2.2.13 *Given $0 < \beta < \beta_0$ and an analytic Killing field X on M satisfying $\|X\| < \beta$, we have $\exp X \in G(\iota, \beta)$.*

Proof By (2.15), we have

$$\tau := \sup\{s > 0 : \exp t X \in G(\iota, \beta), \ \forall |t| \leq s\} > 0.$$

We only need to show that $\tau \geq 1$. Suppose $\tau < 1$. We deduce a contradiction. Choose $c > 0$ in (2.15) and $0 < t_1 < c, (\beta + 1)^{-1}$. Let $s > 0$ be such that $|t| \leq s$ implies $\exp t X \in G(\iota, \beta)$. We show that $\exp t X \in G(\iota, \beta)$ for $0 < t \leq s + t_1$. If $0 < t \leq t_1$, we have $\|\widetilde{\exp t X} - \mathbf{1}\|_{\varphi(C_p)} \leq \|X\| |t| < \beta$. If $t_1 < t \leq s + t_1$, then $0 < t - t_1 \leq s$ gives $\exp(t - t_1) X \in G(\iota, \beta)$ and we have

$$\|\widetilde{\exp t X} - \mathbf{1}\|_{\varphi(C_p)} = \|\widetilde{\exp(t - t_1) X \exp t_1 X} - \mathbf{1}\|_{\varphi(C_p)}$$
$$\leq \|\widetilde{\exp(t - t_1) X \exp t_1 X} - \widetilde{\exp t_1 X}\|_{\varphi(C_p)} + \|\widetilde{\exp t_1 X} - \mathbf{1}\|_{\varphi(C_p)}$$
$$\leq \|\widetilde{\exp t_1 X}\|_{\varphi(C_p)} \|\widetilde{\exp(t - t_1) X} - \mathbf{1}\|_{\varphi(C_p)} + \|\widetilde{\exp t_1 X} - \mathbf{1}\|_{\varphi(C_p)}$$
$$\leq \|X\| t_1 \beta + \|X\| t_1 = \beta t_1 (\beta + 1) < \beta.$$

It follows that $\tau + t_1 \leq \tau$, which is impossible. $\qquad \square$

Let $\|X\| < \beta_0$ and $|t| < 1$. Then $\|X\| < \beta < \beta_0$ for some β implies that $\|t X\| = |t| \|X\| < \beta$ and $\exp t X \in G(\iota, \beta)$. Hence, as noted before, we have

$$\exp t X(C_p) \subset B_p. \qquad (2.16)$$

Theorem 2.2.14 *The Lie algebra* $\mathfrak{g}_a(M)$ *of analytic Killing fields on* M, *equipped with the norm defined in* (2.14), *is a Banach space.*

Proof Let (X_n) be a Cauchy sequence in $\mathfrak{g}_a(M)$. We may assume, by omitting the first few terms, that $\|X_n\| < \beta_0$ for all n and hence $\exp t X_n(C_p) \subset B_p$.

The Cauchy sequence $(\widetilde{X}_n)$ of bounded analytic functions on $\varphi(B_p)$ converges in the supremum norm to a bounded analytic function $Y : \varphi(B_p) \longrightarrow V$.

Consider the smooth map

$$f : (-1, 1) \times \varphi(B_p) \longrightarrow V$$

defined by

$$f(t, y) = \widetilde{X}_n(y).$$

For each $a \in C_p$, we have

$$f(t, \widetilde{\exp t X_n}(\varphi(a))) = \widetilde{X}_n(\widetilde{\exp t X_n}(\varphi(a))) = \frac{d}{dt}\widetilde{\exp t X_n}(\varphi(a)).$$

Let $u(t) = \widetilde{\exp t X_m}(\varphi(a))$ for $t \in (-1, 1)$. Then we have

$$\|u'(t) - f(t, u(t))\|$$
$$= \|\widetilde{X}_m(\widetilde{\exp t X_m}(\varphi(a))) - \widetilde{X}_n(\widetilde{\exp t X_m}(\varphi(a)))\| \le \|X_m - X_n\|.$$

By comparison of solutions of the differential equation $x' = f(t, x)$ in [33, 10.5.1.1], we have

$$\|\widetilde{\exp t X_m}(\varphi(a)) - \widetilde{\exp t X_n}(\varphi(a))\| \le \|X_m - X_n\|\frac{e^{k|t|} - 1}{k}, \qquad (2.17)$$

where, by the Cauchy inequality (2.1), we have

$$\sup_n \|\widetilde{X}'_n\|_{\varphi(B_p)} \le k$$

for some $k > 0$. Hence

$$\|\widetilde{\exp t X_m} - \widetilde{\exp t X_n}\|_{\varphi(C_p)} \longrightarrow 0 \quad \text{as} \quad m, n \to \infty$$

for all $|t| < 1$. Therefore the sequence $(\widetilde{\exp t X_n})$ norm converges to an analytic function $\widetilde{g}_p(t)$ on $\varphi(C_p)$, for each $|t| < 1$.

Let ρ_{C_p} be the metric defined by the admissible neighbourhood C_p, as in (2.10). Then we have

$$\rho_{C_p}(\exp t X_n, \exp t X_m) \longrightarrow 0 \quad \text{as} \quad m, n \to \infty$$

for $|t| < 1$. Applying Lemma 2.2.12, one can find an admissible neighbourhood C_q in a chart $(\mathcal{U}_q, \varphi_q, V)$ at every point $q \in M$ such that $(\widetilde{\exp t X_n})$ is a Cauchy sequence on $\varphi_q(C_q)$ and hence norm converges to an analytic function $\widetilde{g}_q(t)$: $\varphi_q(C_q) \longrightarrow V$, for each $|t| < 1$. It is readily seen that the analytic function $g(t) : M \longrightarrow M$ defined by

$$g_t(q) = \varphi_q^{-1} \widetilde{g}_q(t) \varphi_q(q) \qquad (q \in M),$$

where $\varphi_p = \varphi$, is well defined. Moreover, we have

$d(g_t(q), g_t(y))$
$$\leq d(g_t(q), \exp t X_n(q))) + d(\exp t X_n(q), \exp t X_n(y)) + d(\exp X_n(y), g_t(y))$$
$$= d(g_t(q), \exp t X_n(q)) + d(q, y) + d(\exp X_n(y), g_t(y)) \longrightarrow d(q, y).$$

The same arguments imply that the sequence $(\exp -t X_n)$ converges locally uniformly to a contractive analytic function $h_t : M \longrightarrow M$ for $|t| < 1$, and $g_t h_t(q) = q$ for all $q \in M$. It follows that $g_t \in G_a(M)$ for $|t| < 1$.

Since $\exp(s + t)X_n = (\exp s X_n)(\exp t X_n)$, we have $g_{s+t} = g_s g_t$ whenever both sides are defined. Since $(-1, 1)$ generates $\mathbb{R}$, we can extend the map $t \in (-1, 1) \mapsto g_t \in G_a(M)$ continuously to a one-parameter group $t \in \mathbb{R} \mapsto g_t \in G_a(M)$ of bihomorphic isometries on M. Hence there is a Killing field X in $\mathfrak{g}_a(M)$ such that $g_t = \exp t X$ for all $t \in \mathbb{R}$.

By (2.17), we have, for $0 < |t| < 1$,

$$\left\| \frac{\widetilde{\exp t X} - \mathbf{1}}{t} - \frac{\widetilde{\exp t X_n} - \mathbf{1}}{t} \right\|_{\varphi(C_p)} = \lim_{m \to \infty} \left\| \frac{\widetilde{\exp t X_m} - \mathbf{1}}{t} - \frac{\widetilde{\exp t X_n} - \mathbf{1}}{t} \right\|_{\varphi(C_p)}$$

$$\leq \overline{\lim}_{m \to \infty} \|X_m - X_n\| \frac{e^{k|t|} - 1}{k|t|}.$$

Let $t \longrightarrow 0$. Then $\|\widetilde{X} - \widetilde{X}_n\|_{\varphi(C_p)} \longrightarrow 0$ as $n \to \infty$. By the principle of analytic continuation, we have $\widetilde{X} = Y$ on $\varphi(B_p)$ and hence

$$\|X_n - X\| = \|\widetilde{X}_n - \widetilde{X}\|_{\varphi(B_p)} = \|\widetilde{X}_n - Y\|_{\varphi(B_p)} \longrightarrow 0. \qquad \square$$

Corollary 2.2.15 *The Lie algebra* $\mathfrak{g}_a(M)$ *of analytic Killing fields on* M *is a real Banach Lie algebra.*

Proof Let $Z = [X, Y]$ for $X, Y \in \mathfrak{g}_a(M)$. Then for $a \in C_p$, we have

$$\widetilde{Z}(\varphi(a)) = \widetilde{Y}'(\varphi(a))\widetilde{X}(\varphi(a)) - \widetilde{X}'(\varphi(a))\widetilde{Y}(\varphi(a)),$$

which gives, via the Cauchy inequality,

$$\|\widetilde{Z}\|_{\varphi(C_p)} \le \|\widetilde{Y}'\|_{\varphi(C_p)}\|\widetilde{X}\|_{\varphi(C_p)} + \|\widetilde{X}'\|_{\varphi(C_p)}\|\widetilde{Y}\|_{\varphi(C_p)}$$

$$\le \frac{2}{R}\|\widetilde{Y}\|_{\varphi(B_p)}\|\widetilde{X}\|_{\varphi(B_p)},$$

where R is the distance between $\varphi(C_p)$ and the topological boundary of $\varphi(B_p)$. Since the norms $\|\cdot\|_{\varphi(C_p)}$ and $\|\cdot\|_{\varphi(B_p)}$ are equivalent on $\mathfrak{g}_a(M)$, we have shown that $\mathfrak{g}_a(M)$ is a Banach Lie algebra. $\qquad\square$

Lemma 2.2.16 *The map* $X \in \mathfrak{g}_a(M) \mapsto \exp X \in G_a(M)$ *is injective in a neighbourhood of* $0 \in \mathfrak{g}_a(M)$.

Proof There exists $r_0 > 0$ such that $\varphi(B_p)$ contains the open ball $\{v \in V : \|v\| < r_0\}$. Let $\mathfrak{N} = \{X \in \mathfrak{g}_a(M) : \|X\| < \min(r_0, \beta_0/2)\}$. For each $X \in \mathfrak{N}$, we have $\widetilde{X}(\varphi(B_p)) \subset \varphi(B_p)$ and also, by (2.16),

$$\widetilde{\exp t X}(\varphi(C_p)) \subset \varphi(B_p)$$

for $|t| \le 1$. For each $a \in C_p$, we have

$$\frac{d}{dt}\widetilde{\exp t X}(\varphi(a)) = \widetilde{X}(\widetilde{\exp t X}(\varphi(a)))$$

and it follows that

$$\frac{d^n}{dt^n}\widetilde{\exp t X}(\varphi(a)) = \widetilde{X}^n(\widetilde{\exp t X}(\varphi(a))) \qquad (n = 1, 2, \ldots).$$

Therefore we have the power series expansion for the analytic map

$$\widetilde{\exp t X}(\varphi(a)) = \sum_{n=0}^{\infty} \frac{t^n}{n!} \frac{d^n}{dt^n}\bigg|_{t=0} \widetilde{\exp t X}(\varphi(a)) = \sum_{n=0}^{\infty} \frac{t^n}{n!}\widetilde{X}^n(\widetilde{\exp X}(\varphi(a))).$$

In particular,

$$\widetilde{\exp X}(\varphi(a)) = \sum_{n=0}^{\infty} \frac{1}{n!}\widetilde{X}^n(\widetilde{\exp X}(\varphi(a))) \quad (X \in \mathfrak{N}, a \in C_p).$$

Let $H^\infty(\varphi(B_p), V)$ be the Banach space of bounded analytic maps from $\varphi(B_p)$ to V, equipped with the supremum norm. Define a differentiable map $F : \mathfrak{N} \longrightarrow H^\infty(\varphi(B_p), V)$ by

$$F(X) = \widetilde{\exp X} - \widetilde{X} - \mathbf{1} \qquad (X \in \mathfrak{N}).$$

Then we have $F(0) = 0 = F'(0)$ and one can find a convex neighbourhood $\mathfrak{N}_0 \subset \mathfrak{N}$ such that $\|F'(X)\| \le 1/2$ for all $X \in \mathfrak{N}_0$. Given $X, Y \in \mathfrak{N}_0$ with

$\exp X = \exp Y$, the Mean Value Theorem implies that

$$\|\widetilde{X} - \widetilde{Y}\|_{\varphi(B_p)}$$
$$= \|F(X) - F(Y)\| \leq \|X - Y\| \sup_{0 \leq \lambda \leq 1} \|F'(\lambda X + (1 - \lambda)Y)\| \leq \|X - Y\|/2$$

and hence $\|X - Y\| = \|\widetilde{X} - \widetilde{Y}\|_{\varphi(B_p)} = 0$. This proves injectivity of exp on $\mathfrak{N}_0$. $\qquad\square$

Theorem 2.2.17 *Let M be a connected manifold modelled on a complex Hilbert space, with a Riemannian metric. Then the biholomorphic isometry group $G_a(M)$ of M can be topologized to a real Banach Lie group with Lie algebra $\mathfrak{g}_a(M)$ formed by the analytic Killing fields on M.*

Proof We show that $G_a(M)$ can be equipped with an analytic structure modelled on the real Banach Lie algebra $\mathfrak{g}_a(M)$. By Birkhoff [8], there is a neighbourhood $\mathfrak{N}$ of $0 \in \mathfrak{g}_a(M)$ such that the Campbell–Baker–Hausdorff series

$$H(X, Y) = X + Y + \frac{1}{2}[X, Y] + \frac{1}{12}([X, [X, Y]] - [Y, [X, Y]]) + \cdots, \tag{2.18}$$

written in terms of iterates of $[X, Y]$, converges to an analytic map $H : \mathfrak{N} \times \mathfrak{N} \longrightarrow \mathfrak{g}_a(M)$ satisfying

$$\exp H(X, Y) = (\exp X)(\exp Y).$$

By Lemma 2.2.16, we may assume the restriction $\exp|_{\mathfrak{N}} : \mathfrak{N} \longrightarrow G_a(M)$ is injective. Since $H(0, 0) = 0$, we can find open symmetric neighbourhoods $\mathfrak{V} \subset \mathfrak{W} \subset \mathfrak{N}$ of $0 \in \mathfrak{g}_a(M)$ such that

$$H(\mathfrak{V} \times \mathfrak{V}) \subset \mathfrak{W} \quad \text{and} \quad H(\mathfrak{W} \times \mathfrak{W}) \subset \mathfrak{N}.$$

Let $\mathcal{U} = \exp(\mathfrak{W}) \subset G_a(M)$ and let

$$\varphi = (\exp|_{\mathfrak{W}})^{-1} : \mathcal{U} \longrightarrow \mathfrak{W} \subset \mathfrak{g}_a(M)$$

so that $\iota \in \mathcal{U}$ and $\varphi(\exp X) = X$ for $X \in \mathfrak{W}$. For each $g \in G_a(M)$, we define an injective map

$$\varphi_g : g\mathcal{U} \longrightarrow \mathfrak{W} \subset \mathfrak{g}_a(M)$$

by

$$\varphi_g(h) = \varphi(g^{-1}h) \qquad (h \in g\mathcal{U}).$$

If $g\mathcal{U} \cap h\mathcal{U} \neq \emptyset$, then $h^{-1}g \in \mathcal{U}$ and $h^{-1}g = \exp X$ for some $X \in \mathfrak{W}$. Hence for each $Y \in \varphi_g(g\mathcal{U} \cap h\mathcal{U})$, we have

$$\varphi_h \varphi_g^{-1}(Y) = \varphi(h^{-1}g \exp Y) = \varphi(\exp X)(\exp Y) = H(X, Y)$$

and therefore the composed map

$$\varphi_h \varphi_g^{-1} : \varphi_g(g\mathcal{U} \cap h\mathcal{U}) \longrightarrow \varphi_h(g\mathcal{U} \cap h\mathcal{U})$$

is analytic. It follows that $G_a(M)$ can be given a topology such that $\{(g\mathcal{U}, \varphi_g, \mathfrak{g}(M))\}$ is an analytic structure on $G_a(M)$ where $\varphi_\iota = \varphi$.

It remains to show that the group operations on $G_a(M)$ are analytic. Let $\mathcal{V} = \exp \mathfrak{V}$, which is a neighbourhood of $\iota \in G_a(M)$. Then the map $(g, h) \in \mathcal{V} \times \mathcal{V} \mapsto gh^{-1} \in G_a(M)$ has the local representation

$$(X, Y) \in \mathfrak{V} \times \mathfrak{V} \mapsto H(X, -Y) \in \mathfrak{W},$$

which is analytic. For each $g \in G_a(M)$, the left translation $\ell_g : h\mathcal{U} \longrightarrow gh\mathcal{U}$ on a chart $(h\mathcal{U}, \varphi_h)$ is given by $\ell_g = \varphi_{gh}^{-1} \varphi_h$ and hence analytic.

Finally, each $g \in G_a(M)$ induces a Lie algebra isomorphism $g_* : \mathfrak{g}_a(M) \longrightarrow \mathfrak{g}_a(M)$ by (2.2) and the conjugation $h \in G_a(M) \mapsto ghg^{-1} \in G_a(M)$ is analytic in a neighbourhood of $\iota \in G_a(M)$ via the local representation

$$h = \exp X \in \exp(\mathfrak{W} \cap g_*^{-1}(\mathfrak{W})) \longrightarrow g(\exp X)g^{-1} \in G$$

$$\downarrow \qquad\qquad\qquad \uparrow \exp$$

$$X \in \mathfrak{W} \cap g_*^{-1}(\mathfrak{W}) \xrightarrow{\;g_*\;} \quad g_* X \in \mathfrak{W}$$

where the commutativity of the diagram follows from (2.4). □

Corollary 2.2.18 *Let K be a closed subgroup of the Banach Lie group $G_a(M)$ in Theorem 2.2.17. Then K can be topologized to a real Banach Lie group with Lie algebra*

$$\mathfrak{k} = \{X \in \mathfrak{g}_a(M) : \exp t X \in K, \forall t \in \mathbb{R}\}.$$

Proof By Upmeier [111, 6.8], $\mathfrak{k}$ is a closed real subalgebra of $\mathfrak{g}_a(M)$ and hence a real Banach Lie algebra. Repeating the previous arguments, using K and $\mathfrak{k}$ in place of $G_a(M)$ and $\mathfrak{g}_a(M)$, one can equip K with a real Banach manifold structure, modelled on $\mathfrak{k}$. □

Remark 2.2.19 Corollary 2.2.18 is a special case of the fact that a closed subgroup K of a real Banach Lie group G with Lie algebra $\mathfrak{g}$ can always be topologized to a real Banach Lie group with Lie algebra $\mathfrak{k} = \{X \in \mathfrak{g} : \exp t X \in K, \forall t \in \mathbb{R}\}$ (cf. [111, 7.8]).

2.3 Jordan algebras and Riemannian symmetric spaces

We now discuss Riemannian manifolds which arise naturally in Jordan algebras. This provides a setting to show the useful relationships between the geometric structures of these manifolds and the underlying Jordan algebraic structures.

The important class of Riemannian manifolds closely related to Jordan algebras and Jordan triple systems is the class of *symmetric spaces*. In finite dimensions, these manifolds were classified by É. Cartan.

Definition 2.3.1 A connected Riemannian manifold M is called a *(Riemannian) symmetric space* if there is a symmetry s_p at every point $p \in M$; in other words, if every point p is an isolated fixed point of an involutive isometry $s_p : M \longrightarrow M$.

An isometry s_p is called *involutive* if the composite s_p^2 is the identity map. A fixed point p of s_p is *isolated* if it is the only fixed point in a neighbourhood of p. The following lemma, together with Proposition 2.2.9, shows that there can only be one symmetry s_p at p.

Lemma 2.3.2 *In a Riemannian symmetric space M, the differential ds_p of a symmetry s at $p \in M$ is minus the identity map on T_pM.*

Proof Since s^2 is the identity map on M, the composite $(ds_p)^2$ is the identity map on the tangent space T_pM, and hence ds_p induces an eigenspace decomposition of T_pM:

$$T_pM = \{X_p : ds_p(X_p) = X_p\} \oplus \{X_p : ds_p(X_p) = -X_p\}.$$

Let $ds_p(X_p) = X_p$. Then we must have $X_p = 0$ for otherwise, in a normal neighbourhood of p, there is a unique geodesic $\alpha(t)$ with $\alpha(0) = p$ and $\alpha'(0) = X_p$ which implies that $s \circ \alpha = \alpha$, since $s \circ \alpha$ is a geodesic satisfying the same conditions. This contradicts the fact that p is an isolated fixed point of s. Hence

$$T_pM = \{X_p : ds_p(X_p) = -X_p\}.$$

$\square$

Example 2.3.3 A real Hilbert space V is a Riemannian symmetric space. The symmetry at a point $p \in V$ is the map $s_p(x) = 2p - x$.

Example 2.3.4 Let M be a Riemannian manifold. If there is a symmetry s_p at some point $p \in M$ and if $G(M)$ acts transitively on M, then M is a symmetric space. Indeed, given $q \in M$ with $q = f(p)$ for some $f \in G(M)$, the map $s_q = f \circ s_p \circ f^{-1}$ is a symmetry at q. We note, however, that transitivity of the action of $G(M)$ on M alone is not sufficient for M to be a symmetric space [96].

Example 2.3.5 The unit sphere $S(V) = \{(\lambda, v) \in \mathbb{R} \times V : \lambda^2 + \|v\|^2 = 1\}$ in a Hilbert space $\mathbb{R} \times V$ is a symmetric space in the induced Riemannian metric. The isometry group $O(\mathbb{R} \times V)$ acts transitively on $S(V)$ and the symmetry s_p at the north pole $p = (1, 0)$ is given by

$$s_p(\lambda, v) = (\lambda, -v).$$

Lemma 2.3.6 *Let M be a Riemannian symmetric space. Then the isometry group $G(M)$ acts transitively on M.*

Proof Given a geodesic $\gamma : [0, 1] \longrightarrow M$, the symmetry $s_{\gamma(\frac{1}{2})}$ at the midpoint $\gamma(\frac{1}{2})$ reverses γ, sending the endpoints $\gamma(0)$ and $\gamma(1)$ to each other. Since M is connected, any two points $p, q \in M$ can be joined by the union of a finite number of geodesics, and hence a finite composite of symmetries would send p to q. $\qquad\square$

Riemannian symmetric spaces associated with real Jordan algebras, in the form of symmetric cones and manifolds of idempotents, were first studied in Helwig [52], Hirzebruch [55], Koecher [76, 79] and Vinberg [113]. Koecher and Vinberg [76, 79, 113] have shown that the category of finite-dimensional formally real Jordan algebras is equivalent to the category of symmetric cones in finite-dimensional Euclidean spaces. They showed that the latter are exactly the interiors of positive cones in formally real Jordan algebras. Hirzebruch [55] has shown that the manifold of minimal idempotents in a finite-dimensional *simple* formally real Jordan algebra is a Riemannian symmetric space. We now discuss the infinite-dimensional extension of these results.

By Theorem 1.1.14, finite-dimensional formally real Jordan algebras are characterised by the existence of a positive definite trace form $\langle x, y \rangle =$ Trace $(x \square y)$, which provides a natural topology for the aforementioned manifolds. In the absence of a trace form in infinite dimension, an appropriate generalisation of finite-dimensional formally real Jordan algebras is the class of JB-algebras, which presents a proper topological setting to extend Hirzebruch's result. Another generalisation suitable for geometry and symmetric cones is the class of JH-algebras.

A real Jordan algebra $\mathcal{B}$ is called a *JB-algebra* if it is also a Banach space and the norm satisfies

$$\|ab\| \leq \|a\|\|b\|, \quad \|a^2\| = \|a\|^2, \quad \|a^2\| \leq \|a^2 + b^2\|$$

for all $a, b \in \mathcal{B}$.

A JB-algebra $\mathcal{A}$ is called a *JBW-algebra* if it is the dual of a Banach space, in which case the predual of $\mathcal{A}$ is unique, the weak* topology on $\mathcal{A}$ is unambiguous and $\mathcal{A}$ must have an identity, denoted by **1**. A JBW-algebra is called a *JBW-factor* if its *centre*

$$Z = \{z \in \mathcal{A} : z(ab) = (za)b, \quad \forall\, a, b \in \mathcal{A}\}$$

is trivial, that is, $Z = \{\alpha\mathbf{1} : \alpha \in \mathbb{R}\}$. We refer to Hanche-Olsen and Størmer [47] and Topping [109] for basic results of JB-algebras and JBW-algebras.

An important class of JB-algebras is furnished by C*-algebras. We recall that a C*-algebra is representable as a norm-closed subalgebra $\mathcal{A}$ of the algebra $L(H)$ of bounded linear operators on a complex Hilbert space H such that $a \in \mathcal{A}$ implies $a^* \in \mathcal{A}$, where a^* denotes the adjoint of the operator $a : H \longrightarrow H$. If H is a *real* Hilbert space, then $\mathcal{A}$ satisfying the same conditions is called a *real* C*-*algebra*. We regard a C*-algebra over a complex Hilbert space as a *real* C*-algebra by restricting to the real scalar field. The Banach space

$$\mathcal{A}_{sa} = \{a \in \mathcal{A} : a^* = a\}$$

of self-adjoint operators in a real C*-algebra $\mathcal{A}$ is a JB-algebra in the special Jordan product

$$a \circ b = \frac{1}{2}(ab + ba),$$

where the Jordan product $\circ$ is the same as the associative product on the right if $\mathcal{A}$ is commutative.

The complexification $\mathcal{A}_c = \mathcal{A} \oplus i\mathcal{A}$ of a real C*-algebra $\mathcal{A}$ is a C*-algebra, and if $\mathcal{A}$ is commutative, then so is $\mathcal{A}_c$, which must be of the form $C_0(S)$, the C*-algebra of complex continuous functions vanishing at infinity on a locally compact Hausdorff space S. The self-map τ on $\mathcal{A} \oplus i\mathcal{A} = C_0(S)$ defined by

$$\tau(a + ib) = \overline{a} + i\overline{b} \qquad (a, b \in \mathcal{A})$$

is a complex linear *-automorphism and is of period 2, that is, τ^2 is the identity map. The dual map τ^* restricts to a homeomorphism $\sigma : S \longrightarrow S$ of period 2 and we have the identification

$$\mathcal{A} = \{f \in C_0(S) : f \circ \sigma = \overline{f}\}, \tag{2.19}$$

where the involution f^* is the complex conjugation $\overline{f}$ of f.

Bilinearity implies that the multiplication $(a, b) \in \mathcal{A}^2 \mapsto ab \in \mathcal{A}$ in a JB-algebra $\mathcal{A}$ is analytic (cf. Example 2.1.5), and hence the Jordan triple product $(a, b, c) \mapsto \{a, b, c\} = (ab)c + a(bc) - b(ac)$ is also analytic.

A real Jordan algebra $\mathcal{H}$ is called a *JH-algebra* in Nomura [93] if it is also a real Hilbert space in which the inner product $\langle \cdot, \cdot \rangle$ is *associative*; that is,

$$\langle ab, c \rangle = \langle b, ac \rangle \qquad (a, b, c \in \mathcal{H}).$$

A finite-dimensional JH-algebra is called *Euclidean* in Faraut and Koranyi [38]. Evidently, a finite-dimensional formally real Jordan algebra, equipped with the trace norm as an inner product, is Euclidean. In fact, all three concepts introduced above are equivalent in finite dimensions.

Lemma 2.3.7 *Let $\mathcal{A}$ be a finite-dimensional real Jordan algebra. The following conditions are equivalent:*

(i) $\mathcal{A}$ *is formally real.*
(ii) $\mathcal{A}$ *is Euclidean and has an identity* **1**.
(iii) $\mathcal{A}$ *is a JB-algebra.*

Proof See Braun and Koecher [13, p. 320] and Hanche-Olsen and Størmer [47, pp. 77, 84]. We simply note that, in a finite-dimensional formally real Jordan algebra $\mathcal{A}$, the positive cone $\{a^2 : a \in \mathcal{A}\}$ induces a partial ordering $\leq$ in $\mathcal{A}$ and a norm

$$\|a\| := \inf\{\lambda > 0 : \lambda \mathbf{1} \leq a \leq \lambda \mathbf{1}\}$$

which makes $\mathcal{A}$ into a JB-algebra. $\qquad \square$

Example 2.3.8 An infinite-dimensional formally real Jordan algebra need not be a JB-algebra. The *disc algebra* $\mathcal{D}$ on the open unit disc U in the complex plane consists of continuous functions on the closure $\overline{U}$ which are holomorphic on U. Let

$$\mathcal{A} = \{f \in \mathcal{D} : f(U \cap \mathbb{R}) \subset \mathbb{R}\},$$

which is a real closed subalgebra of $\mathcal{D}$ and is an associative Jordan algebra. It is formally real and we also have $\|f^2\| = \|f\|^2$ for all $f \in \mathcal{A}$, but

$$\| \sin^2 nz + \cos^2 nz \| = 1 \not\geq \| \sin^2 nz \|$$

for large n.

Example 2.3.9 The exceptional Jordan algebra $H_3(\mathbb{O})$ in Example 1.1.5 is a formally real Jordan algebra. Hence it is a JB-algebra with the norm described in the proof of Lemma 2.3.7. Following [117], we equip

$$H_3(\mathcal{O}) = H_3(\mathbb{O}) + i H_3(\mathbb{O})$$

(cf. Example 1.2.8) with the norm

$$\|(a_{ij})\|_c = \inf\{\alpha > 0 : (a_{ij}) \in \alpha C\},$$

which is the Minkowski functional of the convex hull C of the set

$$\left\{ \exp iz = \sum_{n=0}^{\infty} \frac{i^n z^n}{n!} : z \in H_3(\mathcal{O}) \right\}.$$

$(H_3(\mathcal{O}), \|\cdot\|_c)$ is a Banach space complexification of $H_3(\mathbb{O})$, and we will always assume this norm structure of $H_3(\mathcal{O})$. As a subtriple of $H_3(\mathcal{O})$, the Hermitian Jordan triple $M_{12}(\mathcal{O})$ also inherits this norm.

Let $\mathcal{A}$ be a JH-algebra with identity $\mathbf{1}$. Then for each $a \in \mathcal{A}$, we have

$$\|a\|^2 = \langle a, a \rangle = \langle \mathbf{1}, a^2 \rangle \leq \|\mathbf{1}\|\|a^2\|.$$

The box operator $a \,\square\, a : \mathcal{A} \longrightarrow \mathcal{A}$ is norm continuous, since it is weakly continuous by associativity of the inner product,

$$\langle (a \,\square\, a)x_\alpha, y \rangle = \langle x_\alpha, (a \,\square\, a)y \rangle \longrightarrow \langle x, (a \,\square\, a)y \rangle = \langle (a \,\square\, a)x, y \rangle,$$

whenever a net (x_α) converges weakly to $x \in \mathcal{A}$. Actually, it follows from Lemma 3.5.5 that there is a constant $c > 0$ such that $\|a \,\square\, a\| \leq c\|a\|^2$ for all $a \in \mathcal{A}$. Hence

$$\|a^2\| = \|(a \,\square\, a)\mathbf{1}\| \leq \|a \,\square\, a\|\|\mathbf{1}\| \leq c\|\mathbf{1}\|\|a\|^2. \tag{2.20}$$

Now we consider symmetric cones.

Definition 2.3.10 Let V be a real Hilbert space with inner product $\langle \cdot, \cdot \rangle$. An open cone $\Omega \subset V$ is called *symmetric* if it satisfies the following conditions:

(i) (self-duality) $\Omega = \{v \in V : \langle v, x \rangle > 0 \ \forall x \in \overline{\Omega}\backslash\{0\}\}$;
(ii) (homogeneity) given $x, y \in \Omega$, there is a continuous linear isomorphism $h : V \longrightarrow V$ such that $h(\Omega) = \Omega$ and $h(x) = y$.

Example 2.3.11 For $n > 2$, the *Lorentz cone* $\Lambda_n \subset \mathbb{R}^n$, defined as follows, is symmetric:

$$\Lambda_n = \{(x_1, \ldots, x_n) \in \mathbb{R}^n : x_1 > 0 \text{ and } x_1^2 - x_2^2 - \cdots - x_n^2 > 0\}.$$

Example 2.3.12 Let V be the Hilbert space of all real symmetric $m \times m$ matrices, equipped with the inner product

$$\langle A, B \rangle = \mathrm{Trace}\,(AB).$$

The cone Ω consisting of positive definite symmetric $m \times m$ real matrices is symmetric in V.

Example 2.3.13 Let $\mathcal{A}$ be a finite-dimensional formally real Jordan algebra. The interior Ω of the cone $\{a^2 : a \in \mathcal{A}\}$ is a symmetric cone with respect to the inner product $\langle x, y \rangle = \mathrm{Trace}\,(x \,\square\, y)$. This example is a special case of Lemma 2.3.17.

A symmetric cone Ω in a finite-dimensional Hilbert space V can be given a Riemannian metric g which turns it into a Riemannian symmetric space. The metric g can be constructed via a Jordan algebra, and indeed, one only needs to do this for the positive cone of a finite-dimensional formally real Jordan algebra, by the following result of Koecher [76] and Vinberg [113], which, together with Example 2.3.13, establishes the one–one correspondence between symmetric cones and formally real Jordan algebras in finite dimensions.

Theorem 2.3.14 *Let Ω be a symmetric cone in a finite-dimensional real Hilbert space V. Then V can be equipped with a formally real Jordan product such that Ω is the interior of its closure $\overline{\Omega}$ and*

$$\overline{\Omega} = \{x^2 : x \in V\}.$$

We will construct a Riemannian metric on the interior of the positive cone of a unital JH-algebra, and thereby extend Koecher's result [79] to the infinite-dimensional setting. An important example of a unital JH-algebra is the spin factor $H \oplus \mathbb{R}$, which is a direct sum of a real Hilbert space $(H, \langle \cdot, \cdot \rangle_H)$ and $\mathbb{R}$, with Jordan product

$$(x \oplus \zeta) \circ (y \oplus \eta) = (\eta x + \zeta y) \oplus (\langle x, y \rangle_H + \zeta \eta)$$

and the associative inner product

$$\langle x \oplus \zeta, y \oplus \eta \rangle = \langle x, y \rangle_H + \zeta \eta.$$

We note that the spin factor $H \oplus \mathbb{R}$ is also a JB-algebra with the equivalent norm

$$\|(x \oplus \zeta)\| = \|x\|_H + |\zeta|.$$

$H \oplus \mathbb{R}$ has the identity $0 \oplus 1$.

Example 2.3.15 Let $\mathcal{M}$ be a von Neumann algebra with a faithful semifinite normal trace τ. Then $L_2(\mathcal{M}, \tau)$ is a two-sided $\mathcal{M}$-module, where $L_2(\mathcal{M}, \tau)$ is the completion of the inner product space

$$N_\tau = \{a \in \mathcal{M} : \tau(a^*a) < \infty\}$$

with the inner product

$$\langle a, b \rangle = \tau(ab^*) \qquad (a, b \in N_\tau).$$

The multiplication in N_τ is continuous with respect to the inner product norm and therefore extends to an associative binary product in $L_2(\mathcal{M}, \tau)$. It follows that $L_2(\mathcal{M}, \tau)$ is a Jordan algebra in the special Jordan product

$$a \circ b = \frac{1}{2}(ab + ba).$$

Moreover, $L_2(\mathcal{M}, \tau)$ is a JH-algebra. If $\mathcal{M} = L(H)$, the von Neumann algebra of bounded linear operators on a complex Hilbert space H, then the canonical normal trace τ on $\mathcal{M}$ is faithful semifinite and $L_2(\mathcal{M}, \tau)$ is the algebra of Hilbert–Schmidt operators on H, with the Hilbert–Schmidt norm.

We recall that L_a and Q_a denote respectively the left multiplication and the quadratic operator on a Jordan algebra $\mathcal{A}$, defined by an element $a \in \mathcal{A}$. Let $\mathcal{A}$ have an identity $\mathbf{1}$. An element $a \in \mathcal{A}$ is called *invertible* if there exists $a^{-1} \in \mathcal{A}$ such that $aa^{-1} = \mathbf{1}$ and $(a^2)a^{-1} = a$. This is equivalent to saying that the operator Q_a is invertible with inverse $Q_a^{-1} = Q_{a^{-1}}$.

Let $p \in \mathcal{A}$ be an idempotent with Peirce decomposition

$$\mathcal{A} = \mathcal{A}_0(p) \oplus \mathcal{A}_1(p) \oplus \mathcal{A}_2(p)$$

and the corresponding *Peirce projections* $P_k(p) : \mathcal{A} \longrightarrow \mathcal{A}_k(p)$, given by

$$P_0(p) = Q_{1-p}, \quad P_1(p) = 4L_p - 4L_p^2, \quad P_2(p) = Q_p.$$

By Peirce arithmetic in Theorem 1.2.44, we have

$$\mathcal{A}_0(p)\mathcal{A}_2(p) = \{0\} \quad \text{and} \quad \mathcal{A}_1(p)\mathcal{A}_1(p) \subset \mathcal{A}_0(p) \oplus \mathcal{A}_2(p).$$

An idempotent in a JB-algebra is more often called a *projection* in the literature, which we follow henceforth, and we use the same term for JH-algebras. Given a JH-algebra $\mathcal{A}$ and a projection $p \in \mathcal{A}$, we have $\langle p, a^2 \rangle \geq 0$ for all $a \in \mathcal{A}$, since

$$\langle p, a^2 \rangle = \langle pa, a \rangle = \frac{1}{2}\|P_1(p)(a)\|^2 + \|P_2(p)(a)\|^2.$$

Also, two projections p and q are orthogonal with respect to the inner product if, and only if, the Jordan product $pq = 0$.

The spectral decomposition of elements in a finite-dimensional formally real Jordan algebra, proved in Theorem 1.1.14, can be extended to JH-algebras. Indeed, applying similar arguments in the proof of Proposition 3.5.19 to the closed subalgebra $\mathcal{A}(x)$ generated by an element x in a JH-algebra $\mathcal{A}$, one can find a maximal family of orthogonal projections in $\mathcal{A}(x)$ and hence a sequence

$\{e_k\}$ of orthogonal projections such that

$$x = \sum_{k=1}^{\infty} \lambda_k e_k \qquad (\lambda_k \in \mathbb{R}),$$

where we have $\lambda_k = \langle x, e_k \rangle$, and if $\lambda_k \geq 0$ for all k, then

$$x = \sum_{k=1}^{\infty} (\sqrt{\lambda_k})^2 e_k = \left(\sum_{k=1}^{\infty} \sqrt{\lambda_k} e_k \right)^2,$$

which is in the positive cone

$$\mathcal{A}^+ = \{x^2 : x \in \mathcal{A}\}.$$

This remark implies that each $a = x^2 \in \mathcal{A}^+$ has the form

$$a = \sum_{k=1}^{\infty} \lambda_k^2 e_k$$

for some sequence $\{e_k\}$ of orthogonal projections. It follows that, for every $y \in \mathcal{A}$,

$$\langle L_a y, y \rangle = \sum_{k=1}^{\infty} \lambda_k^2 \langle e_k, y^2 \rangle \geq 0.$$

The positive cone $\mathcal{A}^+$ is not only norm-closed in the Hilbert space $\mathcal{A}$, but also weakly closed, as shown in the proof of Lemma 2.3.17.

Definition 2.3.16 Let $\mathcal{A}$ be a JH-algebra with identity $\mathbf{1}$. We denote by $\Omega(\mathcal{A}) = int\, \mathcal{A}^+$ the interior of the positive cone $\mathcal{A}^+$, and call it the *open cone* of $\mathcal{A}$.

We exhibit a typical example of an infinite-dimensional symmetric cone.

Lemma 2.3.17 *The open cone $\Omega(\mathcal{A})$ of a JH-algebra $\mathcal{A}$ with identity is a symmetric cone.*

Proof To prove self-duality, we first show that

$$\mathcal{A}^+ = \{y \in \mathcal{A} : \langle y, a \rangle \geq 0, \forall a \in \mathcal{A}^+\},$$

which implies immediately that $\mathcal{A}^+$ is weakly closed. Given $y \in \mathcal{A}^+$ with

$$y = \sum_{k=1}^{\infty} \alpha_k p_k \qquad (\alpha_k \geq 0)$$

for some mutually orthogonal projections $\{p_k\}$, we have, for $a = x^2 \in \mathcal{A}^+$,

$$\langle y, a \rangle = \sum_{k=1}^{\infty} \alpha_k \langle p_k, x^2 \rangle \geq 0.$$

Conversely, if $y = \sum_k \beta_k q_k$ is such that $\langle y, x^2 \rangle \geq 0$ for all $x \in \mathcal{A}$, where q_k's are mutually orthogonal projections, then

$$0 \leq \langle y, q_k^2 \rangle = \langle y, q_k \rangle = \beta_k \langle q_k, q_k \rangle$$

implies that $\beta_k \geq 0$ for all k and $y \in \mathcal{A}^+$.

Hence, if $y \in int \, \mathcal{A}^+$ and $a \in \mathcal{A}^+ \backslash \{0\}$, then we have $\langle z, a \rangle \geq 0$ for all z in some neighbourhood N of y. For any $\varepsilon > 0$, we can choose $w \in \mathcal{A}$ such that $y + w \in N$ and $\langle w, a \rangle \leq \varepsilon$. This gives

$$\langle y, a \rangle + \varepsilon \geq \langle y, a \rangle + \langle w, a \rangle = \langle y + w, a \rangle \geq 0$$

and it follows that $\langle y, a \rangle > 0$.

On the other hand, $int \, \mathcal{A}$ contains the set $S = \{y \in \mathcal{A} : \langle y, a \rangle > 0, \forall a \in \mathcal{A}^+ \backslash \{0\}\}$, which is open. To see that $\mathcal{A} \backslash S$ is closed, let (y_n) be a sequence norm converging to $y \in \mathcal{A}$ and $\langle y_n, a_n^2 \rangle \leq 0$ for some $a_n \in \mathcal{A}$ with $\|a_n\| = 1$. We show that $y \in \mathcal{A} \backslash S$. Note that the sequence (a_n^2) is bounded by (2.20) and hence it has a subnet (a_β^2) which converges weakly to some $a \in \mathcal{A}$. We have $a \in \mathcal{A}^+$ and also $a \neq 0$, since

$$1 = \|a_\beta\|^2 = \langle a_\beta, a_\beta \rangle = \langle \mathbf{1}, a_\beta^2 \rangle \longrightarrow \langle \mathbf{1}, a \rangle.$$

Moreover,

$$\langle y, a \rangle = (\langle y, a \rangle - \langle y, a_\beta^2 \rangle) + (\langle y, a_\beta^2 \rangle - \langle y_\beta, a_\beta^2 \rangle) + \langle y_\beta, a_\beta^2 \rangle$$
$$\leq (\langle y, a \rangle - \langle y, a_\beta^2 \rangle) + (\langle y, a_\beta^2 \rangle - \langle y_\beta, a_\beta^2 \rangle) \longrightarrow 0 \quad \text{as} \quad \beta \to 0,$$

which implies $y \in \mathcal{A} \backslash S$.

For homogeneity, we observe that, given $a^2, b^2 \in \Omega(\mathcal{A})$, the element a is invertible and $Q_b Q_{a^{-1}} : \mathcal{A} \longrightarrow \mathcal{A}$ is a continuous linear isomorphism, sending Ω to itself, and a^2 to b^2. $\qquad \square$

We note from the above lemma that $\mathbf{1} \in \Omega(\mathcal{A}) = \{y \in \mathcal{A} : \langle y, a \rangle > 0, \forall a \in \mathcal{A}^+ \backslash \{0\}\}$. Each element x in the open cone $\Omega(\mathcal{A})$ is invertible in $\mathcal{A}$. In fact, if $N \subset \mathcal{A}^+$ is a neighbourhood of x, then $x - N$ is a neighbourhood of 0 and hence $\mathbf{1} \in \alpha(x - N)$ for some $\alpha > 0$. It follows that $x = a + \alpha^{-1} \mathbf{1}$ for some $a \in N \subset \mathcal{A}^+$. Let $\mathcal{A}(a, \mathbf{1})$ be the smallest closed subalgebra of $\mathcal{A}$ containing a and $\mathbf{1}$, which is just the closed real linear span of polynomials in a. Then $\mathcal{A}(a, \mathbf{1})$ is a JH-algebra and the operator $L_x = L_a + \alpha^{-1} I : \mathcal{A}(a, \mathbf{1}) \longrightarrow \mathcal{A}(a, \mathbf{1})$ is invertible, since $\langle L_a y, y \rangle \geq 0$ for all y in the Hilbert space $\mathcal{A}(a, \mathbf{1})$. It follows that x is invertible with inverse $x^{-1} = L_x^{-1}\mathbf{1} \in \mathcal{A}(a, \mathbf{1})$ by power associativity.

As in the case of JB-algebras, Jordan multiplication and Jordan triple product are analytic maps on JH-algebras $\mathcal{A}$. We show next that the inverse map $x \in \Omega(\mathcal{A}) \mapsto x^{-1} \in \mathcal{A}$ is a smooth map.

Lemma 2.3.18 *Let $\mathcal{A}$ be a JH-algebra with identity and open cone $\Omega(\mathcal{A})$. The inverse map $f : x \in \Omega(\mathcal{A}) \mapsto x^{-1} \in \mathcal{A}$ is smooth and has derivative*

$$f'(a) = -Q_a^{-1} : \mathcal{A} \longrightarrow \mathcal{A}$$

at $a \in \Omega(\mathcal{A})$.

Proof Let $F : \mathcal{A} \longrightarrow L(\mathcal{A}, \mathcal{A})$ be defined by

$$F(x) = Q_x \qquad (x \in \mathcal{A}).$$

Then $F(a + h)(z) - F(a)(z) - 2\{a, z, h\} = \{a + h, z, a + h\} - \{a, z, a\} - 2\{a, z, h\} = \{h, z, h\}$ implies that F is differentiable and the derivative $F'(a) : \mathcal{A} \longrightarrow L(\mathcal{A}, \mathcal{A})$ is given by

$$F'(a)(h) = 2\{a, \cdot, h\}.$$

Observe that $F(x) \circ f(x) = x$ for all $x \in \mathcal{A}$; in other words, $F(\cdot) \circ f$ is the identity map $\iota_\mathcal{A}$ on $\mathcal{A}$. Differentiating $F(\cdot) \circ f$ at $a \in \Omega(\mathcal{A})$ gives

$$F'(a)(\cdot)f(a) + F(a) \circ f'(a) = \iota_\mathcal{A}.$$

Hence

$$2\{a, f(a), x\} + \{a, f'(a)(x), a\} = x \qquad (x \in \mathcal{A}),$$

which simplifies to

$$\{a, f'(a)(x), a\} = -x,$$

since $a(a^{-1}x) = a^{-1}(ax)$. Therefore we obtain

$$f'(a)(x) = -Q_{a^{-1}}(x) = -Q_a^{-1}(x) \qquad (x \in \mathcal{A}).$$

Since the derivative $f'(a) = -F(a^{-1})$ is the composite of the inverse map with $-F$, which can be differentiated repeatedly, we see that f is smooth.   □

Theorem 2.3.19 *Let $\mathcal{A}$ be a JH-algebra with identity $\mathbf{1}$. Then the open cone $\Omega(\mathcal{A})$ is a Riemannian symmetric space.*

Proof Since $\Omega(\mathcal{A})$ is open in $\mathcal{A}$, it is a submanifold of $\mathcal{A}$. Also $\Omega(\mathcal{A})$ is connected because it is convex. For each $\omega \in \Omega(\mathcal{A})$, we define a symmetric bilinear form on the tangent space $T_\omega \Omega(\mathcal{A}) = \mathcal{A}$ by

$$\langle u, v \rangle_\omega = \langle Q_{\omega^{-1}}(u), v \rangle = \langle \{\omega^{-1}, u, \omega^{-1}\}, v \rangle \qquad (u, v \in \mathcal{A}),$$

which is completely positive definite, since $\omega^{-1} = x^{-2}$ for some $x \in \mathcal{A}$, and we have

$$\langle Q_{\omega^{-1}}(u), u \rangle = \langle Q_{x^{-2}}(u), u \rangle = \langle Q_{x^{-1}}(u), Q_{x^{-1}}(u) \rangle \geq 0,$$

and also, the operator $Q_{\omega^{-1}} : \mathcal{A} \longrightarrow \mathcal{A}$ is invertible. Hence $g_\omega(u, v) = \langle \{\omega^{-1}, u, \omega^{-1}\}, v \rangle$ defines a Riemannian metric on $\Omega(\mathcal{A})$.

To show that $\Omega(\mathcal{A})$ is a Riemannian symmetric space, it suffices to show that the identity $\mathbf{1}$ is an isolated fixed point of an involutive isometry on $\Omega(\mathcal{A})$, by the homogeneous property that, given $x = a^2 \in \Omega(\mathcal{A})$, the linear isomorphism $\theta = Q_{a^{-1}} : \mathcal{A} \longrightarrow \mathcal{A}$ restricts to an isometry on $\Omega(\mathcal{A})$ with respect to g, sending x to $\mathbf{1}$:

$$
\begin{aligned}
&g_{\theta(\omega)}(d\theta_\omega(u), d\theta_\omega(v)) \\
&= g_{\theta(\omega)}(\theta(u), \theta(v)) \\
&= \langle \{\theta(\omega)^{-1}, \theta(u), \theta(\omega)^{-1}\}, \theta(v) \rangle \\
&= \langle \{\{a, \omega^{-1}, a\}, \{a^{-1}, u, a^{-1}\}, \{a, \omega^{-1}, a\}\}, \{a^{-1}, v, a^{-1}\} \rangle \\
&= \langle \{\omega^{-1}, \{a, \{a^{-1}, u, a^{-1}\}, a\}, \omega^{-1}\}, v \rangle \\
&= \langle \{\omega^{-1}, u, \omega^{-1}\}, v \rangle
\end{aligned}
$$

by associativity of the inner product.

Finally, the symmetry at $\mathbf{1}$ is given by the involutive isometry $\sigma : \Omega(\mathcal{A}) \longrightarrow \Omega(\mathcal{A})$, defined by $\sigma(\omega) = \omega^{-1}$, where

$$
\begin{aligned}
g_{\sigma(\omega)}(d\sigma_\omega(u), d\sigma_\omega(v)) &= g_{\sigma(\omega)}(-Q_{\omega^{-1}}(u), -Q_{\omega^{-1}}(v)) \\
&= \langle \{\omega, -Q_{\omega^{-1}}(u), \omega\}, -Q_{\omega^{-1}}(v) \rangle \\
&= \langle u, \{\omega^{-1}, v, \omega^{-1}\} \rangle \\
&= \langle \{\omega^{-1}, u, \omega^{-1}\}, v \rangle.
\end{aligned}
$$

$\square$

To extend Hirzebruch's result mentioned previously to infinite dimensions, one needs to identify the class of JB-algebras which generalises the *simple* finite-dimensional formally real Jordan algebras. This is actually the class of JBW-factors defined earlier. Hence our goal is to show that the rank-n projections in a JBW-factor form a Riemannian symmetric space. These manifolds can be regarded as an infinite-dimensional analogue of the Grassmann manifolds which are manifolds of subspaces of a particular dimension of a vector space.

We first develop some algebraic results for projections in JB-algebras. A JB-algebra may contain only the trivial projection 0; for instance, the algebra $C_0(\mathbb{R})$ of real continuous functions on $\mathbb{R}$ vanishing at infinity contains only the projection 0. However, a JBW-algebra contains an abundance of projections [47, 4.2.3].

A nonzero projection p in a JB-algebra $\mathcal{A}$ is called *minimal* if $\{p, \mathcal{A}, p\} = \mathbb{R}p$, where $\{\cdot, \cdot, \cdot\}$ denotes the canonical Jordan triple product. Given a *nonzero* projection p in a JBW-algebra $\mathcal{A}$, we say that p has *infinite rank* if there are infinitely many mutually orthogonal nonzero projections in $\{p, \mathcal{A}, p\}$; otherwise, p is said to have *finite rank*, and the unique maximal cardinality of mutually orthogonal nonzero projections in $\{p, \mathcal{A}, p\}$ is defined to be the *rank* of p, denoted by rank(p), in which case, p is a sum of orthogonal minimal projections $p_1, \ldots, p_n$ with $n = \text{rank}(p)$. The minimal projections are exactly the rank-1 projections. We regard 0 as a finite rank projection with rank(0) $= 0$. The *rank* of a JBW-algebra $\mathcal{A}$, rank($\mathcal{A}$), is defined to be the rank of the identity.

Lemma 2.3.20 *Let $\mathcal{A}$ be a unital JB-algebra and let $p \in \mathcal{A}$ be a minimal projection with Peirce decomposition*

$$\mathcal{A} = \mathcal{A}_0(p) \oplus \mathcal{A}_1(p) \oplus \mathcal{A}_2(p).$$

Then for every $x \in \mathcal{A}_1(p)\backslash\{0\}$, we have $x^2 \in \mathcal{A}_0(p) \oplus \mathcal{A}_2(p)$, and the Jordan subalgebra $\mathcal{A}(p, x)$ in $\mathcal{A}$ generated by p and x has dimension 3.

Proof Note that $\{p, x, p\} = 2p(px) - px = 0$ for $x \in \mathcal{A}_1(p)$. By the Shirshov–Cohn Theorem, $\mathcal{A}(p, x)$ is special and we have $x = 2(px) = x.p + p.x$, where the product on the right is that of a containing associative algebra. This gives $x^2 = x^2.p + x.p.x$ and, by minimality, we have $p.x^2 = p.x^2p + p.x.p.x = p.x^2.p = \alpha p$ for some $\alpha \in \mathbb{R}$. Likewise $x^2.p = \alpha p$ and hence $px^2 = \alpha p = \{p, x^2, p\}$. Moreover, $x^2(px) = (x^2p)x$ gives $x^3 = \alpha x$. Hence $\mathcal{A}(p, x)$ is the linear span of $\{p, x, x^2\}$, which can be seen readily to be linearly independent, using the above facts. $\square$

An element s in a unital JB-algebra $\mathcal{A}$ is called a *symmetry* if $s^2 = \mathbf{1}$, in which case s is a tripotent in $\mathcal{A}$ with $\|s\| = 1$, and $\{s, \{s, a, s\}, s\} = a$ for all $a \in \mathcal{A}$. Hence the quadratic operator $Q_s : \mathcal{A} \longrightarrow \mathcal{A}$ is an involutive Jordan automorphism and an isometry (cf. [47, 2.8.6, 3.4.3]). Two projections p and q in $\mathcal{A}$ are called *Jordan equivalent* if they are exchanged by a symmetry s, that is, $p = \{s, q, s\}$, which is equivalent to $q = \{s, p, s\}$. We note that any two minimal projections in a JBW-factor are Jordan equivalent, by the comparison theorem for projections [47, 5.2.13].

Remark 2.3.21 A symmetry element *in a JB-algebra* should not be confused with a *symmetry* $s_p : M \longrightarrow M$ for a symmetric space M in Definition 2.3.1.

Lemma 2.3.22 *Let p, q be two Jordan equivalent orthogonal projections in a unital JB-algebra $\mathcal{A}$. Then there is an element $x \in \mathcal{A}_1(p) \cap \mathcal{A}_1(q)$ such that $x^2 = p + q$.*

Proof Let $q = \{t, p, t\}$ for some symmetry $t \in \mathcal{A}$. Let $s = 2q - \mathbf{1}$. Then s is a symmetry and we have $\{s, \{t, p, t\}, s\} = q$. We define $x = 2\{p, t, s\}$. Following the computation in [47, p. 125], one finds $x^2 = p + q$. Further, we have

$$x = 2\{p, t, 2q - \mathbf{1}\} = 4\{p, t, \{t, p, t\}\} - 2\{p, t, \mathbf{1}\}$$
$$= 4pt - 2pt = 2pt,$$

which gives $px = 2p(pt) = pt + \{p, t, p\}$, where, by orthogonality of p and q, we have

$$\{p, t, p\} = \{p, t, \{t, q, t\}\}$$
$$= \{\{p, t, t\}, q, t\} - \{t, \{t, p, q\}, t\} + \{t, q, \{p, t, t\}\}$$
$$= 2\{\{p, t, t\}, q, t\} = 2\{p, q, t\} = 0.$$

Therefore we obtain $px = \frac{1}{2}x$, that is, $x \in \mathcal{A}_1(p)$. Since $qt = \{t, p, t\}t = pt$, we also have $x \in \mathcal{A}_1(q)$. $\qquad\square$

Similar to the Murray–von Neumann classification of von Neumann algebras, the JBW-factors are classified into the following types:

type I_2: spin factors $H \oplus \mathbb{R}$,
type I_3: $H_3(\mathbb{O})$,
type I_n: $L(H)_{sa}$ ($\dim H = n \in \mathbb{N} \cup \{\infty\}\backslash\{2, 3\}$),
type II: semifinite and continuous,
type III: purely infinite,

where $L(H)_{sa}$ is the JBW-algebra of self-adjoint bounded linear operators on a Hilbert space H over $\mathbb{R}$, $\mathbb{C}$ or $\mathbb{H}$, with the special Jordan product

$$a \circ b = \frac{1}{2}(ab + ba)$$

(cf. [47, 5.3.8, 7.5.11]). The type II and type III factors do not contain any minimal projection and hence, neither do they contain nonzero finite rank projections [47, 5.1.5].

Given a finite-dimensional (hence type I) JBW-factor $\mathcal{A}$ of dimension n, we define $\lambda_1 : \mathcal{A} \longrightarrow \mathbb{R}$ to be the trace

$$\lambda_1(x) = \frac{\text{rank}(\mathcal{A})}{n} \text{Trace } L_x$$

so that $\lambda_1(p) = 1$ for every minimal projection p in $\mathcal{A}$.

If $\mathcal{A}$ is an infinite-dimensional type I JBW-factor, then $\mathcal{A}$ is of type I_2 or type I_∞. In the former case, say $\mathcal{A} = H \oplus \mathbb{R}$, we define $\lambda_2 : \mathcal{A} \longrightarrow \mathbb{R}$ by

$$\lambda_2(x \oplus \zeta) = 2\zeta.$$

In the type I_∞ case, $\mathcal{A} = L(H)_{sa}$ and we define $\lambda_\infty : \mathcal{A} \longrightarrow \mathbb{R} \cup \{\infty\}$ by

$$\lambda_\infty(x) = \begin{cases} \text{Trace } x & \text{if Trace } x < \infty \\ \infty & \text{otherwise,} \end{cases}$$

where $\text{Trace } x = \sum_\alpha \langle xe_\alpha, e_\alpha \rangle$ for some orthonormal basis $\{e_\alpha\}$ in H. We have $\lambda_\infty(p) = 1$ for every minimal projection p and $\lambda_\infty(x) < \infty$ for each x in the Peirce 1-space $\mathcal{A}_1(p)$ since $x = px + xp - 2pxp$.

In a type I_2 JBW-factor, an element $p = x \oplus \zeta \in H \oplus \mathbb{R}$ is a minimal projection if, and only if, $\|x\|_H = \frac{1}{2} = \zeta$. Hence we also have $\lambda_2(p) = 1$ for a minimal projection p in $H \oplus \mathbb{R}$.

Given a type I JBW-factor $\mathcal{A}$, we now define a function $\lambda : \mathcal{A} \longrightarrow \mathbb{R} \cup \{\infty\}$, called the *canonical trace*, by

$$\lambda = \begin{cases} \lambda_1 & \text{if } \dim \mathcal{A} < \infty, \\ \lambda_2 & \text{if } \mathcal{A} \text{ is an infinite-dimensional spin factor,} \\ \lambda_\infty & \text{if } \mathcal{A} \text{ is of type } I_\infty. \end{cases} \qquad (2.21)$$

We note from the above classification that the Jordan product in a JBW-factor is denoted by $\circ$. In fact, they are all special Jordan algebras except $H_3(\mathbb{O})$. It is readily verified that the canonical trace λ is *associative* in the sense that

$$\lambda((x \circ y) \circ z) = \lambda(x \circ (y \circ z))$$

if $\lambda(x) < \infty$. We also note that $\lambda(\{x, y, x\}) = \lambda(x^2 \circ y)$ if $\lambda(x) < \infty$.

A JB-algebra $\mathcal{A}$ is equipped with a partial ordering defined by the positive cone $\mathcal{A}^+ = \{a^2 : a \in \mathcal{A}\}$. A linear functional $\mu : \mathcal{A} \longrightarrow \mathbb{R}$ is called *positive* if $x \in \mathcal{A}^+$ implies $\mu(x) \geq 0$. We call μ *associative* if

$$\mu((xy)z) = \mu(x(yz)) \qquad (x, y, z \in \mathcal{A}).$$

A positive functional μ on $\mathcal{A}$ is automatically continuous, and if $\mathcal{A}$ has an identity $\mathbf{1}$, then $\|\mu\| = \mu(\mathbf{1})$ [47, 1.2.2].

Lemma 2.3.23 *Let $\mathcal{A}$ be a JB-algebra and let $\mu : \mathcal{A} \longrightarrow \mathbb{R}$ be an associative positive linear functional. Then for each $a \in \mathcal{A}^+$, we have*

$$|\mu(xa)| \leq \|x\|\mu(a) \qquad (x \in \mathcal{A}).$$

Proof We may assume that $\mathcal{A}$ has an identity $\mathbf{1}$ since, otherwise, the natural extension of μ to $\mathcal{A} \oplus \mathbb{R}$ is still associative and positive. By associativity, we have $\mu(\{x, y, x\}) = \mu(x^2 y)$ for all $x, y \in \mathcal{A}$. The linear functional $\psi(x) = \mu(xa)$ is positive since $x \in \mathcal{A}^+$ implies

$$\mu(xa) = \mu((x^{1/2})^2 a) = \mu(\{x^{1/2}, a, x^{1/2}\}) \geq 0$$

as $a \in \mathcal{A}^+$. Hence we have

$$|\mu(xa)| = |\psi(x)| \leq \|x\|\|\psi\| = \|x\|\psi(\mathbf{1}) = \|x\|\mu(a). \qquad \square$$

As noted earlier, a JBW-factor may not contain any minimal projection.

Proposition 2.3.24 *Let $\mathcal{A}$ be a JBW-factor and let M be the subspace of minimal projections in $\mathcal{A}$. Given $p \in M$ and given $x \in \mathcal{A}_1(p)$ satisfying $\lambda(x^2) = 2$, we have*

$$M \cap \mathcal{A}(p, x) = \left\{ (\cos 2\theta)p + \left(\frac{1}{2}\sin 2\theta \right) x + \frac{1}{2}(1 - \cos 2\theta) x^2 : \theta \in \mathbb{R} \right\}.$$

Proof Since $\mathcal{A}$ contains a minimal projection, it is of type I. Let $\lambda : \mathcal{A} \longrightarrow \mathbb{R} \cup \{\infty\}$ be the canonical trace defined in (2.21). We first note that $\lambda(x) = 0$ since $\lambda(x) = 2\lambda(p \circ x) = 2\lambda(p \circ (p \circ x)) = \lambda(p \circ x) = \frac{1}{2}\lambda(x)$. As in the proof of Lemma 2.3.20, we have $p \circ x^2 = \alpha p$ for some $\alpha \in \mathbb{R}$. Since $\lambda(p \circ x^2) = \lambda((p \circ x) \circ x) = \frac{1}{2}\lambda(x^2) = 1$, we have $\alpha = 1$.

Now let $q = \zeta p + \eta x + \kappa x^2 \in M \cap \mathcal{A}(p, x)$. Then $\{p, q, p\} = \zeta p + \kappa\{p, x^2, p\} = (\zeta + \kappa)p$ implies $0 \leq \zeta + \kappa \leq 1$. Also $1 = \lambda(q) = \zeta + 2\kappa$ implies $-1 \leq -\kappa \leq \zeta \leq 1 - \kappa$. On the other hand, we have

$$\zeta p + \eta x + \kappa x^2 = (\zeta p + \eta x + \kappa x^2)^2$$
$$= (\zeta^2 + 2\zeta\kappa)p + (\zeta\eta + 2\eta\kappa)x + (\eta^2 + \kappa^2)x^2,$$

which implies $\kappa = \eta^2 + \kappa^2 \geq 0$. Therefore $|\zeta| \leq 1$ and $\zeta = \cos 2\theta$ for some $\theta \in \mathbb{R}$, which gives $\kappa = \frac{1}{2}(1 - \cos 2\theta)$ and $\eta = \frac{1}{2}\sin 2\theta$.

Conversely, given any

$$z = (\cos 2\theta)p + \left(\frac{1}{2}\sin 2\theta \right) x + \frac{1}{2}(1 - \cos 2\theta) x^2$$

for some $\theta \in \mathbb{R}$, it is evident that $z^2 = z$ by the above arguments. Since $\lambda(z) = 1$, it follows that z is a minimal projection and hence $z \in M \cap \mathcal{A}(p, x)$. $\qquad \square$

Corollary 2.3.25 *Let M be the space of minimal projections in a JBW-factor* $\mathcal{A}$. *Then M is path-connected.*

Proof By definition, the empty set is path-connected. Fix $p \in M$. We show that any other $q \in M$ is of the form

$$q = (\cos 2\theta)p + \left(\frac{1}{2}\sin 2\theta\right)x + \frac{1}{2}(1 - \cos 2\theta)x^2$$

for some $\theta \in \mathbb{R}$ and hence q is joined to p by a continuous path. Note that p and q are Jordan equivalent. If q and p are orthogonal, then by Lemma 2.3.22, we have $q = x^2 - p \in M \cap \mathcal{A}(p, x)$ for some $x \in \mathcal{A}_1(p)$ and we are done, by Proposition 2.3.24.

Suppose q and p are not orthogonal. Then the Peirce-1 component $q_1 = P_1(p)(q) = 2(p \circ q - P_2(p)(q))$ is in the Jordan algebra $\mathcal{A}(p, q)$ generated by p and q. Therefore we have $\mathcal{A}(p, q_1) \subset \mathcal{A}(p, q)$, where $\dim \mathcal{A}(p, q) = 3$, since $p \circ q \neq 0$. We have $q_1 \neq 0$ for otherwise, $p \circ q = P_2(p)(q) = \alpha p$ for some $\alpha \in \mathbb{R}$ which is impossible, since p and q are two distinct minimal projections. It follows from Lemma 2.3.20 that $\dim \mathcal{A}(p, q_1) = 3$. Hence $\mathcal{A}(p, q_1) = \mathcal{A}(p, q)$ and $q \in \mathcal{A}(p, q_1)$. By Proposition 2.3.24, q is joined to p by a continuous path. $\qquad\square$

Remark 2.3.26 This result is false for JBW-algebras. In fact, it is even false for the abelian algebra $\mathbb{R}^2$ in which the space of minimal projections consists of two points $\{(1, 0), (0, 1)\}$, which is not connected.

Let $\mathcal{A}$ be a JBW-algebra. For each $a \in \mathcal{A}$, the smallest weak* closed Jordan subalgebra $W(a) \subset \mathcal{A}$ containing a is an associative JBW-algebra which is isometric and Jordan isomorphic to the algebra of real continuous functions on an extremely disconnected compact space [47, 4.1.11]. This isomorphism enables us to apply functional calculus to $W(a)$.

Given two projections p and q in a JBW-algebra $\mathcal{A}$, their supremum $p \vee q$ is the range projection $r(p + q)$ of $p + q$ [47, 4.2.8]. For a positive element $a \in \mathcal{A}$, functional calculus implies that its range projection $r(a)$ is the weak* limit of the sequence of Jordan products $\left((a + \frac{1}{m})^{-1}a\right)$, where $(a + \frac{1}{m})^{-1}$ is the inverse of $a + \frac{1}{m}$ in the JBW-algebra $W(a)$ generated by a. By continuity of the inverse and Jordan product, we see that if (a_k) is a sequence of positive elements norm converging to some $a \in \mathcal{A}$, then $(r(a_k))$ weak* converges to $r(a)$. In particular, if $\mathcal{A}$ is finite-dimensional, then this convergence is equivalent to norm convergence.

Corollary 2.3.27 *The subspace $\mathcal{P}_n$ of rank-n projections in a JBW-factor $\mathcal{A}$ is path-connected.*

Proof There is nothing to prove for $\mathcal{P}_0 = \{0\}$. Let $n \geq 1$ and let $p, q \in \mathcal{P}_n$ with $p \neq q$. Then p and q are rank-n projections in the finite-dimensional JBW-factor $\{(p \vee q), \mathcal{A}, (p \vee q)\}$. Each is an orthogonal sum of n minimal projections:

$$p = p_1 + \cdots + p_n, \quad q = q_1 + \cdots + q_n.$$

By Corollary 2.3.25, each p_k is joined to q_k by a continuous path $\{p_k(\theta)\}$ of minimal projections, with parametrization $\theta \in [0, 1]$. By the previous remark, the path

$$p(\theta) = p_1(\theta) \vee \cdots \vee p_n(\theta) = r(p_1(\theta) + \cdots + p_n(\theta))$$

is a continuous path of rank-n projections with $p(0) = p$ and $p(1) = q$. $\quad\square$

We now study the differential structures of the space $\mathcal{P}$ of projections in a JB-algebra $\mathcal{A}$ where the Jordan product of two elements a and b is written ab. Given a projection p in $\mathcal{A}$ with Peirce decomposition

$$\mathcal{A} = \mathcal{A}_0(p) \oplus \mathcal{A}_1(p) \oplus \mathcal{A}_2(p)$$

and given $v \in \mathcal{A}_0(p)$, we define a linear map $p_v : \mathcal{A} \longrightarrow \mathcal{A}$ by

$$p_v = 4[L_v, L_p],$$

where $[\cdot, \cdot]$ denotes the Lie brackets. The usual exponential, $\exp p_v : \mathcal{A} \longrightarrow \mathcal{A}$, is a Jordan algebra automorphism; in particular, $(\exp p_v)(z)$ is a projection if and only if z is such.

Lemma 2.3.28 *Let p be a projection in a JB-algebra $\mathcal{A}$. Then for each nonzero projection $q \in \mathcal{A}_2(p) \oplus \mathcal{A}_0(p)$, we have $\|q - p\| \geq 1$ if $q \neq p$.*

Proof Write $q - p = z_2 \oplus z_0 \in \mathcal{A}_2(p) \oplus \mathcal{A}_0(p)$. Then z_0 and z_2 cannot both be 0 and we have

$$\begin{aligned}
p + z_2 + z_0 = q = q^2 &= (p + z_2 + z_0)^2 \\
&= p + z_2^2 + z_0^2 + 2pz_2 + 2pz_0 \\
&= p + z_2^2 + z_0^2 + 2z_2,
\end{aligned}$$

which gives $z_0 = (z_2^2 + z_2) + z_0^2$ and $z_0^2 = (z_2^2 + z_2)z_0 + z_0^3$.

Since $\mathcal{A}_2(p)\mathcal{A}_0(p) = \{0\}$, we have $z_0^2 = z_0^3 \in \mathcal{A}_0(p)$. Therefore $z_2^2 + z_2 = 0$ and $z_0 = z_0^2$. It follows that, if $z_0 \neq 0$,

$$\|q - p\|^2 = \|(q - p)^2\| = \|z_2^2 + z_0^2\| = \|z_0 - z_2\| \geq \|z_0\| = 1,$$

since $-z_2 \geq 0$. If $z_0 = 0$, we also have $\|q - p\| \geq 1$. $\quad\square$

We now show that the projections in a JB-algebra form a Banach manifold.

Proposition 2.3.29 *Let $\mathcal{A}$ be a JB-algebra. The subspace $\mathcal{P}$ of projections in $\mathcal{A}$ is a submanifold of $\mathcal{A}$.*

Proof Let $p \in \mathcal{P}$ and let

$$V = \mathcal{A}_1(p) \quad \text{and} \quad W = \mathcal{A}_2(p) \oplus \mathcal{A}_0(p).$$

We define an analytic map $\varphi : V \times W \longrightarrow \mathcal{A}$ by

$$\varphi(v, w) = (\exp p_v)(w).$$

We have $\varphi(0, p) = p$ and the derivative $\varphi'(0, p) : V \times W \longrightarrow \mathcal{A} = V \oplus W$ is given by

$$\varphi'(0, p)(v, w) = v + w$$

and is therefore an isomorphism. Hence, by the inverse mapping theorem, φ is bianalytic on an open set $O_1 \times O_2$ in $V \times W$ containing $(0, p)$. Let

$$N = \{w \in W : \|w - p\| < 1\}$$

and let $\mathcal{U} = \varphi(O_1 \times N)$. Then $\mathcal{U}$ is an open neighbourhood of p in $\mathcal{A}$ and we have

$$\mathcal{P} \cap \mathcal{U} = \varphi(O_1 \times \{p\}).$$

Indeed, given $(v, p) \in O_1 \times \{p\}$, we have $\varphi(v, p) = (\exp p_v)(p)$, which is a projection in $\mathcal{U}$. Conversely, for $q \in \mathcal{P} \cap \mathcal{U}$ with $q = \varphi(v, w)$ and $(v, w) \in O_1 \times N$, we have $q = (\exp p_v)(w)$, which implies that w is a projection in W. Since $\|w - p\| < 1$, we must have $w = p$, by Lemma 2.3.28. Hence $(\mathcal{U}, \varphi^{-1}, V \times W)$ is a chart at p such that

$$\varphi^{-1}(\mathcal{P} \cap \mathcal{U}) = (V \times \{p\}) \cap \varphi^{-1}(\mathcal{U})$$

and we have proved that $\mathcal{P}$ is a submanifold of $\mathcal{A}$. $\square$

Now we focus on projections in JBW-algebras.

Proposition 2.3.30 *Let $\mathcal{A}$ be a JBW-algebra. Then the subspace $\mathcal{P}_f$ of finite rank projections in $\mathcal{A}$ is an open subset of the manifold $\mathcal{P}$ of projections in $\mathcal{A}$. Also, the subspace $\mathcal{P}_\infty$ of infinite rank projections in $\mathcal{A}$ is open in $\mathcal{P}$.*

Proof The openness of $\mathcal{P}_f$ follows from the fact that, for each $p \in \mathcal{P}_f$, the set

$$\{q \in \mathcal{P} : \|q - p\| < 1\}$$

is an open subset of $\mathcal{P}_f$ because $\|q - p\| < 1$ implies that q and p are Jordan equivalent, by Topping [109, proposition 7] and by considering the special JBW-algebra generated by p and q, if necessary.

Likewise $\mathcal{P}_\infty$ is open in $\mathcal{P}$. $\square$

Remark 2.3.31 If two projections p and q in $\mathcal{A}$ can be joined by a continuous path $\{p(\theta)\}_{\theta \in [0,1]}$ of projections in $\mathcal{A}$, then p and q must be Jordan equivalent, since the path can be subdivided into shorter paths such that $\| p(\theta) - p(\theta') \| < 1$ on each of them and hence $p(\theta)$ is Jordan equivalent to $p(\theta')$.

The Banach manifolds $\mathcal{P}_f$ and $\mathcal{P}_\infty$ need not be connected, and $\mathcal{P}_\infty$ need not have a Riemannian structure. However, these structures occur in JBW-factors.

Theorem 2.3.32 *Let $\mathcal{A}$ a JBW-algebra. Then the subspace $\mathcal{P}_n$ of projections of rank n in $\mathcal{A}$ is a submanifold of $\mathcal{P}$, for $n \in \mathbb{N} \cup \{0\}$. Further, if $\mathcal{A}$ is a JBW-factor, then $\mathcal{P}_n$ is a Riemannian symmetric space and the tangent space $T_p \mathcal{P}_n$ of $\mathcal{P}_n$ at each $p \in \mathcal{P}_n$ identifies with the Peirce 1-space $\mathcal{A}_1(p)$.*

Proof As in the proof of Proposition 2.3.30, $\mathcal{P}_n$ is an open subset of $\mathcal{P}$ and hence the first assertion follows.

Now let $\mathcal{A}$ be a JBW-factor with Jordan product $\circ$ and let $\mathcal{P}_n \neq \emptyset$ for some $n \neq 0$. Then $\mathcal{A}$ must be of type I. Let $p \in \mathcal{P}_n$ and let

$$\alpha : (-c, c) \longrightarrow \mathcal{P}_n \subset \mathcal{A}$$

be a smooth curve with $\alpha(0) = p$. The derivative $\alpha'(0) : \mathbb{R} \longrightarrow \mathcal{A}$ satisfies

$$\alpha'(0) = 2\alpha(0) \circ \alpha'(0),$$

since $\alpha(t)^2 = \alpha(t)$. In particular, $\alpha'(0)(1) \in \mathcal{A}_1(p)$. On the other hand, given $v \in \mathcal{A}_1(p)$, we can define a smooth curve $\beta : (-c, c) \longrightarrow \mathcal{P}_n$ by

$$\beta(t) = \exp(4t[L_v, L_p])(p).$$

Then $\beta(0) = p$ and the derivative $\beta'(0) : \mathbb{R} \longrightarrow \mathcal{A}$ is given by

$$\beta'(0)(t) = 4t[L_v, L_p](p)$$

and we have $\beta'(0)(1) = v$. It follows that the tangent space $T_p \mathcal{P}_n$ identifies with $\{\alpha'(0)(1) : p = \alpha(0)$ for a smooth curve $\alpha\} = \mathcal{A}_1(p)$.

To see that $\mathcal{P}_n$ has a Riemannian structure, we let, for each $p \in \mathcal{P}_n$,

$$\lambda_p : \mathcal{A}_1(p) \longrightarrow \mathbb{R}$$

be the restriction of the canonical trace $\lambda : \mathcal{A} \longrightarrow \mathbb{R} \cup \{\infty\}$ defined in (2.21), where

$$\lambda(v) = 2\lambda(p \circ v) \leq 2\lambda(p)\|v\| = 2n\|v\| \qquad (v \in \mathcal{A}_1(p))$$

by Lemma 2.3.23. On the tangent space $\mathcal{A}_1(p)$, we can define an inner product

$$\langle \cdot, \cdot \rangle_p : \mathcal{A}_1(p) \longrightarrow \mathbb{R}$$

by

$$\langle u, v \rangle_p = \lambda_p (u \circ v).$$

The inner product norm $|v|_p = \lambda_p(v^2)^{1/2}$ is equivalent to the JBW-algebra norm on $\mathcal{A}_1(p)$. Indeed, we have, by Lemma 2.3.23 again,

$$\|v\|^2 = \|v^2\| \le |v|_p^2 = \lambda_p(v^2) = 2\lambda_p((p \circ v) \circ v) = 2\lambda_p(p \circ v^2) \le 2n\|v^2\|.$$

In particular, the inner product $\langle \cdot, \cdot \rangle_p$ is complete. Since any two projections $p, q \in \mathcal{P}_n$ are exchanged by a symmetry $s \in \mathcal{A}$, the tangent space $\mathcal{A}_1(q)$ identifies with $\mathcal{A}_1(p)$ via the map $v \in \mathcal{A}_1(q) \mapsto \{s, v, s\} \in \mathcal{A}_1(p)$, which preserves the JBW-algebra norm as well as the inner product: $\lambda_q(u \circ v) = \lambda(u \circ v) = \lambda_p(\{s, u \circ v, s\})$. Hence in a local chart $(\mathcal{U}, \varphi, \mathcal{A}_1(p))$ at p, we have

$$\langle u, v \rangle_q = \lambda(u \circ v) \qquad (q \in \mathcal{U}; u, v \in \mathcal{A}_1(p)).$$

It follows that the inner product $\langle \cdot, \cdot \rangle_p$ depends smoothly on $p \in \mathcal{P}_n$ and defines a Riemannian metric.

By Corollary 2.3.27, $\mathcal{P}_n$ is connected.

Given $p \in \mathcal{P}_n$, the element $\mathbf{1} - 2p$ is a symmetry in $\mathcal{A}$ and the map $\sigma : \mathcal{A} \longrightarrow \mathcal{A}$ defined by

$$\sigma(a) = \{\mathbf{1} - 2p, a, \mathbf{1} - 2p\}$$

is a Jordan automorphism of $\mathcal{A}$. Its restriction $\sigma_p : \mathcal{P}_n \longrightarrow \mathcal{P}_n$ is an isometry with p as an isolated fixed point. This proves that $\mathcal{P}_n$ is a symmetric space. $\quad\square$

Corollary 2.3.33 *In a JBW-factor, the connected components of the manifold $\mathcal{P}_f$ of finite rank projections are exactly the manifolds*

$$\{\mathcal{P}_n\}_{n=0}^k \qquad (k \in \mathbb{N} \cup \{\infty\}),$$

where $\mathcal{P}_0 = \{0\}$ and $k = \infty$ if and only if the factor is of type I_∞.

Proof In a type *II* or *III* factor, we have $\mathcal{P}_f = \{0\}$. For a type *I* factor, we only need to observe that two projections in a connected component, which is now path-connected, must be of the same rank, since they can be joined by a continuous path of projections $\{p(\theta)\}$, which can be subdivided into shorter paths such that $\|p(\theta) - p(\theta')\| < 1$ on each of them, and it follows that these projections are all Jordan equivalent. $\quad\square$

2.4 Jordan triples and Riemannian symmetric spaces

We introduce in this section a class of Jordan triple systems which corresponds to a large class of Riemannian symmetric spaces. The correspondence is established via the Tits–Kantor–Koecher Lie algebras. To include infinite-dimensional manifolds, we need to consider topological structures for Jordan triple systems and their TKK Lie algebras.

If V is a *normed* Jordan triple system, that is, V is a Jordan triple system which is a normed vector space, we denote by $L(V)$ the normed vector space of linear continuous self-maps on V. It is a Banach space if V is one also.

A Lie algebra $\mathfrak{g}$ is called a *normed Lie algebra* if $\mathfrak{g}$ is a normed vector space and the Lie product is continuous:

$$\|[X, Y]\| \leq C \|X\| \|Y\| \qquad (X, Y \in \mathfrak{g})$$

for some $C > 0$.

A TKK Lie algebra $\mathfrak{g} = \mathfrak{g}_{-1} \oplus \mathfrak{g}_0 \oplus \mathfrak{g}_1$ with involution θ is said to be *quasi-normed* or *admit a quasi-norm* if $\mathfrak{g}$ is a normed vector space such that the maps $\theta : \mathfrak{g}_{-1} \longrightarrow \mathfrak{g}_1$ and $(a, b) \in \mathfrak{g}_\alpha \times \mathfrak{g}_\beta \mapsto [a, b] \in \mathfrak{g}_{\alpha+\beta}$ are continuous. In this case, the dual involution $\theta^*|_{\mathfrak{g}_{-1}} = -\theta|_{\mathfrak{g}_{-1}}$ is also continuous on $\mathfrak{g}_{-1}$.

Let V be a Jordan triple system and let, as in (1.39),

$$\mathfrak{L}(V) = V \oplus V_{00} \oplus V$$

be the corresponding TKK Lie algebra with the main involution θ. We denote by $\mathfrak{g}(V)$ the symmetric part of $\mathfrak{L}(V)$. For $(h^+, h^-) \in V_{00} \subset V_0 \times V_0$ with $\theta(h^+, h^-) = (h^+, h^-)$, we have $h^+ = h^- \in V_0$, where V_0 is the linear span of $V \square V$. Moreover, given $h^+ = \sum_j a_j \square b_j$, we have $h^- = -\sum_j b_j \square a_j$. If $h^+ = h^-$, then

$$2h^+ = \sum_j (a_j \square b_j - b_j \square a_j).$$

Hence the symmetric part of $\mathfrak{L}(V)$ can be written as

$$\mathfrak{g}(V) = \{(a, k, -a) : a \in V, \theta k = k \in V_{00}\}$$
$$= \{(a, (h, h), -a) : h = \sum_j (a_j \square b_j - b_j \square a_j), a, a_j, b_j \in V\}.$$
$$(2.22)$$

The restriction of θ to $\mathfrak{g}(V)$ is an involution, also denoted by θ. The dual symmetric part of $\mathfrak{L}(V)$ will be denoted by

$$\mathfrak{g}^*(V) = \{(a, (h, h), a) : h = \sum_j (a_j \square b_j - b_j \square a_j), a, a_j, b_j \in V\}.$$

Lemma 2.4.1 *Let V be a Jordan triple system and $\mathfrak{L}(V) = V_{-1} \oplus V_{00} \oplus V_1$ its TKK Lie algebra with symmetric part $\mathfrak{g}(V)$. The following conditions are equivalent:*

(i) $\mathfrak{g}(V)$ *can be normed to become a normed Lie algebra.*

(ii) $\mathfrak{g}^*(V)$ *can be normed to become a normed Lie algebra.*

(iii) V *can be normed to have continuous inner derivations, that is, one can define a norm on V such that the bilinear map $(a, b) \in V \times V \mapsto a \,\square\, b - b \,\square\, a \in L(V)$ is well defined and continuous:*

$$\|a \,\square\, b - b \,\square\, a\| \le c\|a\|\|b\| \qquad (a, b \in V)$$

for some $c > 0$.

Proof (i) $\Longrightarrow$ (iii). Let $\mathfrak{g}(V)$ be a normed Lie algebra with norm $\| \cdot \|_{\mathfrak{g}(V)}$. We equip V with the norm

$$\|a\| = \|(a, 0, -a)\|_{\mathfrak{g}(V)} \qquad (a \in V).$$

Then there is some constant $C > 0$ such that

$$\|[(a, 0, -a), (b, 0, -b)]\|_{\mathfrak{g}(V)} \le C\|a\|\|b\|$$

for all $a, b \in V$. We have, for $a, b, x \in V$,

$$
\begin{aligned}
&\|(a \,\square\, b - b \,\square\, a)(x)\| \\
&= \|((b \,\square\, a - a \,\square\, b)(x), 0, (a \,\square\, b - b \,\square\, a)(x))\|_{\mathfrak{g}(V)} \\
&= \|[(0, \ (b \,\square\, a - a \,\square\, b, \ b \,\square\, a - a \,\square\, b), \ 0), \ (x, 0, -x)]\|_{\mathfrak{g}(V)} \\
&= \|[[(a, 0, -a), (b, 0, -b)], \ (x, 0, -x)]\|_{\mathfrak{g}(V)} \\
&\le C^2\|a\|\|b\|\|x\|.
\end{aligned}
$$

This proves $a \,\square\, b - b \,\square\, a \in L(V)$ and the map $(a, b) \in V^2 \mapsto a \,\square\, b - b \,\square\, a \in L(V)$ is continuous.

  (iii) $\Longrightarrow$ (i). Let V be equipped with a norm $\| \cdot \|_V$ so that the inner derivations are continuous and the bilinear map in (iii) is bounded by c. For $(a, (h, h) - a) \in \mathfrak{g}(V)$ with $h = \sum_j (a_j \,\square\, b_j - b_j \,\square\, a_j)$, we have $h \in L(V)$. Therefore we can equip $\mathfrak{g}(V)$ with the norm

$$\|(a, (h, h) - a)\| = \|a\|_V + \|h\|_{L(V)}$$

for $(a, (h, h), -a) \in \mathfrak{g}(V)$. Then we have

$$\| [a \oplus (h, h) \oplus -a, \, u \oplus (g, g) \oplus -u] \|$$
$$= \|(hu - ga, \, ([h, g] - a \square u + u \square a, \, [h, g] - a \square u + u \square a), \, ga - hu)\|$$
$$\leq \|h\|\|u\| + \|g\|\|a\| + 2\|h\|\|g\| + c\|a\|\|u\|$$
$$\leq (2 + c)\|a \oplus h \oplus -a\|\|u \oplus g \oplus -u\|$$

for some $c > 0$. Hence $\mathfrak{g}(V)$ is a normed Lie algebra with the above norm.
The equivalence of (ii) and (iii) is proved analogously since

$$[a \oplus h \oplus a, \, u \oplus g \oplus u] = (hu - ga, \, (k, k), \, hu - ga),$$

where $k = [h, g] + a \square u - u \square a$. $\qquad\square$

The continuity of the triple product in a normed Jordan triple V requires its
TKK Lie algebra $\mathfrak{L}(V)$ be quasi-normed, as shown next.

Lemma 2.4.2 *Let V be a Jordan triple system and $\mathfrak{L}(V) = V \oplus V_{00} \oplus V$ its
TKK Lie algebra. The following conditions are equivalent:*

(i) *$\mathfrak{L}(V)$ admits a quasi-norm.*
(ii) *V can be normed to have continuous left multiplication; that is, one can
define a norm on V such that the bilinear map $(a, b) \in V \times V \mapsto
a \square b \in L(V)$ is well defined and continuous:*

$$\|a \square b\| \leq c\|a\|\|b\| \qquad (a, b \in V)$$

for some $c > 0$.

Proof (i) $\implies$ (ii). Let $\mathfrak{L}(V)$ be quasi-normed. Then V inherits the norm of
$\mathfrak{L}(V)$. For $\alpha, \beta \in \{0, \pm 1\}$, there are positive constants $c_{\alpha,\beta}$ such that

$$\|[x, y]\|_{\mathfrak{L}(V)} \leq c_{\alpha,\beta}\|x\|_{\mathfrak{L}(V)}\|y\|_{\mathfrak{L}(V)} \quad \text{for} \quad (x, y) \in \mathfrak{L}(V)_\alpha \times \mathfrak{L}(V)_\beta.$$

Given $a, b, x \in V$, we have

$$\|(a \square b)(x)\|_V = \|[[a, \theta b], x]\|_{\mathfrak{L}(V)}$$
$$\leq c_{0,-1}\|[a, \theta b]\|_{\mathfrak{L}(V)}\|x\|_V$$
$$\leq c_{0,-1}c_{-1,1}\|a\|_V\|\theta b\|_V\|x\|_V$$
$$\leq c_{0,-1}c_{-1,1}\|\theta\|\|a\|_V\|b\|_V\|x\|_V.$$

Hence V satisfies condition (ii).

(ii) $\implies$ (i). We have $V_0 \subset L(V)$, and $\mathfrak{L}(V)$ can be equipped with a natural norm

$$\|x \oplus (h^+, h^-) \oplus y\| = \|x\|_V + \|h^+\|_{L(V)} + \|h^-\|_{L(V)} + \|y\|_V$$
$$(x \oplus (h^+, h^-) \oplus y \in \mathfrak{L}(V)).$$

We have

$$\|\theta(x, 0, 0)\|_{\mathfrak{L}(V)} = \|(0, 0, x)\|_{\mathfrak{L}(V)} = \|x\|_V = \|(x, 0, 0)\|_{\mathfrak{L}(V)}$$

and hence $\theta : V_{-1} \longrightarrow V_1$ is continuous. Given $a \in \mathfrak{L}(V)_\alpha$ and $b \in \mathfrak{L}(V)_\beta$ for $\alpha \neq \beta$, we have $[a, b] = \pm a \square b$ if $\alpha, \beta \neq 0$, and for $\alpha\beta = 0$, we have $[a, b] = \pm h(x)$ for some $h = \sum_k u_k \square v_k$ with $x = a$ or b. It follows from condition (ii) that the maps $(a, b) \in \mathfrak{L}(V)_\alpha \times \mathfrak{L}(V)_\beta \mapsto [a, b] \in \mathfrak{L}(V)_{\alpha+\beta}$ are continuous. $\square$

Remark 2.4.3 Condition (ii) is equivalent to saying that V can be normed to have a continuous Jordan triple product.

The symmetric part $\mathfrak{g}_s$ of a quasi-normed TKK Lie algebra $\mathfrak{g}$ inherits the norm of $\mathfrak{g}$. The norm completion $\overline{\mathfrak{g}}_s$ is called the *complete symmetric part* of $\mathfrak{g}$. The Lie product of $\mathfrak{g}_s$ extends to $\overline{\mathfrak{g}}_s$ and the involution θ extends to $\overline{\theta}$ on $\overline{\mathfrak{g}}_s$. The *complete dual symmetric part* $\overline{\mathfrak{g}}_s^*$ is defined likewise.

We see from Theorem 1.3.11 and Lemma 2.4.2 that the category of quasi-normed canonical TKK Lie algebras is equivalent to the category of normed Jordan triples with continuous triple product.

In finite dimensions, Riemannian symmetric spaces correspond to orthogonal involutive Lie algebras. The first objective of this section is to extend this correspondence to an infinite-dimensional setting. This then leads to the second objective of characterising the class of Jordan triples whose TKK Lie algebras admit orthogonal symmetric parts.

In what follows, we shall not distinguish a quadratic form and its associated symmetric bilinear form, but use the same notation for both. We recall that a real quadratic form q on a real vector space V is said to be *positive definite* if $q(v) > 0$ for $v \in V \backslash \{0\}$. It is called *completely positive definite* if the symmetric bilinear form $q(\cdot, \cdot)$ is a complete inner product on V.

We adapt the notion of an *orthogonal involutive Lie algebra* in finite dimensions (cf. [51, p. 213], [63, p. 35] and [11, p. 21]) to our setting below, which can be infinite-dimensional.

Definition 2.4.4 Let $(\mathfrak{g}, \theta)$ be an involutive real Lie algebra with the canonical decomposition $\mathfrak{g} = \mathfrak{k} \oplus \mathfrak{p}$ into ± 1-eigenspaces of θ, with $\mathfrak{k}$ the 1-eigenspace.

Then $\mathfrak{g}$ is called *orthogonal* if there is a completely positive definite quadratic form

$$q : \mathfrak{p} \times \mathfrak{p} \longrightarrow \mathbb{R}$$

which is (infinitesimally) invariant under the isotropy representation $ad\,\mathfrak{k}|_{\mathfrak{p}}$ of $\mathfrak{k}$ on $\mathfrak{p}$; that is,

$$q(ad\,Z(X), Y) + q(X, ad\,Z(Y)) = 0 \qquad (Z \in \mathfrak{k}, X, Y \in \mathfrak{p}),$$

where the second condition is equivalent to

$$q(ad\,Z(X), X) = 0 \qquad (Z \in \mathfrak{k}, X \in \mathfrak{p}).$$

We note that q is invariant under θ, that is, $q(\theta X, \theta X) = q(X, X)$ as $\theta X = -X$.

Given a Riemannian symmetric space M, its isometry group $G(M)$ acts transitively on M by Lemma 2.3.6. Pick a point $p \in M$ and let $K = \{g \in G(M) : g(p) = p\}$, called the *isotropy subgroup at p*. Then the map

$$\rho : gK \in G(M)/K \mapsto g(p) \in M \qquad (2.23)$$

is a well-defined bijection from the left coset space $G(M)/K = \{gK : g \in G\}$ onto M. With a suitable topology on $G(M)$, this bijection becomes a homeomorphism. In fact, if $G(M)$ can be topologized to a Banach Lie group, then the homogeneous space $G(M)/K$ is diffeomorphic to M. We give the details below.

Let $G(M)$ be a real Banach Lie group. Then its Lie algebra can be identified with the Lie algebra $\mathfrak{g}(M)$ of Killing fields on M, via the one-parameter subgroups of $G(M)$. Fix the point $p \in M$ in the following discussion. Let $s_p : M \longrightarrow M$ be the symmetry at p and let $\sigma : G(M) \longrightarrow G(M)$ be the conjugation

$$\sigma(g) = s_p g s_p \qquad (g \in G(M)).$$

Let $\theta = Ad(s_p) = d\sigma_\iota : \mathfrak{g}(M) \longrightarrow \mathfrak{g}(M)$ be the differential of σ at the identity $\iota \in G(M)$. Since $s_p^2 = \iota$, we have $\sigma^2 = \iota$ and $\theta^2 = I$, the identity map on $\mathfrak{g}(M)$. Indeed,

$$I = (d\sigma^2)_\iota = (d\sigma)_{\sigma(\iota)} \circ d\sigma_\iota = d\sigma_\iota \circ d\sigma_\iota = \theta^2.$$

Hence $(\mathfrak{g}(M), \theta)$ is an involutive Lie algebra and θ has eigenvalues ± 1 with the eigenspace decomposition

$$\mathfrak{g}(M) = \mathfrak{k} \oplus \mathfrak{p},$$

where

$$\mathfrak{k} = \{X \in \mathfrak{g}(M) : \theta X = X\} \quad \text{and} \quad \mathfrak{p} = \{X \in \mathfrak{g}(M) : \theta X = -X\}. \quad (2.24)$$

We will see that the involutive Lie algebra $(\mathfrak{g}(M), \theta)$ is orthogonal. We retain the above notation throughout this section.

Lemma 2.4.5 *Let M be a Riemannian symmetric space and $\theta = \mathrm{Ad}(s_p)$ the adjoint representation induced by the symmetry at $p \in M$. Then the 1-eigenspace of θ is equal to*

$$\mathfrak{k} = \{X \in \mathfrak{g}(M) : X_p = 0\} = \{X \in \mathfrak{g}(M) : \exp t X(p) = p, \forall t \in \mathbb{R}\}.$$

Proof In the notation of (2.4), we have $\theta = (s_p)_*$ and $\exp t\theta X = s_p(\exp t X)s_p$ for all $t \in \mathbb{R}$. If $\theta X = X$, then $\exp t X = s_p \exp t X s_p$ and $\exp t X(p) = s_p(\exp t X(p))$ implies $\exp t X(p) = p$ for sufficiently small t since p is an isolated fixed point of s_p. It follows that

$$X_p = \left.\frac{d}{dt}\right|_{t=0} \exp t X(p) = 0.$$

On the other hand, given $X_p = 0$, we have

$$\left.\frac{d}{dt}\right|_{t=s} \exp t X(p) = \left.\frac{d}{dt}\right|_{t=0} \exp(s + t)X(p)$$

$$= \left.\frac{d}{dt}\right|_{t=0} \exp s X \circ \exp t X(p)$$

$$= d(\exp s X)_p(X_p) = 0$$

for all $s \in \mathbb{R}$. Therefore $\exp t X$ is constant in t and we have $\exp t X(p) = p$ for all $t \in \mathbb{R}$.

In turn, the last condition implies $\theta X = X$. Indeed, by Lemma 2.3.2, the differential $(ds_p)_p$ is minus the identity on T_pM, which gives

$$d(s_p(\exp t X)s_p)_p = (ds_p)_p \circ d(\exp t X)_p \circ (ds_p)_p = d(\exp t X)_p$$

and it follows from Proposition 2.2.9 that $\exp t X = s_p(\exp t X)s_p$ for all $t \in \mathbb{R}$. Hence

$$\theta X(\cdot) = \left.\frac{d}{dt}\right|_{t=0} \exp t\theta X(\cdot) = \left.\frac{d}{dt}\right|_{t=0} s_p(\exp t X)s_p(\cdot)$$

$$= \left.\frac{d}{dt}\right|_{t=s} \exp t X(\cdot) = X(\cdot). \qquad \square$$

By Remark 2.2.19 and Lemma 2.4.5, the isotropy group K is a real Banach Lie group with Lie algebra $\mathfrak{k}$. Since $\mathfrak{g}(M) = \mathfrak{k} \oplus \mathfrak{p}$, the inclusion map $\iota_K : K \hookrightarrow G(M)$ is an immersion as its differential $(d\iota_K)_\iota : \mathfrak{k} \longrightarrow \mathfrak{g}(M)$ has complemented image $\mathfrak{k}$. Hence K is a submanifold of $G(M)$ and a Banach Lie subgroup of $G(M)$. It follows that the left coset space $G(M)/K$ is a Banach manifold and the quotient map $\pi : G(M) \longrightarrow G(M)/K$ is a submersion.

By the *identity component* of a topological group G, we mean the connected component of G containing the identity. It is a closed normal subgroup of G and will always be denoted by G^0.

Lemma 2.4.6 *Let $G_\sigma(M) = \{g \in G(M) : \sigma(g) = g\}$ and let $G_\sigma(M)^0$ be the identity component of $G_\sigma(M)$. Then the isotropy group K satisfies*

$$G_\sigma(M)^0 \subset K \subset G_\sigma(M).$$

Proof Given $g \in K$, we have $\sigma(g)(p) = s_p \circ g \circ s_p(p) = p$ and $d\sigma(g)_p(X_p) = d(s_p \circ g \circ s_p)_p(X_p) = ds_p \circ dg \circ ds_p(X_p) = dg_p(X_p)$ by Lemma 2.3.2. It follows from Proposition 2.2.9 that $\sigma(g) = g$.

By Remark 2.2.19, $G_\sigma(M)$ is a real Banach Lie group with Lie algebra $\mathfrak{g}_\sigma \subset \mathfrak{g}(M)$. The map σ restricts to the identity map on $G_\sigma(M)$ and hence $\theta X = d\sigma_\iota(X) = X$ for all $X \in \mathfrak{g}_\sigma$. By Lemma 2.4.5, we have $\exp tX(p) = p$ for all $t \in \mathbb{R}$. Thus the exponential map exp maps homeomorphically a neighbourhood of $\mathfrak{g}_\sigma$ onto a neighbourhood V of the identity in $G_\sigma(M)$ with $V \subset K$. Therefore $G_\sigma(M)^0 \subset \bigcup_{n \geq 1} V^n \subset K$.   □

For $p \in M$, the evaluation map $\epsilon : g \in G(M) \mapsto g(p) \in M$ is a submersion since its differential at $\iota \in G(M)$ is the evaluation map $X \in \mathfrak{g}(M) \mapsto X(p) \in T_pM$ of which the kernel $\{X \in \mathfrak{g}(M) : X(p) = 0\} = \mathfrak{k}$ is complemented in $\mathfrak{g}(M)$, by Lemma 2.4.5.

Since submersions are open maps by Remark 2.1.18, the following commutative diagram implies that the bijection $\rho : G(M)/K \longrightarrow M$ in (2.23) is a homeomorphism:

$$
\begin{array}{ccc}
G(M)/K & \xrightarrow{\rho} & M \\
& \pi \nwarrow \quad \nearrow \epsilon & \\
& G(M) &
\end{array}
\qquad (2.25)
$$

Further, ρ is a diffeomorphism since the submersions π and ϵ have local left inverses, by Lemma 2.1.17. For instance, the submersion π has a local left inverse f_π on some neighbourhood $\mathcal{V}$ of each point in $G(M)/K$, which implies $\rho|_\mathcal{v} = \epsilon \circ f_\pi$ is smooth. Likewise, ρ^{-1} is smooth. Therefore we can identify the manifold M with the homogeneous space $G(M)/K$.

Taking differentials of the commutative diagram in (2.25) gives

$$
\begin{array}{ccc}
T_K\,G(M)/K & \xrightarrow{\;d\rho_K\;} & T_pM \\[4pt]
{}_{d\pi_\iota}\nwarrow & & \nearrow_{d\epsilon_\iota} \\[4pt]
& \mathfrak{g}(M) &
\end{array}
$$

where the evaluation map $d\epsilon_\iota$ restricts to a real linear isomorphism

$$
X \in \mathfrak{p} \mapsto X(p) \in T_pM, \tag{2.26}
$$

where $\mathfrak{p}$ is the (-1)-eigenspace of $\theta : \mathfrak{g}(M) \longrightarrow \mathfrak{g}(M)$.

Remark 2.4.7 Let $G(M)^0$ be the identity component of $G(M)$. Then $G(M)^0$ also acts transitively on M. Indeed, $G(M)^0$ is a clopen set in $G(M)$ which is locally connected, and the above commutative diagram implies that $\epsilon(G(M)^0) = \{g(p) : g \in G(M)^0\}$ is a non-empty clopen set in M and hence equals M, by connectedness of M. If we let $K_0 = \{g \in G(M)^0 : g(p) = p\}$ be the isotropy subgroup of $G(M)^0$ at $p \in M$, then likewise M is also diffeomorphic to the homogeneous space $G(M)^0/K_0$.

Example 2.4.8 Let $\mathcal{P}_n$ be the Riemannian symmetric space of rank-n projections in the JBW-factor $\mathcal{A} = L(H)_{sa}$ in Corollary 2.3.27, where $L(H)$ is the von Neumann algebra of bounded linear operators on a complex Hilbert space H. The real Banach Lie group

$$
U = U(H) = \{u \in L(H) : uu^* = u^*u = \mathbf{1}\}
$$

of unitary operators on H can be represented as a group of isometries of $\mathcal{P}_n$, where $q \in \mathcal{P}_n \mapsto uqu^* \in \mathcal{P}_n$ is an isometry for each $u \in U$. The action

$$
(u, q) \in U \times \mathcal{P}_n \mapsto uqu^* \in \mathcal{P}_n
$$

on $\mathcal{P}_n$ by U is transitive. Let $\mathfrak{u}$ be the Lie algebra of U, which consists of skew-Hermitian operators on H. Fix some $p \in \mathcal{P}_n$, the evaluation map $\epsilon : U \longrightarrow \mathcal{P}_n$, given by $\epsilon(u) = upu^*$, is a submersion. Indeed, its differential

$$
d\epsilon_1 : \mathfrak{u} \longrightarrow T_p\mathcal{P}_n
$$

is given by

$$
d\epsilon_1(X) = Xp - pX \qquad (X \in \mathfrak{u})
$$

with kernel $\mathfrak{k} = \{X \in \mathfrak{u} : Xp = pX\} = \{X \in \mathfrak{u} : (\mathbf{1} - 2p)X(\mathbf{1} - 2p) = X\}$. The Lie algebra $\mathfrak{u}$ has the decomposition

$$
\mathfrak{u} = \mathfrak{k} \oplus \{X \in \mathfrak{u} : X = Xp + pX\},
$$

where $\mathfrak{p} = \{X \in \mathfrak{u} : X = Xp + pX\}$ is the (-1)-eigenspace of the involution $\theta(X) = (1 - 2p)X(1 - 2p)$ on $\mathfrak{u}$. We have the real linear isomorphism

$$X \in \mathfrak{p} \mapsto Xp - pX \in \{X \in L(H)_{sa} : X = Xp + pX\} = \mathcal{A}_1(p) = T_p \mathcal{P}_n,$$

where each $X \in T_p \mathcal{P}_n$ has pre-image $Xp - pX \in \mathfrak{p}$.

The previous discussion reveals two important features of Riemannian symmetric spaces whose isometries form a Lie group; namely, they are homogeneous spaces (of connected Lie groups) and the Lie algebras of vector fields associated with them are involutive. The latter hinted at a connection with TKK Lie algebras and Jordan triple systems. We are going to determine the class of Riemannian symmetric spaces, among homogeneous spaces, whose Lie algebras are related to TKK Lie algebras. We show that they correspond to a class of normed Jordan triples called JH-triples.

In what follows, G/K denotes a homogeneous space where G is a connected real Banach Lie group with identity e and K a Banach Lie subgroup of G. We first characterize homogeneous spaces G/K which are symmetric spaces, in terms of the Lie algebra of G.

Let $\pi : G \longrightarrow G/K$ be the quotient map, which is a submersion. Let $\mathfrak{g}$ be the Lie algebra of G and $\mathfrak{k}$ the Lie algebra of K, which is a Lie subalgebra of $\mathfrak{g}$ as well as a complemented subspace, since K is a submanifold of G. Let $p = \pi(e) = K$. Then the differential $d\pi_e : \mathfrak{g} \longrightarrow T_p(G/K)$ has kernel $\mathfrak{k}$ and induces to an isomorphism $\mathfrak{g}/\mathfrak{k} \approx T_p(G/K)$.

Denote the natural left action of G on G/K by

$$\tau_g : aK \in G/K \mapsto gaK \in G/K \qquad (g \in G).$$

A Riemannian metric $\langle \cdot, \cdot \rangle$ on G/K (cf. Remark 2.2.1) is called G-*invariant* if the left translation τ_g is an isometry in this metric for all $g \in G$; that is,

$$\langle X_{aK}, Y_{aK} \rangle_{aK} = \langle (d\tau_g)_{aK} X_{aK}, (d\tau_g)_{aK} Y_{aK} \rangle_{gaK} \qquad (a, g \in G).$$

Definition 2.4.9 Let G/K be a homogenous space. We call (G, K) a *Riemannian symmetric pair* if there is an involutive automorphism $\sigma : G \longrightarrow G$ such that

(i) $G_\sigma^0 \subset K \subset G_\sigma$, where G_σ is the closed subgroup of fixed points of σ and G_σ^0 the identity component of G_σ;

(ii) G/K admits a G-invariant Riemannian metric.

In the identification of a Riemannian symmetric space M with the homogeneous space $G(M)/K$, for each $g \in G(M)$, the left translation τ_g :

$G(M)/K \longrightarrow G(M)/K$ is identified with the isometry $g : M \longrightarrow M$. Therefore $(G(M), K)$ is a Riemannian symmetric pair by Lemma 2.4.6. Conversely, if (G, K) is a Riemannian symmetric pair, we will see that G/K is a symmetric space.

Let $\sigma : G \longrightarrow G$ be an involutive automorphism in a Riemannian symmetric pair (G, K). Then the differential $d\sigma_e : \mathfrak{g} \longrightarrow \mathfrak{g}$ is involutive on the Lie algebra $\mathfrak{g}$ of G and hence $\mathfrak{g}$ has a ± 1-eigenspace decomposition

$$\mathfrak{g} = \{X \in \mathfrak{g} : (d\sigma_e)X = X\} \oplus \mathfrak{p},$$

where $\mathfrak{p}$ is the (-1)-eigenspace.

Lemma 2.4.10 *Given a Riemannian symmetric pair (G, K) with involutive automorphism σ on G, the Lie algebra $\mathfrak{k}$ of K is the 1-eigenspace of the differential $d\sigma_e$ and the differential $d\pi_e$ of the quotient map $\pi : G \longrightarrow G/K$ is a linear isomorphism from $\mathfrak{p}$ onto the tangent space $T_p(G/K)$, where $\mathfrak{p}$ is the (-1)-eigenspace of $d\sigma_e$ and $p = \pi(e)$.*

Proof The Lie algebra $\mathfrak{k}$ of K is given by $\mathfrak{k} = \{X \in \mathfrak{g} : \exp tX \in K, \ \forall t \in \mathbb{R}\}$. We need to show

$$\mathfrak{k} = \{X \in \mathfrak{g} : (d\sigma_e)X = X\}.$$

Let $X \in \mathfrak{k}$. Since $K \subset G_\sigma$, we have $\sigma(\exp tX) = \exp tX$ for all $t \in \mathbb{R}$, which implies $(d\sigma_e)X = X$. Conversely, given $(d\sigma_e)X = X$, we have $\sigma(\exp tX) = \exp tX$ for all $t \in \mathbb{R}$, since

$$\frac{d}{dt}\bigg|_{t=0} \sigma(\exp tX) = X = \frac{d}{dt}\bigg|_{t=0} \exp tX.$$

It follows that $\exp tX \in G_\sigma^0 \subset K$ for all $t \in \mathbb{R}$ and hence $X \in \mathfrak{k}$.

Finally, the differential $d\pi_e : \mathfrak{g} \longrightarrow T_p(G/K)$ has kernel $\mathfrak{k}$ and $\mathfrak{g} = \mathfrak{k} \oplus \mathfrak{p}$. Hence $d\pi_e$ restricts to an isomorphism between $\mathfrak{p}$ and $T_p(G/K)$. □

Note that in Lemma 2.4.10 we have

$$[\mathfrak{p}, \mathfrak{p}] \subset \mathfrak{k} \quad \text{and} \quad [\mathfrak{k}, \mathfrak{p}] \subset \mathfrak{p}.$$

Let $Ad : G \longrightarrow \text{Aut}\,\mathfrak{g}$ be the adjoint representation. Given $X, Y \in \mathfrak{g}$, we recall from Lemma 2.1.25 that

$$[X, Y] = ad(X)(Y) = \frac{d}{dt}\bigg|_{t=0} (Ad(\exp tX)Y - Y). \tag{2.27}$$

Lemma 2.4.11 *Let (G, K) be a Riemannian symmetric pair with the decomposition $\mathfrak{g} = \mathfrak{k} \oplus \mathfrak{p}$ into ± 1-eigenspaces of the involution $d\sigma_e$, as in Lemma 2.4.10. Then for $k \in K$ and $Y \in \mathfrak{p}$, we have $Ad(k)Y \in \mathfrak{p}$.*

Proof Since $\sigma(k) = k$, we have, by (2.5),

$$
\begin{aligned}
d\sigma_e(Ad(k)Y) &= \left.\frac{d}{dt}\right|_{t=0} \sigma(\exp t\, Ad(k)Y) = \left.\frac{d}{dt}\right|_{t=0} \sigma(k(\exp t\, Y)k^{-1}) \\
&= \left.\frac{d}{dt}\right|_{t=0} k\sigma(\exp t\, Y)k^{-1} = Ad(k)d\sigma_e(X) \\
&= -Ad(k)Y.
\end{aligned}
$$

Therefore $Ad(k)Y \in \mathfrak{p}$. $\qquad\qquad\square$

Lemma 2.4.12 *Let (G, K) be a Riemannian symmetric pair with the decomposition $\mathfrak{g} = \mathfrak{k} \oplus \mathfrak{p}$ into ± 1-eigenspaces of the involution $\theta = d\sigma_e$, as in Lemma 2.4.10. Then $(\mathfrak{g}, \theta)$ is orthogonal and G/K is a symmetric space.*

Proof Let $p = \pi(e) = K$ as before. We first define a complete inner product q on the (-1)-eigenspace $\mathfrak{p}$ via the isomorphism $d\pi_e : \mathfrak{p} \longrightarrow T_p(G/K)$. Since (G, K) is a Riemannian symmetric pair, there is a G-invariant Riemannian metric $\langle \cdot, \cdot \rangle$ on G/K. Define $q : \mathfrak{p} \times \mathfrak{p} \longrightarrow \mathbb{R}$ by

$$
q(X, Y) = \langle (d\pi_e)X, (d\pi_e)Y \rangle_p \qquad (X, Y \in \mathfrak{p}).
$$

We show that q is $Ad(k)$-invariant for all $k \in K$, Let $k \in K$ and $X \in \mathfrak{p}$. Then we have

$$
\begin{aligned}
(d\pi_e)(Ad(k)X) &= \left.\frac{d}{dt}\right|_{t=0} \exp(t\, Ad(k)X)K = \left.\frac{d}{dt}\right|_{t=0} (k(\exp t\, X)k^{-1})K \\
&= (d\tau_k)_p \left.\frac{d}{dt}\right|_{t=0} (\exp t\, X)K = (d\tau_k)_p(d\pi_e)X.
\end{aligned}
$$

For $X, Y \in \mathfrak{p}$, we have shown earlier $Ad(k)X, Ad(k)Y \in \mathfrak{p}$. It follows that

$$
\begin{aligned}
q(Ad(k)X, Ad(k)Y) &= \langle (d\pi_e)Ad(k)X, (d\pi_e)Ad(k)Y \rangle_p \\
&= \langle (d\tau_k)_p(d\pi_e)X, (d\tau_k)_p(d\pi_e)Y \rangle_p \\
&= \langle (d\pi_e)X, (d\pi_e)Y \rangle_p \\
&= q(X, Y),
\end{aligned}
$$

which is equivalent to ad $\mathfrak{k}$-invariance of q, since for $Z \in \mathfrak{k}$ and $X, Y \in \mathfrak{p}$, we have

$$
\begin{aligned}
q(ad\, Z(X), Y) &= q\left(\frac{d}{dt}\bigg|_{t=0} Ad(\exp t Z)X, Y\right) \\
&= \frac{d}{dt}\bigg|_{t=0} q(Ad(\exp t Z)X, Y) \\
&= \frac{d}{dt}\bigg|_{t=0} q(X, Ad(\exp -t Z)Y) \\
&= q(X, -ad\, Z(Y)).
\end{aligned}
$$

This proves that $(\mathfrak{g}, \theta)$ is orthogonal with respect to q.

Finally, we show that G/K is a symmetric space in the Riemannian metric $\langle \cdot, \cdot \rangle$. Since G acts transitively on G/K by left translations which are isometries, it suffices to exhibit a symmetry at $p = \pi(e)$. Given $K \subset G_\sigma$, we can define an involution $s : G/K \longrightarrow G/K$ by

$$
s(aK) = \sigma(a)K.
$$

Since $G_\sigma^0 \subset K$, it is readily seen that $p = K$ is an isolated fixed point of s. We complete the proof by showing that s is an isometry.

We have

$$
s\tau_{g^{-1}} = \tau_{\sigma(g^{-1})}s
$$

since, for each $aK \in G/K$,

$$
s\tau_{g^{-1}}(aK) = s(g^{-1}aK) = \sigma(g^{-1})\sigma(a)K = \tau_{\sigma(g^{-1})}s(aK).
$$

It follows from the G-invariance of the metric that

$$
\begin{aligned}
\langle X, Y \rangle_{gK} &= \langle d\tau_{g^{-1}}X, d\tau_{g^{-1}}Y \rangle_K \\
&= \langle (ds)_p d\tau_{g^{-1}}X, (ds)_p d\tau_{g^{-1}}Y \rangle_K \\
&= \langle d\tau_{\sigma(g^{-1})}(ds)_{gK} X, d\tau_{\sigma(g^{-1})}(ds)_{gK} Y \rangle_{\sigma(g^{-1})s(gK)} \\
&= \langle (ds)_{gK} X, (ds)_{gK} Y \rangle_{s(gK)}.
\end{aligned}
$$

$\square$

Example 2.4.13 Every Banach Lie group is a homogeneous space since translations act transitively on the group. In fact, a connected real Banach Lie group G is a symmetric space if it admits a left and right invariant Riemannian metric. On the product $G \times G$ of such a group G, we can define an involutive automorphism $\sigma : (g, h) \in G \times G \mapsto (h, g) \in G \times G$. Then the fixed-point set is the diagonal

$$
G_\sigma = \{(g, g) : g \in G\} \subset G \times G
$$

and $G = G \times G / G_\sigma$. Hence $(G \times G, G_\sigma)$ is a Riemannian symmetric pair. In particular, abelian or compact connected Lie groups are symmetric spaces.

We have the following Lie algebraic characterization of Riemannian symmetric spaces in any dimension.

Theorem 2.4.14 *Let G/K be a homogeneous space, where G is a simply connected Banach Lie group and K is connected. The following conditions are equivalent:*

(i) *(G, K) is a Riemannian symmetric pair.*
(ii) *The Lie algebra $\mathfrak{g}$ of G is involutive and orthogonal, and the Lie algebra $\mathfrak{k}$ of K is the 1-eigenspace of the involution.*

In the preceding case, the involution of $\mathfrak{g}$ is the differential $(d\sigma)_e$ at the identity $e \in G$, where $\sigma : G \longrightarrow G$ is the given involution in (i).

Proof (i) $\Longrightarrow$ (ii). This is proved in Lemma 2.4.12.

(ii) $\Longrightarrow$ (i). The involution θ of $\mathfrak{g}$ induces a local involutive automorphism of G, which can then be extended to an involutive automorphism σ of G by simply connectedness (cf. [20, p.49]). The exponential map exp sends a neighbourhood of $\mathfrak{k}$ onto a neighbourhood V of the identity in K with $\sigma(k) = k$ for all $k \in V$. Since K is connected, we have $K = \bigcup_{n \geq 1} V^n \subset G_\sigma$ and $G_\sigma^0 = K$.

Let $\mathfrak{g} = \mathfrak{k} \oplus \mathfrak{p}$ be the eigenspace decomposition with respect to θ. Orthogonality provides a completely positive definite quadratic form q on $\mathfrak{p}$, invariant under $ad\,\mathfrak{k}$. Via the identification $\mathfrak{p} \approx T_p G/K$ where $p = \pi(e)$, the quadratic form q induces an $Ad\,K$-invariant complete inner product $\langle \cdot, \cdot \rangle_p$ on $T_p G/K$, which extends to a G-invariant metric $\langle \cdot, \cdot \rangle$ of G/K:

$$\langle X, Y \rangle_{gK} = \langle d\tau_{g^{-1}} X, d\tau_{g^{-1}} Y \rangle_p \qquad (g \in G \text{ and } X, Y \in T_{gK} G/K),$$

where τ_g is the left translation by g. This proves that (G, K) is a Riemannian symmetric pair.

The last assertion is evident from the proof. $\qquad \square$

Remark 2.4.15 In the above proof of (i) $\Longrightarrow$ (ii), it is not necessary to assume simple connectedness of G, nor connectedness of K.

Our next task is to give a Jordan algebraic characterization of symmetric spaces on which the vector fields form a TKK Lie algebra. In view of Theorem 2.4.14, this amounts to identifying the Jordan triple structures in those orthogonal involutive Lie algebras having TKK Lie algebraic structures.

Lemma 2.4.16 *Let V be a real Jordan triple and let $\mathfrak{L}(V)$ be its TKK Lie algebra, with the main involution θ. The following conditions are equivalent:*

(i) *The symmetric part $(\mathfrak{g}(V), \theta)$ of $\mathfrak{L}(V)$ is orthogonal.*

(ii) *The dual symmetric part $(\mathfrak{g}^*(V), \theta^*)$ of $\mathfrak{L}(V)$ is orthogonal.*

(iii) *V admits a complete inner product $\langle \cdot, \cdot \rangle$ satisfying*

$$\langle (a \,\square\, b)x, x \rangle = \langle x, (b \,\square\, a)x \rangle \qquad (a, b, x \in V).$$

Proof (i) $\Longrightarrow$ (iii). Let $(\mathfrak{g}(V), \theta)$ be orthogonl with eigenspace decomposition

$$\begin{aligned}
\mathfrak{g}(V) &= \mathfrak{k} \oplus \mathfrak{p} \\
&= \{(0, h, 0) : \theta h = h\} \oplus \{(a, 0, -a) : a \in V\},
\end{aligned}$$

where $h \in V_{00}$ is of the form $h = (h^+, h^-)$ with $h^+ = h^- = \sum_j a_j \,\square\, b_j - b_j \,\square\, a_j$ for some $a_j, b_j \in V$. Let q be a completely positive definite quadratic form on $\mathfrak{p}$, invariant under $ad\, \mathfrak{k}$. Then V is endowed with the following complete inner product:

$$\langle x, y \rangle := q((x, 0, -x), (y, 0, -y)).$$

Let $a, b \in V$ and let $h = (a \,\square\, b - b \,\square\, a,\ a \,\square\, b - b \,\square\, a)$ Then $\theta h = h$ and the $ad\, \mathfrak{k}$-invariance of q gives

$$\begin{aligned}
0 &= q([(0, h, 0), (x, 0, -x)], (x, 0, -x)) \\
&= q((\{abx\} - \{bax\}, 0, \{bax\} - \{abx\}), (x, 0, -x)) \\
&= \langle (a \,\square\, b)x, x \rangle - \langle (b \,\square\, a)x, x \rangle.
\end{aligned}$$

(iii) $\Longrightarrow$ (i). Let $\mathfrak{k} \subset \mathfrak{g}(V)$ be the 1-eigenspace of θ and $\mathfrak{p}$ the (-1)-eigenspace. We define a completely positive definite quadratic form $q_v : \mathfrak{p} \times \mathfrak{p} \longrightarrow \mathbb{R}$ by the complete inner product $\langle \cdot, \cdot \rangle$ of V:

$$q_v((a, 0, -a), (u, 0, -u)) = \langle a, u \rangle \qquad ((a, 0, -a), (u, 0, -u) \in \mathfrak{p}).$$

Thus q_v is $ad\, \mathfrak{k}$-invariant. Indeed, let $Z = (0, h, 0) \in \mathfrak{k}$ with $h = (h^+, h^-)$ and $h^+ = h^-$. Then we have

$$\begin{aligned}
q_v([Z, (u, 0, -u)], (u, 0, -u)) &= q_v((h^+ u, 0, -h^- u), (u, 0, -u)) \\
&= \langle h^+ u, u \rangle,
\end{aligned}$$

which vanishes. Indeed, let $h^+ = \sum_j (x_j \,\square\, y_j - y_j \,\square\, x_j) = 0$, say. Then

$$\begin{aligned}
\langle h^+ u, u \rangle &= \sum_j \langle (x_j \,\square\, y_j)u, u \rangle - \langle (y_j \,\square\, x_j)u, u \rangle \\
&= \sum_j \langle (x_j \,\square\, y_j)u, u \rangle - \langle u, (x_j \,\square\, y_j)u \rangle = 0.
\end{aligned}$$

Therefore $\mathfrak{g}(V)$ is orthogonal with respect to q_v.

The equivalence of (ii) and (iii) is proved similarly. □

This lemma motivates the following definition.

Definition 2.4.17 A Jordan triple system V is called a *Jordan Hilbert triple* if V is a Hilbert space in which the inner product $\langle \cdot, \cdot \rangle$ satisfies

$$\langle (a \,\square\, b)x, x \rangle = \langle x, (b \,\square\, a)x \rangle \qquad (a, b, x \in V). \tag{2.28}$$

According to this definition, Lemma 2.4.16 says that a real Jordan triple V is a Jordan Hilbert triple if, and only if, the Lie algebra $\mathfrak{g}(V)$ is orthogonal. Condition (2.4.19) can be rephrased in terms of inner derivations. For a Jordan triple system V and $a, b \in V$, denote by

$$d(a, b) : V \longrightarrow V$$

the inner derivation $d(a, b) = a \,\square\, b - b \,\square\, a$. A simple calculation shows that (2.4.19) is equivalent to saying that the inner derivations $d(a, b)$ are skew-symmetric:

$$\langle d(a, b)x, y \rangle = \langle x, d(b, a)y \rangle \qquad (a, b, x, y \in V).$$

Inner derivations are automatically continuous in a Jordan Hilbert triple.

Lemma 2.4.18 *Let V be a Jordan Hilbert triple and $a, b \in V$. The inner derivation $d(a, b) : V \longrightarrow V$ is continuous.*

Proof Since

$$\langle d(a, b)x, y \rangle = \langle x, d(b, a)y \rangle \qquad (x, y \in V),$$

the linear operator $d(a, b)$ is weakly continuous on V and therefore norm continuous. □

In view of Lemma 2.4.1 and Lemma 2.4.2, given a normed Jordan triple V, we always equip the Lie algebras $\mathfrak{g}(V)$, $\mathfrak{g}^*(V)$ and $\mathfrak{L}(V)$ with the following norms:

$$\|(a, (h, h), -a)\|_{\mathfrak{g}(V)} = \|a\|_V + \|h\|_{L(V)}$$
$$\|(a, (h, h), a)\|_{\mathfrak{g}^*(V)} = \|a\|_V + \|h\|_{L(V)}$$
$$\|(x, (h^+, h^-), y)\|_{\mathfrak{L}(V)} = \|x\|_V + \|h^+\|_{L(V)} + \|h^-\|_{L(V)} + \|y\|_V.$$

With these norms in the context of normed Jordan triples, Lemma 2.4.1 and Lemma 2.4.2 can be stated as follows. A normed Jordan triple V admits a continuous map $(a, b) \in V \times V \mapsto d(a, b) \in L(V)$ if, and only if, $\mathfrak{g}(V)$ is a normed Lie algebra. A normed Jordan triple V admits a continuous map

$(a, b) \in V \times V \mapsto a \square b \in L(V)$ if, and only if, the TKK Lie algebra $\mathfrak{L}(V)$ is a quasi-normed Lie algebra.

Definition 2.4.19 A real Jordan Hilbert triple V is called a *JH-triple* if the bilinear map

$$d : (a, b) \in V^2 \mapsto d(a, b) \in L(V)$$

is continuous; in other words, if there is a constant $c > 0$ such that

$$\|d(a, b)\| \leq c\|a\|\|b\| \qquad (a, b, \in V).$$

Theorem 2.4.20 *Let V be a real normed Jordan triple and $(\mathfrak{L}(V), \theta)$ its TKK Lie algebra, with symmetric part $\mathfrak{g}(V) = \mathfrak{k} \oplus \mathfrak{p}$, where $\mathfrak{p} = \{(u, 0, -u) : u \in V\}$ is the (-1)-eigenspace of θ on $\mathfrak{g}(V)$. The following conditions are equivalent:*

(i) *V is a JH-triple.*
(ii) *$(\mathfrak{g}(V), \theta)$ is a normed orthogonal Lie algebra.*
(iii) *$(\mathfrak{g}^*(V), \theta^*)$ is a normed orthogonal Lie algebra.*

In this case, the quadratic form q on $\mathfrak{p}$ for orthogonality is related to the inner product $\langle \cdot, \cdot \rangle$ of V by

$$q(u, 0, -u) = \langle u, u \rangle \qquad (u \in V).$$

Proof This follows from Lemma 2.4.1 and Lemma 2.4.16. $\qquad\qquad \square$

We see from this result and Theorem 1.3.11 that the category of TKK Lie algebras with normed orthogonal symmetric part is equivalent to the category of JH-triples. Our final task is to show that they correspond to a large class of Riemannian symmetric spaces.

Definition 2.4.21 A JH-triple V is said to be *continuous* if its triple product is continuous, which is equivalent to the condition

$$\|a \square b\| \leq c\|a\|\|b\| \qquad (a, b \in V)$$

for some constant $c > 0$.

Theorem 2.4.22 *Let V be a real normed Jordan triple and $(\mathfrak{L}(V), \theta)$ its TKK Lie algebra, with symmetric part $\mathfrak{g}(V)$. The following conditions are equivalent:*

(i) *V is a continuous JH-triple.*
(ii) *$(\mathfrak{L}(V), \theta)$ is quasi-normed and $(\mathfrak{g}(V), \theta)$ is a normed orthogonal Lie algebra.*

(iii) $(\mathfrak{L}(V), \theta)$ *is quasi-normed and* $(\mathfrak{g}^*(V), \theta^*)$ *is a normed orthogonal Lie algebra.*

Proof This follows from Lemma 2.4.2 and Theorem 2.4.20. □

Let V be a JH-triple with TKK Lie algebra $\mathfrak{L}(V) = V \oplus V_{00} \oplus V$ and involution θ. For $h = \sum_j a_j \,\square\, b_j \in V_0 \subset L(V)$ and $h^\natural = \sum_j b_j \,\square\, a_j$, we have

$$h + h^\natural = 0 \quad \text{if, and only if,} \quad 2h = \sum_j (a_j \,\square\, b_j - b_j \,\square\, a_j).$$

Let

$$V_{00}^\theta = \{k \in V_{00} : \theta k = k\} = \{(h, h) \in V_0^2 : h + h^\natural = 0\},$$

which is a Lie subalgebra of V_0 and a subspace of $L(V)$, where the last equality can be seen from (2.22). The symmetric part $\mathfrak{g}(V)$ is contained in $V \oplus V_{00}^\theta \oplus V$. The dual symmetric part $\mathfrak{g}^*(V)$, with the dual involution $\theta^*(a, k, a) = (-a, k, -a)$, is also contained in $V \oplus V_{00}^\theta \oplus V$.

The symmetric part $\mathfrak{g}(V)$ is equipped with the norm

$$\|(u, (h, h), -u)\| = \langle u, u \rangle^{1/2} + \|h\| \qquad ((u, (h, h), -u) \in \mathfrak{g}(V)).$$

Let $\overline{V_{00}^\theta}$ be the closure of V_{00}^θ in the ℓ^∞-sum $L(V) \oplus_\infty L(V)$. Then

$$\overline{\mathfrak{g}}(V) = \{(u, (h, h), -u) : u \in V, (h, h) \in \overline{V_{00}^\theta}\}$$

is the completion of $\mathfrak{g}(V)$. The Lie product of $\mathfrak{g}(V)$ extends naturally to $\overline{\mathfrak{g}}(V)$. With the extended involution

$$\overline{\theta}(u, k, -u) = (-u, k, u) \qquad ((u, k, -u) \in \overline{\mathfrak{g}}(V)),$$

$(\overline{\mathfrak{g}}(V), \overline{\theta})$ is a real involutive Banach Lie algebra with involution $\overline{\theta}$ and eigenspace decomposition

$$\overline{\mathfrak{g}}(V) = \{(0, (h, h), 0) : (h, h) \in \overline{V_{00}^\theta}\} \oplus \mathfrak{p}, \tag{2.29}$$

where $\mathfrak{p} = \{(u, 0, -u) : u \in V\}$ is also the (-1)-eigenspace of θ in $\mathfrak{g}(V)$.

Likewise $\overline{\mathfrak{g}^*}(V) = \{(u, (h, h), u) : u \in V, (h, h) \in \overline{V_{00}^\theta}\}$ is the completion of the dual symmetric part $\mathfrak{g}^*(V)$, which is a real involutive Banach Lie algebra with the extended involution $\overline{\theta^*}(u, k, u) = (-u, k, -u)$ for $(u, k, u) \in \overline{\mathfrak{g}^*}(V)$.

Given $(h, h) \in V_{00}^\theta \subset L(V) \times L(V)$, we have $\langle hu, u \rangle = 0$ for all $u \in V$, by the proof of (iii) $\Longrightarrow$ (i) in Lemma 2.4.16. Hence h is skew-symmetric with respect to the inner product of V:

$$\langle hx, y \rangle + \langle x, hy \rangle = 0 \qquad (x, y \in V).$$

In other words, $h^* = -h = h^\natural$, where h^* is the adjoint of h in $L(V)$.

Remark 2.4.23 If V is finite-dimensional in Theorem 2.4.20, then one can define an inner product $\langle \cdot, \cdot \rangle_{\mathfrak{k}}$ on $\mathfrak{k} = \{(0, (h, h), 0) : (h, h) \in V_{00}^{\theta}\}$ by the trace

$$\langle (0, (h, h), 0), (0, (g, g), 0) \rangle_{\mathfrak{k}} = \text{Trace} (hg^*),$$

which is $ad\ \mathfrak{k}$-invariant since

$$\langle [(0, (h, h), 0), (0, (g, g), 0)], (0, (g, g), 0) \rangle_{\mathfrak{k}} = \text{Trace} ([h, g]g^*) = 0,$$

where $[h, g]g^* = hgg^* - ghg^* = -hg^2 + ghg$ by skew symmetry of g. It follows that the positive definite quadratic form $\widetilde{q} : \mathfrak{g}(V) \times \mathfrak{g}(V) \longrightarrow \mathbb{R}$ defined by

$$\widetilde{q}((a, (h, h), -a), (a, (h, h), -a))$$
$$= \langle (0, (h, h), 0), (0, (h, h), 0) \rangle_{\mathfrak{k}} + q(a, 0, -a)$$

is invariant under θ and $ad\ \mathfrak{k}$.

Lemma 2.4.24 *Let* $(V, \langle \cdot, \cdot \rangle)$ *be a JH-triple and* $\mathfrak{g}(V) = \mathfrak{k} \oplus \mathfrak{p}$ *the symmetric part of the TKK Lie algebra* $(\mathfrak{L}(V), \theta)$. *Then the completion* $(\overline{\mathfrak{g}}(V), \overline{\theta})$ *is orthogonal with respect to the inner product of* $\mathfrak{p}$.

Proof Let $\overline{\mathfrak{g}}(V) = \overline{\mathfrak{k}} \oplus \mathfrak{p}$ be the eigenspace decomposition with respect to $\overline{\theta}$, where

$$\overline{\mathfrak{k}} = \{(0, (h, h), 0) : (h, h) \in \overline{V_{00}^{\theta}}\}.$$

Let $\langle \cdot, \cdot \rangle_{\mathfrak{p}}$ be the inner product of $\mathfrak{p}$. For $(h, h) \in \overline{V_{00}^{\theta}}$ with $h = \lim_n h_n$ and $(h_n, h_n) \in V_{00}^{\theta}$, we have $\langle hu, u \rangle = \lim_n \langle h_n u, u \rangle = 0$ for all $u \in V$. Hence for $Z = (0, (h, h), 0) \in \overline{\mathfrak{k}}$, we have

$$\langle [Z, (u, 0, -u)], (u, 0, -u) \rangle_{\mathfrak{p}} = \langle (hu, 0, -hu), (u, 0, -u) \rangle_{\mathfrak{p}} = \langle hu, u \rangle = 0.$$
$$\square$$

Remark 2.4.25 Evidently, the completion $\overline{\mathfrak{g}}^*(V)$ of the dual symmetric part is also orthogonal in the preceding lemma. We omit the details to avoid repetition.

Theorem 2.4.20 provides a key to the correspondence between JH-triples and Riemannian symmetric spaces. We first derived some results to pave the way towards such a correspondence. We begin by determining the centre of the complete symmetric part $\overline{\mathfrak{g}}(V)$.

Lemma 2.4.26 *Let* V *be a JH-triple and* $\mathfrak{g}(V)$ *the symmetric part of its TKK Lie algebra* $(\mathfrak{L}(V), \theta)$. *The centre of* $\overline{\mathfrak{g}}(V)$ *is given by*

$$\mathfrak{z}(\overline{\mathfrak{g}}(V)) = \{(a, 0, -a) : a \square x = x \square a, \forall x \in V\}.$$

Proof Let $X = a \oplus (h_0, h_0) \oplus -a \in \overline{\mathfrak{g}}(V)$. If $X \in \mathfrak{z}(\overline{\mathfrak{g}}(V))$, then for $x \oplus (g, g) \oplus -x \in \mathfrak{g}(V)$, we have

$$0 = [a \oplus (h_0, h_0) \oplus -a, \; x \oplus (g, g) \oplus -x]$$
$$= (h_0 x - ga, \; ([h_0, g] - a \square x + x \square a, \; [h_0, g] - a \square x + x \sqcup a), \; ga - h_0 x).$$

Choose $g = 0$; then $h_0 x = ga = 0$ for all $x \in V$ and hence $h_0 = 0$ and $a \square x = x \square a$. The arguments can be reversed, since $a \square x = x \square a$ for all $x \in V$ implies $ga = 0$ for all $(g, g) \in V_{00}^{\theta}$. □

The above lemma leads to the definition of the following closed subspace of a JH-triple V:

$$Z(V) = \{a \in V : a \square x = x \square a, \forall x \in V\}, \tag{2.30}$$

where $Z(V)$ is closed by continuity of the bilinear inner derivation map $d : (a, x) \in V^2 \mapsto d(a, x) \in L(V)$. As V is a Hilbert space, we have the direct sum decomposition

$$V = Z(V) \oplus Z(V)^{\perp}.$$

Lemma 2.4.27 *Given a JH-triple V, the subspace $Z(V)$ is an abelian subtriple of V.*

Proof We need only show that $Z(V)$ is a subtriple of V, for which it suffices to show that $a \in Z(V)$ implies $a^3 = \{a, a, a\} \in Z(V)$, by the polarization formulae for $a, b, x \in Z(V)$:

$$6\{b, a, b\} = (a + b)^3 + (a - b)^3 - 2a^3$$
$$2\{a, x, b\} = \{a + b, x, a + b\} - \{a, x, a\} - \{b, x, b\}.$$

Let $a \in Z(V)$. We have

$$\{a, x, \{a, a, y\}\} = \{\{a, x, a\}, a, y\} - \{a, \{a, x, a\}, y\} + \{a, a, \{a, x, y\}\}$$
$$= \{a, a, \{a, x, y\}\}$$
$$= \{a^3, x, y\} - \{a, \{a, a, x\}, y\} + \{a, x, \{a, a, y\}\}$$
$$= \{a^3, x, y\},$$

where $\{a, \{a, a, x\}, y\} = \{a, y, \{a, a, x\}\} = \{a, a, \{a, y, x\}\}$. Hence

$$(a^3 \square x)(y) = \{a, a, \{x, a, y\}\}$$
$$= \{\{a, a, x\}, a, y\} - \{x, a^3, y\} + \{x, a, \{a, a, y\}\}$$
$$= 2\{a^3, x, y\} - \{x, a^3, y\}$$
$$= (x \square a^3)(y)$$

and $a^3 \in Z(V)$. □

Lemma 2.4.28 *Let V be a JH-triple and let $\mathfrak{z}(\overline{\mathfrak{g}})$ be the centre of $\overline{\mathfrak{g}}(V)$. Then we have the decomposition $\overline{\mathfrak{g}}(V) = \mathfrak{z}(\overline{\mathfrak{g}}) \oplus \mathfrak{z}(\overline{\mathfrak{g}})^o$, where*

$$\mathfrak{z}(\overline{\mathfrak{g}})^o = \left\{ (x, (h, h), -x) : x \in Z(V)^\perp \text{ and } (h, h) \in \overline{V_{00}^\theta} \right\}$$

is a Lie subalgebra of $\overline{\mathfrak{g}}(V)$ with trivial centre.

Proof The direct sum decomposition follows from Lemma 2.4.26 and the decomposition $V = Z(V) \oplus Z(V)^\perp$. We note that $\overline{V_{00}^\theta} \subset L(V) \times L(V)$, and by a note before Remark 2.4.23, for each $(g, g) \in \overline{V_{00}^\theta}$, the linear map $g \in L(V)$ is skew-symmetric with respect to the inner product of V. Given $x \in Z(V)^\perp$ and $(g, g) \in \overline{V_{00}^\theta}$, we have

$$\langle gx, a \rangle = -\langle x, ga \rangle = 0$$

for all $a \in Z(V)$, where $ga = 0$ as in the proof of Lemma 2.4.26. Hence $gx \in Z(V)^\perp$.

Let $X = (x, (h, h), -x), Y = (y, (g, g), -y) \in \mathfrak{z}(\overline{\mathfrak{g}})^o$. We have

$$[X, Y] = (hy - gx, ([h, g] - x \,\square\, y + y \,\square\, x, [h, g] - x \,\square\, y + y \,\square\, x), \; gx - hy),$$

where $hy - gx \in Z(V)^\perp$ implies $[X, Y] \in \mathfrak{z}(\overline{\mathfrak{g}})^o$.

Further, if $[X, Y] = 0$, then choosing $g = 0$, we have $hy = 0$ for all $y \in Z(V)^\perp$ and hence $h = 0$ since $ha = 0$ for all $a \in Z(V)$. It follows that $x \,\square\, y = y \,\square\, x$ for all $y \in Z(V)^\perp$. But $x \,\square\, a = a \,\square\, x$ for all $a \in Z(V)$. Hence $x \in Z(V) \cap Z(V)^\perp$ and $x = 0$. This proves the triviality of the centre of $\mathfrak{z}(\overline{\mathfrak{g}})^o$. $\qquad\square$

Definition 2.4.29 Let (G, K) be a Riemannian symmetric pair. The symmetric space G/K is said to be *associated to a TKK Lie algebra* if the involutive Lie algebra $(\mathfrak{g}, \sigma)$ of G is the complete symmetric part $(\overline{\mathfrak{h}}_s, \overline{\theta})$ of a TKK Lie algebra $(\mathfrak{h}, \theta)$, in which case G/K is called a *Jordan symmetric space*.

Remark 2.4.30 A symmetric space G/K is a also Jordan symmetric space if the Lie algebra $(\mathfrak{g}, \sigma)$ of G is the complete *dual* symmetric part $(\overline{\mathfrak{h}}_s^*, \overline{\theta^*})$ of a TKK Lie algebra $(\mathfrak{h}, \theta)$. Indeed, $\overline{\mathfrak{h}}_s^*$ is the complete symmetric part of $(\mathfrak{h}, \theta^*)$, by Remark 1.3.6.

Theorem 2.4.31 *We have the following correspondence between JH-triples and Jordan symmetric spaces.*

(i) *Let V be a JH-triple. Then there is a Riemannian symmetric space G/K such that the Lie algebra of G is the complete symmetric part $\overline{\mathfrak{g}}(V)$ of the TKK Lie algebra $\mathfrak{L}(V)$ of V.*

(ii) *Let G/K be a Jordan symmetric space. Then there is a JH-triple V such that $\mathfrak{L}(V)$ is the canonical part of the TKK Lie algebra associated to G/K. Further, if G/K is associated to a canonical TKK Lie algebra, then the complete symmetric part $\overline{\mathfrak{g}}(V)$ of $\mathfrak{L}(V)$ is the Lie algebra of G.*

Proof (i) Let V be a JH-triple and let $\mathfrak{L}(V)$ be its TKK Lie algebra with involution θ. Then the complete symmetric part $\overline{\mathfrak{g}}(V)$ is a Banach Lie algebra and θ extends to an involution $\overline{\theta}$ on $\overline{\mathfrak{g}}(V)$. By Lemma 2.4.28, we have $\overline{\mathfrak{g}}(V) = \mathfrak{z}(\overline{\mathfrak{g}}) \oplus \mathfrak{z}(\overline{\mathfrak{g}})^o$, where the centre $\mathfrak{z}(\overline{\mathfrak{g}})$ is the Lie algebra of itself as a Lie group, and the Lie algebra $\mathfrak{z}(\overline{\mathfrak{g}})^o$ has a trivial centre and is therefore the Lie algebra of a connected Banach Lie group, by Remark 2.1.27. Taking the universal covering of the group (cf. [111, lemma 6.10]), it follows that $\overline{\mathfrak{g}}(V)$ is the Lie algebra of a simply connected Banach Lie group G, and the 1-eigenspace $\mathfrak{k} = \{(0, (h, h), 0) : (h, h) \in V_{00}^{\theta}\} \subset \overline{\mathfrak{g}}(V)$ of $\overline{\theta}$ is the Lie algebra of a connected Banach Lie subgroup K of G. By Lemma 2.4.24, $(\overline{\mathfrak{g}}(V), \overline{\theta})$ is an orthogonal involutive Banach Lie algebra and by Theorem 2.4.14, G/K is a Riemannian symmetric space.

(ii) Conversely, let G/K be a symmetric space such that the Lie algebra $(\mathfrak{g}, \sigma)$ of G is the complete symmetric part $(\overline{\mathfrak{h}}_s, \overline{\theta})$ of a TKK Lie algebra $(\mathfrak{h}, \theta)$. Then, by Theorem 1.3.11 and the construction after Proposition 1.3.13, there is a Jordan triple V such that

$$\mathfrak{h} = \mathfrak{h}_{-1} \oplus \mathfrak{h}_0 \oplus \mathfrak{h}_1 \supset \mathfrak{L}(V) = V \oplus V_{00} \oplus V$$

with $\mathfrak{h}_{\pm 1} = V$ and $\overline{\mathfrak{h}}_s$ contains the complete symmetric part $\overline{\mathfrak{g}}(V)$ of the TKK Lie algebra $\mathfrak{L}(V)$ which is the canonical part $\mathfrak{h}^c$ of $\mathfrak{h}$.

By Theorem 2.4.14 and Remark 2.4.15, $(\overline{\mathfrak{h}}_s, \overline{\theta})$ is orthogonal with respect to a complete inner product $\langle \cdot, \cdot \rangle_{\mathfrak{p}}$ on the (-1)-eigenspace of $\overline{\theta}$, which equals the (-1)-eigenspace $\mathfrak{p} = \{(a, 0, -a) : a \in \mathfrak{h}_{-1}\}$ of θ in $\mathfrak{h}_s$. Since $\mathfrak{p}$ is also the (-1)-eigenspace of θ in $\mathfrak{g}(V)$, it follows that $(\mathfrak{g}(V), \theta)$ is also orthogonal with respect to $\langle \cdot, \cdot \rangle_{\mathfrak{p}}$. Hence V is a JH-triple by Theorem 2.4.20.

Finally, if $\mathfrak{h}$ is canonical, then $\mathfrak{h} = \mathfrak{L}(V)$ and $(\overline{\mathfrak{h}}_s, \overline{\theta}) = (\overline{\mathfrak{g}}(V), \overline{\theta})$. $\square$

In the correspondence between JH-triples V and Jordan symmetric spaces, the symmetric part and the dual symmetric part of a TKK Lie algebra $\mathfrak{L}(V)$ correspond to a pair of symmetric spaces which are said to be *dual* to each other. Let $(\mathfrak{g}, \sigma)$ be the involutive Lie algebra of a Jordan symmetric space G/K and V a JH-triple such that $\mathfrak{g} = \overline{\mathfrak{g}}(V)$ and $\sigma = \overline{\theta}$, where $\mathfrak{g}(V)$ is the symmetric part of the TKK Lie algebra $(\mathfrak{L}(V), \theta)$ of V. Let $\mathfrak{g} = \mathfrak{k} \oplus \mathfrak{p}$ be the eigenspace decomposition for θ. Then the complete dual symmetric part $(\overline{\mathfrak{g}^*}(V), \overline{\theta^*})$ gives rise to a symmetric space, the *dual* of G/K, whose Lie algebra can be identified with $(\mathfrak{g}', \sigma')$, where

$$g' = \mathfrak{k} \oplus i\mathfrak{p} \quad \text{and} \quad \sigma'(k + ip) = k - ip.$$

Indeed, given $\overline{\mathfrak{g}}(V) = \{(0, k, 0) : \theta k = k \in \overline{V_{00}^\theta}\} \oplus \{(a, 0, -a) : a \in V\}$, we have

$$\mathfrak{g}' = \{(0, k, 0) : \theta k = k \in \overline{V_{00}^\theta}\} \oplus \{(ia, 0, -ia) : a \in V\}.$$

The following Lie algebra isomorphism identifies $\mathfrak{g}'$ with $\overline{\mathfrak{g}*}(V)$:

$$\psi : (ia, k, -ia) \in \mathfrak{g}' \mapsto (a, k, a) \in \overline{\mathfrak{g}*}(V),$$

where $\psi\sigma' = \theta^*\psi$ and θ^* is the dual involution of $\mathfrak{L}(V)$.

We conclude this section with some examples. The structures of JH-triples will be studied in more detail in the next chapter.

Example 2.4.32 A finite-dimensional formally real Jordan algebra V is a Hilbert space with inner product $\langle a, b \rangle = \text{Trace}\,(a \square b)$ which satisfies

$$\langle ab, c \rangle = \langle b, ac \rangle \qquad (a, b, c \in V).$$

Therefore V is a JH-triple in the canonical Jordan triple product

$$\{a, b, c\} = (ab)c + a(bc) - b(ac).$$

In fact, any JH-algebra, which can be infinite-dimensional, is a JH-triple in the canonical Jordan triple product.

Example 2.4.33 We have shown in Theorem 1.2.34 that a finite-dimensional Jordan triple V has a positive definite trace form $\text{Tr}\,(x \square y)$ if, and only if, it is semisimple and positive. With the trace form as inner product, such a Jordan triple is a JH-triple. These Jordan triples are also called *compact* in Loos [82] because they correspond to a class of Riemannian symmetric spaces, called the *symmetric R-spaces*, which are compact. The Lie algebras $\mathfrak{L}$ for these symmetric spaces, constructed in [82], are TKK Lie algebras. Given a compact Jordan triple $(V, \{\cdot, \cdot, \cdot\})$ which is non-degenerate, the corresponding Lie algebra $\mathfrak{L}$ constructed in Loos [82] is the TKK Lie algebra $\mathfrak{L}((V, -2\{\cdot, \cdot, \cdot\}))$ of the Jordan triple $(V, -2\{\cdot, \cdot, \cdot\})$. The Lie algebra of the corresponding symmetric R-space is the dual symmetric part $\mathfrak{g}^*$ of $\mathfrak{L}((V, -2\{\cdot, \cdot, \cdot\}))$, with dual involution $\theta^*(x, h, y) = (-y, -h^\natural, -x)$. Let β be the Killing form of $\mathfrak{L}((V, -2\{\cdot, \cdot, \cdot\}))$ and let $a \square a$ be the box operator with respect to the triple product $\{\cdot, \cdot, \cdot\}$. By Koecher [78, p. 56], the restriction of β on the (-1)-eigenspace $\mathfrak{p} = \{(a, 0, a) : a \in V\}$ of $\mathfrak{g}^*$ is given by

$$\beta((a, 0, a), (a, 0, a)) = -8 \operatorname{Trace}(a \,\square\, a)$$

which is negative definite.

Example 2.4.34 A JH^*-*triple*, as defined in Kaup [71], is a Hermitian Jordan triple V which is a complex Hilbert space where the triple product is continuous and every operator $a \,\square\, a : V \longrightarrow V$ is self-adjoint. The latter condition is equivalent to the *associativity* of the inner product:

$$\langle (a \,\square\, b)x, y \rangle = \langle x, (b \,\square\, a)y \rangle \qquad (a, b, x, y \in V). \tag{2.31}$$

JH*-triples can be regarded as real Jordan triples with real inner product $\operatorname{Re} \langle \cdot, \cdot \rangle$, in which case they are continuous JH-triples. The inner product $\operatorname{Re} \langle \cdot, \cdot \rangle$ is also *associative* in the sense of (2.31).

Example 2.4.35 The inner product of a JH-triple need not be associative in the sense of (2.31), even if it is continuous and finite-dimensional. A simple example is the real Hilbert space $\mathbb{H}$ of quaternions, with basis $\{\mathbf{1}, \mathbf{i}, \mathbf{j}, \mathbf{k}\}$. The triple product $2\{a, b, c\} = abc + cba$ in $\mathbb{H}$ satisfies

$$\langle (a \,\square\, b)x, x \rangle = \langle x, (b \,\square\, a)x \rangle \qquad (a, b, x \in \mathbb{H}).$$

To see this, we can represent $\mathbb{H}$ by 4×4 matrices in $M_4(\mathbb{R})$ and observe that the inner product satisfies

$$\langle a, x \rangle = \frac{1}{4} \operatorname{Trace}(ax^t) = \frac{1}{4} \operatorname{Trace}(a^t x),$$

where x^t denotes the transpose of $x \in M_4(\mathbb{R})$ and $xx^t = x^t x$ for $x \in \mathbb{H}$. Hence

$$4\langle (a \,\square\, b)x, x \rangle = \operatorname{Trace}(abxx^t + xbax^t) = \operatorname{Trace}(x^t xab + baxx^t)$$
$$= \operatorname{Trace}(x^t(xab + bax)) = 4\langle x, (b \,\square\, a)x \rangle.$$

We also have $\|\{a, b, c\}\| \leq \|a\| \|b\| \|c\|$, since $\|xy\| = \|x\| \|y\|$ for all $x, y \in \mathbb{H}$. Hence $\mathbb{H}$ is a continuous JH-triple. However, the inner product is not associative, since

$$\langle (\mathbf{1} \,\square\, \mathbf{k})\mathbf{k}, \mathbf{1} \rangle \neq \langle \mathbf{k}, (\mathbf{k} \,\square\, \mathbf{1})\mathbf{1} \rangle.$$

We note that $\mathbb{H}$ is not a flat Jordan triple; for instance, $(\mathbf{i} \,\square\, \mathbf{j})(\mathbf{i}) \neq (\mathbf{j} \,\square\, \mathbf{i})(\mathbf{i})$.

It has been shown in Kaup [71] that JH*-triples correspond to the Hermitian symmetric spaces which are Riemannian symmetric spaces with a Hermitian structure and will be discussed in the next section. In fact, JH*-triples are exactly those JH-triples which admit compatible complex structures.

A *complex structure* on a real vector space V is a linear map $J : V \longrightarrow V$ such that $-J^2$ is the identity map on V. If V admits a complex structure J, then we view it as a complex vector space by defining the complex scalar multiplication $iv := Jv$ for $v \in V$. Conversely, every complex vector space, regarded as a real space, admits a natural complex structure $J : x \mapsto ix$.

Proposition 2.4.36 *A JH-triple* $(V, \langle \cdot, \cdot \rangle)$ *admits the structure of a JH*-triple if, and only if, there is a complex structure* $J : V \longrightarrow V$ *satisfying*

$$\langle Jx, Jy \rangle = \langle x, y \rangle \quad and \quad J(x \square y) = (x \square y)J$$

for $x, y \in V$.

Proof We only need to show the necessity. If a JH-triple V satisfies the given condition, then it becomes a complex Hilbert space with the complex inner product

$$\ll x, y \gg \ = \ \langle x, y \rangle - i \langle Jx, y \rangle,$$

and also it becomes a Hermitian Jordan triple with the Jordan triple product

$$\{\{x, y, z\}\} = \ \{x, y, z\} + i\{x, Jy, z\}.$$

Substituting $x + y$ and $x + iy$ for z in the identity

$$\ll (a \square b)z, z \gg \ = \ \ll z, (b \square a)z \gg$$

yields $\ll (a \square b)x, y \gg \ = \ \ll x, (b \square a)y \gg$. Hence $(V, \{\{\cdot, \cdot, \cdot\}\}, \ll \cdot, \cdot \gg)$ is a JH*-triple. $\qquad \square$

2.5 Jordan triples and symmetric domains

In this section, we study Jordan structures in complex Banach manifolds. Let (M, g) be a Riemannian manifold. An *almost complex structure* J on M is a map $p \in M \mapsto J_p \in L(T_pM)$ such that J_p is a complex structure on the tangent space T_pM and for each smooth vector field X on M, the vector field JX is smooth, where $JX(p) = J_p(X_p)$ for $p \in M$. The almost complex structure J is called a *Hermitian structure* if the Riemannian metric g is *Hermitian* with respect to J, in the sense that

$$g_p(J_pu, J_pv) = g_p(u, v) \qquad (u, v \in T_pM, p \in M).$$

Under these conditions, one can define a *Hermitian metric h* on M by

$$h_p(u, v) = g_p(u, v) - i g_p(J_p u, v) \qquad (u, v \in T_p M).$$

Example 2.5.1 Let M be a complex manifold of dimension $n < \infty$, with an atlas $\{(\mathcal{U}_\varphi, \varphi, V_\varphi)\}$. Let $p \in M$. If $p \in \mathcal{U}_\varphi$, the tangent space $T_p M$ identifies with $V_\varphi = \mathbb{C}^n = \mathbb{R}^{2n}$ via φ and has a basis $\{x_k^\varphi, y_k^\varphi : k = 1, \ldots, n\}$. We can define an almost complex structure J on M by putting

$$J_p x_k^\varphi = y_k^\varphi, \qquad J_p y_k^\varphi = -x_k^\varphi \qquad (k = 1, \ldots, n),$$

which can be seen to be independent of the choice of local charts $(\mathcal{U}_\varphi, \varphi, V_\varphi)$. Indeed, if $(\mathcal{U}_\psi, \psi, V_\psi)$ is another local chart at p and $T_p M \simeq V_\psi$ has a basis $\{x_k^\psi, y_k^\psi : k = 1, \ldots, n\}$, the change of basis

$$x_k^\psi = \sum_j a_{kj} x_j^\varphi + \sum_j b_{kj} y_j^\varphi$$

$$y_k^\psi = \sum_j c_{kj} x_j^\varphi + \sum_j d_{kj} y_j^\varphi$$

is effected by the derivative

$$(\psi \varphi^{-1})'(\varphi(p)) : V_\varphi \longrightarrow V_\psi,$$

which is the $2n \times 2n$ matrix

$$\begin{pmatrix} a_{kj} & b_{kj} \\ c_{kj} & d_{kj} \end{pmatrix},$$

where $a_{kj} = d_{kj}$ and $b_{kj} = -c_{kj}$ by the Cauchy–Riemann equations. It follows that $J_p(x_k^\psi) = y_k^\psi$ and $J_p(y_k^\psi) = -x_k^\psi$. We call J the *canonical* almost complex structure.

The following definition is an extension of the concept of a finite-dimensional Hermitian symmetric space (cf. [51, p. 372]).

Definition 2.5.2 A connected manifold M modelled on a complex Hilbert space is called a *Hermitian symmetric space* if it is equipped with a Hermitian structure and every point $p \in M$ is an isolated fixed point of an involutive holomorphic isometry $s_p : M \longrightarrow M$.

The correspondence between Hermitian symmetric spaces and JH*-triples shown in Kaup [71] can be seen from the perspective of TKK Lie algebras. Let M be a Hermitian symmetric space. By definition, M is a Riemannian symmetric space and has an analytic structure. By Theorem 2.2.17, the group $G_a(M)$ of biholomorphic isometries of M forms a real Banach Lie group with Lie algebra $\mathfrak{g}_a(M)$. The manifold M is diffeomorphic to the homogeneous space $G_a(M)/K$, where K is the isotropy group at a point $p \in M$. The isometry $s_p : M \longrightarrow M$ gives rise to an involution $\theta = Ads_p : \mathfrak{g}_a(M) \longrightarrow \mathfrak{g}_a(M)$. Let $\mathfrak{g}_a(M) = \mathfrak{k} \oplus \mathfrak{p}$ be the eigenspace decomposition with respect to θ, with $\mathfrak{p}$ the (-1)-eigenspace. By Theorem 2.4.14, the involutive Lie algebra $\mathfrak{g}_a(M)$ is orthogonal.

By (2.26), the eigenspace $\mathfrak{p}$ identifies with the tangent space $T_p M$ as real vector spaces, via the evaluation map

$$X \in \mathfrak{p} \mapsto X(p) \in T_p M.$$

With the Hermitian structure of M, the tangent space $T_p M$ is a complex Hilbert space. Given $X \in \mathfrak{p}$, the vector $i X(p) \in T_p M$ is the evaluation of a vector field $JX \in \mathfrak{p}$. This gives a complex structure

$$J : \mathfrak{p} \longrightarrow \mathfrak{p}$$

satisfying $\theta J = J\theta$ and

$$[JX, Y](p) = i[X, Y](p) \qquad (X \in \mathfrak{p}, Y \in \mathfrak{g}_a(M)).$$

If $Y \in \mathfrak{k}$, then $[X, Y] \in \mathfrak{p}$ and hence

$$J[X, Y] = [JX, Y]. \tag{2.32}$$

We note that $X \in \mathfrak{g}_a(M)$ does not imply $iX \in \mathfrak{g}_a(M)$. In fact, $\mathfrak{g}_a(M)$ is *totally real*. We show this for $\mathfrak{k}$.

Lemma 2.5.3 *In the eigenspace decomposition* $\mathfrak{g}_a(M) = \mathfrak{k} \oplus \mathfrak{p}$, *we have* $\mathfrak{k} \cap i\mathfrak{k} = \{0\}$.

Proof Let $Y \in \mathfrak{k}$ and $Y = iZ \in i\mathfrak{k}$. We show $Y = 0$. For $s + it \in \mathbb{C}$, the map $(\exp sY)(\exp tZ) = \exp(s - it)Y : M \longrightarrow M$ is a biholomorphic isometry and therefore the map

$$F : s + it \in \mathbb{C} \longrightarrow (d \exp(s - it)Y)_p \in L(T_p M)$$

is a bounded holomorphic map, where $L(T_p M)$ is the complex Banach space of continuous linear self-maps on $T_p M$. By Liouville theorem, $\varphi \circ F : \mathbb{C} \longrightarrow \mathbb{C}$ is constant for each continuous linear functional φ on $L(T_p M)$. Hence F is constant and $(d \exp Y)_p$ is the identity map. Since $\exp Y(p) = p$, Proposition 2.2.9 implies that $\exp Y$ itself is an identity map on M, that is, $Y = 0$. $\qquad \square$

Lemma 2.5.4 *Let* $J : \mathfrak{p} \longrightarrow \mathfrak{p}$ *be the complex structure just described. Then we have*

$$[JX, JY] = [X, Y] \qquad (X, Y \in \mathfrak{p}).$$

Proof Since $[JX, Y](p) = i[X, Y](p)$, we have $[JX, Y] - i[X, Y] \in \mathfrak{k}$, as $\mathfrak{k}$ is the kernel of the evaluation map. Likewise, we have $[X, JY] + i[JX, JY] \in \mathfrak{k}$. It follows that

$$i[JX, JY] - i[X, Y] \in \mathfrak{k} \cap i\mathfrak{k}$$

and hence $[JX, JY] - [X, Y] = 0$. $\square$

Let $\mathfrak{p}_c$ be the complexification of $\mathfrak{p}$ and extend J to a complex linear map on $\mathfrak{p}_c$, also denoted by J. Let

$$\mathfrak{p}_+ = \{X \in \mathfrak{p}_c : JX = iX\}, \quad \mathfrak{p}_- = \{X \in \mathfrak{p}_c : JX = -iX\}$$

be the $\pm i$-eigenspaces of J such that

$$\mathfrak{p}_c = \mathfrak{p}_+ \oplus \mathfrak{p}_-$$

and hence the complexification $\mathfrak{g}_c$ of $\mathfrak{g}_a(M)$ has a decomposition

$$\mathfrak{g}_c = \mathfrak{p}_+ \oplus \mathfrak{k}_c \oplus \mathfrak{p}_-,$$

where

$$X \in \mathfrak{p} \mapsto X - iJX \in \mathfrak{p}_+$$

is a complex linear isomorphism. Considered as a real Lie algebra, $\mathfrak{g}_c$ has an involution given by

$$\sigma X = \theta \overline{X},$$

where the natural extension of θ to $\mathfrak{g}_c$ is still denoted by θ. We have $\sigma(\mathfrak{p}_+) = \mathfrak{p}_-$ and

$$[\mathfrak{k}_c, \mathfrak{p}_{\pm 1}] \subset \mathfrak{p}_{\pm 1}, \qquad [\mathfrak{p}_\pm, \mathfrak{p}_\pm] = 0, \tag{2.33}$$

by Lemma 2.5.4 and (2.32). We also have $[\mathfrak{p}_+, \mathfrak{p}_-] \subset \mathfrak{k}_c$. Hence $(\mathfrak{g}_c, \sigma)$, as a real Lie algebra, is a TKK Lie algebra and the space $\mathfrak{p}_+ \approx \mathfrak{p}$ is a Jordan triple with Jordan triple product

$$\{X, Y, Z\} = [[X, \sigma Y], Z] \qquad (X, Y, Z \in \mathfrak{p}_+). \tag{2.34}$$

Moreover, $\mathfrak{g}_a(M)$ can be identified with the complete symmetric part of $\mathfrak{g}_c$ by the map

$$(Y, X) \in \mathfrak{g}_a(M) = \mathfrak{k} \oplus \mathfrak{p} \mapsto (X - iJX, \ Y, \ X + iJX) \in \mathfrak{g}_c$$

where $X + iJX = -\sigma(X - iJX)$. Hence $\mathfrak{p}$ is a JH-triple and also a JH*-triple by Proposition 2.4.36.

Conversely, let V be a JH*-triple and let

$$\mathfrak{L}(V) = V_{-1} \oplus V_{00} \oplus V_1 \qquad (V_{\pm 1} = V)$$

be its TKK Lie algebra with the main involution θ. Define

$$\sigma X = -X \quad \text{for} \quad X \in V_{\pm 1}, \quad \sigma X = X \quad \text{for} \quad X \in V_{00}$$

and $\overline{X} = \sigma \theta X$ for $X \in \mathfrak{L}(V)$. Let $\mathfrak{g} = \{X \in \mathfrak{L}(V) : \overline{X} = X\}$ be the real form of $X \mapsto \overline{X}$. Then σ is an involution on $\mathfrak{g}$. Since $X \in \mathfrak{g}$ if, and only if, $X = \sigma \theta X$, in which case X is of the form $(a, k, -a)$ with $a \in V$ and $\theta k = k \in V_{00}$, it follows that $\sigma|_{\mathfrak{g}} = \theta|_{\mathfrak{g}}$ and $(\mathfrak{g}, \sigma)$ is the symmetric part of $\mathfrak{L}(V)$ with orthogonal completion $\overline{\mathfrak{g}} = \mathfrak{k} \oplus \mathfrak{p}$, where $\mathfrak{p} = \{(a, 0, -a) : a \in V\}$ inherits the Hermitian structure from V and $\overline{\mathfrak{g}}$ gives rise to a Hermitian symmetric space.

Example 2.5.5 Let $U = \{z \in \mathbb{C} : |z| < 1\}$ be the open unit disc in the complex plane. It is a Riemannian manifold in the Poincaré metric

$$\langle u, v \rangle_z = \mathrm{Re} \, \frac{u \overline{v}}{(1 - |z|^2)^2} \qquad (z \in U, u, v \in T_z D = \mathbb{C}),$$

with the canonical Hermitian structure $Jv = iv$ and the corresponding Hermitian metric

$$h_z(u, v) = \frac{u \overline{v}}{(1 - |z|^2)^2}.$$

The automorphism group of biholomorphic maps on U is the group

$$\mathrm{Aut}\, U = \{\alpha g_a : |\alpha| = 1, a \in U\},$$

where $g_a : U \longrightarrow U$ is the Möbius transformation

$$g_a(z) = \frac{z + a}{1 + \bar{a}z} \qquad (z \in U)$$

induced by a and we have $g_a^{-1} = g_{-a}$ (cf. [98, theorem 12.6]). Evidently $\mathrm{Aut}\, U$ acts transitively on U since $g_a(0) = a$. The Riemannian distance on U is the Poincaré distance

$$\delta(z, w) = \inf_\gamma \left\{ \int_0^1 \frac{|\gamma'(t)|}{1 - |\gamma(t)|^2} dt \right\} = \tanh^{-1} \left| \frac{z - w}{1 - z\overline{w}} \right|, \qquad (2.35)$$

where the infimum is taken over all piecewise smooth curves $\gamma : [0, 1] \to U$ from z to w (cf. [31, p. 42]). By the Schwarz–Pick lemma, a holomorphic map $f : U \longrightarrow U$ satisfies the inequality

$$\left| \frac{f(z) - f(w)}{1 - f(z)\overline{f(w)}} \right| \leq \left| \frac{z - w}{1 - z\overline{w}} \right|$$

and hence Aut U is the group of biholomorphic isometries on U. With the Poincaré metric, U is a Hermitian symmetric space. Each point $a \in D$ has a symmetry

$$s_a(z) = g_a \circ s_0 \circ g_{-a}(z) = \frac{2a - (1 + |a|^2)z}{1 + |a|^2 - 2\bar{a}z} \qquad (z \in U),$$

where $s_0(z) = -z$ is the symmetry at 0.

If we equip U with the Euclidean metric, then the group of biholomorphic isometries of U consists of rotations, and hence U is not a Hermitian symmetric space in the Euclidean metric.

Lemma 2.5.6 *Let $f : U \longrightarrow U$ be holomorphic and satisfy*

$$|f(z) - f(w)| = |z - w| \qquad (z, w \in U).$$

Then f is a rotation.

Proof Consider the function $F(z) = \frac{f(z) - f(0)}{z}$, which has a removable singularity at 0 and satisfies $|F(z)| = 1$ for $z \neq 0$. By the maximum modulus principle, F is a constant function, say $F(z) = e^{i\alpha}$ for some $\alpha \in \mathbb{R}$. It follows that

$$f(z) = e^{i\alpha} z + f(0) \qquad (z \in U).$$

If $f(0) = re^{i\beta} \neq 0$ for some $r \in (0, 1)$, then we have

$$1 > |f((1 - r)e^{i(\beta - \alpha)})| = |(1 - r)e^{i\beta} + re^{i\beta}| = 1,$$

which is impossible. Therefore $f(z) = e^{i\alpha} z$ for all $z \in U$. $\qquad \square$

Example 2.5.7 The upper half-plane $\Pi^+ = \{z \in \mathbb{C} : \operatorname{Im} z > 0\}$ is a Hermitian symmetric space in the hyperbolic metric

$$\langle v, v \rangle_z = \frac{|v|^2}{(\operatorname{Im} z)^2} \qquad (z \in \Pi^+, v \in \mathbb{C})$$

and has the canonical complex structure. In fact, the group $G_a(\Pi^+)$ of biholomorphic isometries identifies with the special linear group

$$SL(2, \mathbb{R}) = \left\{ \begin{pmatrix} a & b \\ c & d \end{pmatrix} : a, b, c, d \in \mathbb{R}, \det \begin{pmatrix} a & b \\ c & d \end{pmatrix} = 1 \right\},$$

where $g = \begin{pmatrix} a & b \\ c & d \end{pmatrix}$ gives the isometry

$$z \in \Pi^+ \mapsto g \cdot z := \frac{az + b}{cz + d} \in \Pi^+.$$

This action of $SL(2, \mathbb{R})$ on Π^+ is transitive, since

$$\begin{pmatrix} a & b \\ 0 & \frac{1}{a} \end{pmatrix} \cdot i = ab + ia^2.$$

The symmetry at i is

$$s_i = \begin{pmatrix} 0 & 1 \\ -1 & 0 \end{pmatrix}.$$

Let $K = \{g \in SL(2, \mathbb{R}) : g \cdot i = i\}$ be the isotropy subgroup at i. Since $\begin{pmatrix} a & b \\ c & d \end{pmatrix} \cdot i = i$ if and only if $a = d$, $b = -c$ and $a^2 + b^2 = 1$, we see that

$$K = \left\{ \begin{pmatrix} \cos t & \sin t \\ -\sin t & \cos t \end{pmatrix} : t \in \mathbb{R} \right\} = SO(2).$$

The symmetric space Π^+ is biholomorphic to the open unit disc U via the Cayley transform

$$z \in \Pi^+ \mapsto \frac{z - i}{z + i} \in U,$$

and $SL(2, \mathbb{R})$ is isomorphic to $\mathrm{Aut}\, U$ via

$$\begin{pmatrix} a & b \\ c & d \end{pmatrix} \in SL(2, \mathbb{R}) \mapsto \alpha g_\beta \in \mathrm{Aut}\, U,$$

where

$$\alpha = \frac{b - c - (a + d)i}{c - b - (a + d)i}, \quad \beta = \frac{b + c + (a - d)i}{c - b + (a + d)i}.$$

In finite dimensions, it is well known [16, 48] that the Hermitian symmetric spaces of non-compact type are the bounded symmetric domains in the space of several complex variables of which the open unit disc U is the simplest example. The concept of a symmetric domain can be extended naturally to the infinite-dimensional setting which is given below.

Definition 2.5.8 Let V be a complex Banach space and let $D \subset V$ be a domain; that is, D is an open connected set. We call D *symmetric*, or a *symmetric domain*, if each point $a \in D$ is an isolated fixed point of an involutive biholomorphic map $s \in \mathrm{Aut}\, D$, in which case s is called a *symmetry* at a.

We note that symmetric domains are Banach manifolds, but in infinite dimension, they need not admit a Riemannian metric. Nevertheless, one can define a metric on a *bounded* domain D, called the *Carathéodory metric*, as follows:

$$\delta_D(z, w) = \sup\{\delta(f(z), f(w)) : f \in H(D, U)\} \qquad (z, w \in D), \quad (2.36)$$

where $H(D, U)$ denotes the vector space of holomorphic maps from D to the open unit disc U in $\mathbb{C}$, and δ is the Poincaré distance in (2.35). In particular, $\delta_U = \delta$ by the Schwarz–Pick lemma, and also, each $g \in \mathrm{Aut}\, D$ is a δ_D-isometry:

$$\delta_D(g(z), g(w)) = \delta_D(z, w) \qquad (z, w, \in D).$$

The metric δ_D is equivalent to the metric on D defined by the norm of the ambient space. If D is symmetric, then Cartan's uniqueness theorem implies that the symmetry s at a is unique since its differential ds_a at a is minus the identity map on V (cf. Lemma 2.3.2). Henceforth we denote the symmetry at a by s_a.

A symmetric domain is called *irreducible* if it is not a Cartesian product of symmetric domains. Finite-dimensional bounded symmetric domains have been classified by É. Cartan [16]. One can view Cartan's classification as an extension of the Riemann mapping theorem in one dimension, which can be stated as follows.

Theorem 2.5.9 *Let D be a finite-dimensional irreducible bounded symmetric domain. Then D is biholomorphic to the open unit ball of one of the following complex Banach spaces of matrices:*

(i) $M_{mn}(\mathbb{C})$
(ii) $S_n(\mathbb{C})$
(iii) $H_n(\mathbb{C})$
(iv) $Sp_n(H)$
(v) $M_{1,2}(\mathcal{O})$
(vi) $H_3(\mathcal{O})$,

where the space $M_{mn}(\mathbb{C})$ is equipped with the operator norm of $L(\mathbb{C}^n, \mathbb{C}^m)$, both $S_n(\mathbb{C})$ and $H_n(\mathbb{C})$ are norm-closed subspaces of $M_n(\mathbb{C})$, consisting of $n \times n$ skew-symmetric and symmetric matrices, respectively, and $Sp_n(H)$ is a complex triple spin factor of dimension $n > 2$, which is a norm-closed subspace of $L(H)$ for some complex Hilbert space H such that $a \in Sp_n(H)$ implies $a^* \in Sp_n(H)$ and $a^2 \in \mathbb{C}\mathbf{1}$.

Classically the first four types of open unit balls in the preceding list are called the *Cartan domains*, while the last two are called the *exceptional domains*. We observe that the six types of finite-dimensional complex Banach spaces in Cartan's classification are all Hermitian Jordan triples, where $S_n(\mathbb{C})$ and $H_n(\mathbb{C})$ are subtriples of $M_{mn}(\mathbb{C})$ and the triple spin factor $Sp_n(H)$ has an equivalent norm of the inner product,

$$\langle a, b \rangle \mathbf{1} = ab^* + b^*a = (a + b^*)^2 - a^2 - (b^*)^2 \in \mathbb{C}\mathbf{1} \qquad (a, b \in Sp_n(H)),$$

and is equipped with the Jordan triple product

$$\{a, b, c\} = \frac{1}{2}(\langle a, b \rangle c + \langle c, b \rangle a - \langle a, c^* \rangle b^*).$$

In Theorem 2.5.9, dim $H < \infty$, although we include the definition of infinite-dimensional complex triple spin factors for later reference. In fact, these finite-dimensional Hermitian Jordan triples, as tangent spaces of Hermitian symmetric manifolds, should all carry the structure of a JH*-triple by the discussion at the beginning of this section. These JH*-structures are described in Proposition 3.5.1.

Our goal in this section is to present an infinite-dimensional version of Cartan's classification, namely, to show that a bounded symmetric domain in a complex Banach space is biholomorphic to the open unit ball of a Hermitian Jordan triple equipped with a complete norm, thereby demonstrating the important relationship of Jordan theory and complex geometry. A natural construction of such a Hermitian Jordan triple for a bounded symmetric domain D is to take the tangent space $T_p D$ at a point $p \in D$ and use the geometry of D to construct a Jordan structure and a norm on $T_p D$ so that D is biholomorphic to the open unit ball of $T_p D$. Although there is no explicit reference to Lie structures in this result, one would, however, expect their relevance in the construction. This is indeed the case. To begin, one makes use of the fact that the automorphism group Aut D has the structure of a real Lie group.

Let D be a *bounded* domain in a complex Banach space V. Then D is a complex Banach manifold modelled on V. Let $\delta = \delta_D$ be the Carathéodory metric defined in (2.36). Then the group

$$G_\delta(D) = \{g \in \text{Aut } D : \delta(g(z), g(w)) = \delta(z, w), \forall z, w \in D\}$$

of biholomorphic δ-isometries is precisely the automorphism group Aut D by the earlier remark. Using the Carathéodory metric δ on D, the concept of an admissible neighbourhood for a Riemannian manifold M with Riemannian distance d, introduced in Section 2.2, can be carried over to D and as in (2.10), one can define a metric ρ_{B_p} on the biholomorphic δ-isometry group $G_\delta(D)$, for a given admissible neighbourhood B_p of $p \in D$. With this metric, $G_\delta(D)$ becomes a topological group.

Each complete holomorphic vector field X on D generates a one-parameter group $\exp t X$ of biholomorphic δ-isometies in $G_\delta(D)$. It follows that the complete holomorphic vector fields aut D of D form a Lie algebra (cf. [111, corollary 11.8]). Replacing M in Theorem 2.2.17 by D and using aut D in place of $\mathfrak{g}_a(M)$, one can repeat the proof of the theorem and conclude that $G_\delta(D)$; that is, Aut D, is a real Banach Lie group with real Banach Lie algebra aut D.

In the absence of a Riemannian metric, the infinitesimal version of the Carathéodory metric δ_D on a *bounded* domain D is the Carathéodory norm on the tangent space $V = T_p D$ defined by

$$\|v\|_p = \sup\{|f'(p)(v)| : f \in H(D, U), f(p) = 0\} \qquad (v \in V),$$

which is equivalent to the original norm $\|\cdot\|$ of V. Indeed, given that D is contained in some ball B of radius $r > 0$ at the origin, with $d(D, \partial B) = R > 0$, Cauchy inequality implies that

$$|f'(p)(v)\| \leq \|f'(p)\|\|v\| \leq \frac{1}{R}\|v\| \qquad (f \in H(D, U), f(p) = 0). \quad (2.37)$$

Hence $\|v\|_p \leq \|v\|/R$. On the other hand, for each continuous linear functional φ of V with $\|\varphi\| \leq 1$, the holomorphic function $\frac{\varphi - \varphi(p)}{2r}$ maps D to U, and p to $0 \in U$. Therefore

$$2r\|v\|_p \geq |(\varphi - \varphi(p)'(p)(v)| = |\varphi(v)|$$

and it follows that

$$2r\|v\|_p \geq \|v\|. \qquad (2.38)$$

In particular, if D is the *open unit ball* of V and if we choose $p = 0 \in D$, then we have $\|v\|_0 = \|v\|$. The Carathéodory norm is invariant under Aut D:

$$\|v\|_p = \|g'(p)(v)\|_{g(p)} \qquad (v \in V, \ g \in \text{Aut } D). \qquad (2.39)$$

A domain D in V is called *homogeneous* if the automorphism group Aut D acts transitively on D.

Lemma 2.5.10 *Let D be a bounded domain in a complex Banach space V and let $a \in D$. The following conditions are equivalent:*

(i) *D is homogeneous.*
(ii) *The evaluation map $X \in \text{aut } D \mapsto X(a) \in V$ is surjective.*
(iii) *The orbit $G(a) = \{g(a) : g \in \text{Aut } D\}$ is open in D.*

Proof (i) $\Rightarrow$ (ii). Let $K = \{g \in \text{Aut } D : g(a) = a\}$ be the isotropy subgroup at a. Since Aut D acts transitively on D and, as discussed in the Riemannian case before, the quotient manifold Aut D/K is diffeomorphic to D via the map $gK \mapsto g(a)$. Hence the map

$$\epsilon : g \in \text{Aut } D \mapsto g(a) \in D$$

is the quotient map, which is a submersion. In particular, the evaluation map $X \mapsto X(a)$, which is the differential $d\epsilon_\iota$ of ϵ at the identity $\iota \in \text{Aut } D$, is surjective.

(ii) $\Rightarrow$ (iii). In the above notation, surjectivity of the evaluation map $d\epsilon_\iota$ implies that the map $\epsilon : g \in \mathrm{Aut}\,D \mapsto g(a) \in D$ is a submersion at the identity ι and therefore open by Remark 2.1.18. Hence the orbit $G(a) = \{g(a) : g \in \mathrm{Aut}\,D\}$ is open.

(iii) $\Rightarrow$ (i). Since D is the disjoint union of all orbits, the orbit $G(a)$ is also closed. By connectedness of D, we have $D = G(a)$. $\qquad\square$

Lemma 2.5.11 *A homogeneous domain D is symmetric if it admits a symmetry at some point. In particular, a homogeneous open unit ball of a complex Banach space is a symmetric domain.*

Proof As in Example 2.3.4, if D is homogeneous, then the symmetry s_a at a point $a \in D$ is given by $s_a = g \circ s_p \circ g^{-1}$, where s_p is the given symmetry at some p, and $g \in \mathrm{Aut}\,D$ satisfies $g(p) = a$.

For the second assertion, we note that an open unit ball has a symmetry $s_0(z) = -z$ at the origin. $\qquad\square$

Now let D be a bounded symmetric domain in a complex Banach space V. It has been shown in Vigué [114, théorèm 3.2.6] that D is homogeneous. As in the case of Riemannian symmetric spaces, the manifold D is diffeomorphic to the homogenous space $\mathrm{Aut}\,D/K$, where K is the isotropy subgroup at a point $p \in D$, with symmetry $s_p : D \longrightarrow D$. The involution $\theta = Ad(s_p) : \mathrm{aut}\,D \longrightarrow \mathrm{aut}\,D$ gives the ± 1-eigenspace decomposition

$$\mathrm{aut}\,D = \mathfrak{k} \oplus \mathfrak{p}, \tag{2.40}$$

where $\mathfrak{p}$ is the (-1)-eigenspace and is real linear isomorphic to the tangent space $T_pD \approx V$ via the evaluation map $X \mapsto X(p)$. Making use of the Carathéodory norm on the tangent space V, we also have $\mathfrak{k} \cap i\mathfrak{k} = \{0\}$ as in Lemma 2.5.3.

As before, let

$$J : \mathfrak{p} \longrightarrow \mathfrak{p}$$

be the complex structure defined by $(JX)(p) = iX(p)$ which satisfies $J\theta = \theta J$. With the complex structure J, $\mathfrak{p}$ is complex isomorphic to V. Analogous to Lemma 2.5.4 and (2.32), we have

$$[JX, JY] = [X, Y] \quad \text{and} \quad J[Z, X] = [Z, JX] \qquad (X, Y \in \mathfrak{p},\ Z \in \mathfrak{k}).$$

Let $\mathfrak{g}_c$ be the complexification of $\mathrm{aut}\,D$. Using the same construction as in the beginning of this section, where the correspondence is shown between Hermitian symmetric spaces and JH*-triples, we obtain the decomposition

$$\mathfrak{g}_c = \mathfrak{p}_+ \oplus \mathfrak{k}_c \oplus \mathfrak{p}_-$$

and a complex linear isomorphism

$$X \in \mathfrak{p} \mapsto X - iJX \in \mathfrak{p}_+$$

such that $\mathfrak{p}_+$ is a Jordan triple with triple product

$$\{X, Y, Z\}_{\mathfrak{p}_+} = \frac{1}{8}[[X, \sigma Y], Z]$$

as in (2.34), apart from the scalar $1/8$ which is added for later application. This gives a Hermitian Jordan triple structure on $\mathfrak{p}$ and V.

Proposition 2.5.12 *Let D be a bounded symmetric domain in a complex Banach space V. Then V admits a Hermitian Jordan triple structure.*

Proof Following the above discussion, it suffices to show that $\mathfrak{p}$, as a complex vector space which is complex isomorphic to V, is a Hermitian Jordan triple. The triple product on $\mathfrak{p}$ is obtained from $\mathfrak{p}_+$ via the isomorphism $X \in \mathfrak{p} \mapsto X - iJX \in \mathfrak{p}_+$.

Let $X, Z \in \mathfrak{p}$. Then the triple product in $\mathfrak{p}_+$ is given by

$$\frac{1}{8}[[X - iJX, \sigma(Z - iJZ)], X - iJX]$$

$$= \frac{1}{8}[[X - iJX, \theta(Z + iJZ)], X - iJX]$$

$$= \frac{1}{8}[[X - iJX, -Z - iJZ], X - iJX]$$

$$= -\frac{1}{4}[[X, Z], X] + \frac{1}{4}[[JX, Z], JX] + i(\frac{1}{4}[[X, Z], JX]$$

$$+ \frac{1}{4}[[JX, Z], X]).$$

Therefore $\mathfrak{p}$ is equipped with the triple product

$$\{X, Y, Z\}_{\mathfrak{p}} = -\frac{1}{4}[[X, Y], Z] + \frac{1}{4}[[JX, Y], JZ], \qquad (2.41)$$

which is Hermitian. Indeed,

$$\{X, iZ, X\}_{\mathfrak{p}} = \{X, JZ, X\}_{\mathfrak{p}} = -\frac{1}{4}[[X, JZ], X] + \frac{1}{4}[[JX, JZ], JX]$$

$$= \frac{1}{4}[[JX, Z], X] + \frac{1}{4}[[X, Z], JX] = -\frac{1}{4}J[[JX, Z], JX]$$

$$+ \frac{1}{4}J[[X, Z], X]$$

$$= -J\{X, Z, X\}_{\mathfrak{p}}.$$

Complex linearity in the outer variables can be verified likewise. $\square$

It would be useful to have an explicit description of the vector fields in $\mathfrak{p}$ for a symmetric domain D. For simplicity, let us assume now that the bounded symmetric domain D happens to be the *open unit ball* of the Banach space V. Naturally we choose $p = 0 \in D$ for the tangent space $T_p D$ and to simplify notation, *we consider a vector field $X \in \text{aut}\, D$ as a holomorphic map*

$$X : D \longrightarrow V.$$

For $X \in \mathfrak{k}$, we have $\exp t X \in K = \{g \in \text{Aut}\, D : g(0) = 0\}$ for all $t \in \mathbb{R}$. It follows from Cartan's uniqueness theorem stated before Proposition 2.2.9 that each $\exp t X : D \longrightarrow D$ is (the restriction of) a linear map and $\| \exp t X \| \le 1$ for all $t \in \mathbb{R}$ implies that $i X$ is (the restriction) of a hermitian linear operator on V. We recall that an element a in a Banach algebra $\mathcal{A}$ is called *hermitian* if it has *real* numerical range $N(a)$, which is equivalent to $\| \exp ita \| = 1$ for all $t \in \mathbb{R}$ (cf. [10, p. 46]). A hermitian continuous linear operator $T : V \longrightarrow V$ is one which is hermitian as an element in the Banach algebra $L(V)$ of continuous linear self-maps on V.

Remark 2.5.13 In the case above, we can and will regard each vector field $X \in \mathfrak{k}$ as i times a hermitian linear map on V and for each $v \in V$, the image $X(v)$ in V is well defined.

For each $\alpha \in D$, there is a unique $X_\alpha \in \mathfrak{p}$ such that $X_\alpha(0) = \alpha$. Let $Y_\alpha = \frac{1}{2}(X_\alpha - iJX_\alpha)$, where $J : \mathfrak{p} \longrightarrow \mathfrak{p}$ is the complex structure defined previously. Then we have $Y_\alpha(0) = \alpha$.

Define a holomorphic map $F : D \longrightarrow D$ by

$$F(z) = \exp Y_z(0) \qquad (z \in D).$$

Evidently $F(0) = 0$. For $v \in V$ and sufficiently small $t \in \mathbb{R}$, we have $tv \in D$ and

$$F'(0)(v) = \frac{d}{dt}\bigg|_{t=0} F(tv) = \frac{d}{dt}\bigg|_{t=0} \exp t Y_v(0) = Y_v(0) = v.$$

Again Cartan's uniqueness theorem implies that F is the identity map; that is,

$$\exp Y_z(0) = z \qquad (z \in D).$$

As in (2.33), we have

$$[Y_\alpha, Y_\beta] = \frac{1}{4}[X_\alpha - iJX_\alpha, X_\beta - iJX_\beta] = 0$$

for all α, $\beta \in D$. Therefore we have from the Campbell–Baker–Hausdorff series (2.18) that

$$(\exp t Y_\beta)(z) = (\exp Y_{t\beta})(\exp Y_z)(0) = \exp(Y_{t\beta+z})(0) = t\beta + z.$$

It follows that

$$Y_\beta(z) = \left.\frac{d}{dt}\right|_{t=0} \exp t Y_\beta = \beta$$

is a constant vector field.

Given $X \in \mathfrak{p}$, the vector field $[X, Y_\beta] \in \mathfrak{k}$ is linear for all Y_β which, together with the fact that Y_β has zero derivative, implies that the derivative of X is linear. Hence we must have

$$X(\cdot) = X(0) + p(\cdot), \tag{2.42}$$

where p is a homogeneous polynomial on V of degree 2.

Remark 2.5.14 In view of (2.42), we can and will regard each vector field $X \in \mathfrak{p}$ on D as a quadratic polynomial on V. Together with Remark 2.5.13, each vector field $X \in \mathrm{aut}\, D = \mathfrak{k} \oplus \mathfrak{p}$ can be identified as a polynomial of degree 2 on V, and the exponential map $\exp X : V \longrightarrow V$ is well defined.

Next, we compute the polynomial p in (2.42).

Lemma 2.5.15 *Let D be the open unit ball of a complex Banach space V such that D is a symmetric domain. Let V be equipped with the induced Hermitian Jordan triple product $\{\cdot, \cdot, \cdot\}$ derived in Proposition 2.5.12 (2.41). Then in the eigenspace decomposition $\mathrm{aut}\, D = \mathfrak{k} \oplus \mathfrak{p}$, each $X \in \mathfrak{p}$ is of the form*

$$X(z) = X(0) - \{z, X(0), z\} \qquad (z \in D).$$

Proof By Proposition 2.5.12, the Jordan triple product of $a = X(0), b = Y(0)$ and $c = Z(0)$ in V is given by

$$\{a, b, c\} = \{X, Y, Z\}_\mathfrak{p}(0).$$

Let $\theta = Ad(s_0)$ be the involution of $\mathrm{aut}\, D$, where $s_0 : D \longrightarrow D$ is the symmetry at 0. Since $\mathfrak{p}$ is the (-1)-eigenspace of θ, we have $X'(0) = 0$ for all $X \in \mathfrak{p}$. Indeed,

$$X'(0) = -\theta X'(0) = -s_0{}'(0)X'(0)s_0{}'(0) = X'(0)s_0{}'(0),$$

which gives

$$X'(0)(v) = X'(0)s_0{}'(0)(v) = X'(0)(-v) = -X'(0)(v) \qquad (v \in V)$$

and therefore $X'(0) = 0$.

Now let $X \in \mathfrak{p}$. By (2.42), we have

$$X(\cdot) = a - p_a(\cdot),$$

where $a = X(0)$ and $p_a(v) = P_a(v, v)$ is a homogeneous polynomial on V of degree 2, P_a being the polar form of p_a.

Pick any $z \in D$ with $z = Y(0)$ and $Y = z - p_z$, where $p_z(v) = P_z(v, v)$ is a homogeneous polynomial of degree 2. We have

$$[Y, X](v) = X'(v)(Y(v)) - Y'(v)(X(v)) = 2P_a(v, Y(v)) - 2P_z(v, X(v)). \tag{2.43}$$

Hence

$$
\begin{aligned}
[[Y, X] Y](0) &= -[Y, X]'(0)(Y(0)) = -2P_a'(0, Y(0))(Y(0)) \\
&\quad + 2P_z'(0, X(0))(Y(0)) = -2P_a(0, Y'(0)(Y(0)) \\
&\quad - 2P_a(Y(0), Y(0)) + 2P_z(0, X'(0)(Y(0))) + 2P_z(Y(0), X(0)) \\
&= -2P_a(z, z) + 2P_z(z, a).
\end{aligned}
$$

Likewise, we have

$$[[JY, X], JY](0) = 2P_a(z, z) + 2P_z(z, a),$$

which gives

$$\{Y, X, Y\}_\mathfrak{p}(0) = -\frac{1}{4}[[Y, X] Y](0) + \frac{1}{4}[[JY, X], JY](0) = P_a(z, z).$$

Therefore

$$X(z) = X(0) - p_a(z) = a - P_a(z, z) = a - \{z, a, z\}. \qquad \square$$

Remark 2.5.16 Since both P_a and $\{\cdot, a, \cdot\}$ are symmetric bilinear maps, we have $P_a(z, w) = \{z, a, w\}$. Also, $P_{ia} = -iP_a$, since the triple product is Hermitian.

From now on, given a complex Banach space V with a symmetric open unit ball, we will always assume the Hermitian Jordan triple structure in V constructed above.

Lemma 2.5.17 *Let D be the open unit ball of V such that D is a symmetric domain. Then the operator $a \square b + b \square a : V \longrightarrow V$ is a hermitian operator for $a, b \in V$. In particular, $a \square a$ is hermitian.*

Proof Let $X, Y \in \mathfrak{p}$ be such that $X(0) = a$ and $Y(0) = ib$. By (2.43), we have

$$
\begin{aligned}
[Y, X](v) &= 2P_a(v, Y(v)) - 2P_{ib}(v, X(v)) \\
&= 2\{v, a, (ib - \{v, ib, v\})\} - 2\{v, ia, (a - \{v, a, v\})\} \\
&= 2\{v, a, ib\} - 2\{v, ib, a\} = 2((ib)\,\square\, a - a\,\square\,(ib))(v).
\end{aligned}
$$

Since $[Y, X] \in \mathfrak{k}$, it follows from Remark 2.5.13 that $i[Y, X]$ is a hermitian linear map on V. In other words,

$$
b\,\square\, a + a\,\square\, b = -\frac{i}{2}(2((ia)\,\square\, a - a\,\square\,(ia))) = -\frac{i}{2}[Y, X]
$$

is hermitian. $\qquad\square$

Let us estimate the norm of $a\,\square\, a = -\dfrac{i}{4}[Y, X]$ in the above proof. We recall from (2.14) that the norm $\|[Y, X]\|_{\text{aut } D}$ of the vector field $[Y, X]$ in the Banach Lie algebra aut D is defined as $\sup\{\|[Y, X](z)\| : z \in B_0\}$, where B_0 is an admissible neighbourhood of $0 \in D$ and hence $rD \subset B_0$ for some $r > 0$. Applying the open mapping theorem to the real continuous linear bijection $Z \in \mathfrak{p} \mapsto Z(0) \in V$, one finds a constant $k > 0$ such that

$$
\|Z\|_{\text{aut } D} \le k\|Z(0)\| \qquad (Z \in \text{aut } D). \tag{2.44}
$$

It follows that

$$
\begin{aligned}
\|a\,\square\, a\| = \sup_{z \in D} \|a\,\square\, a(z)\| &= \sup_{z \in D}\left\|-\frac{i}{4}[Y, X](z)\right\| \\
&\le \frac{1}{4r}\|[Y, X]\|_{\text{aut } D} \le \frac{K}{4r}\|Y\|_{\text{aut } D}\|X\|_{\text{aut } D} \\
&\le \frac{k^2 K}{4r}\|Y(0)\|\|X(0)\| = \frac{k^2 K}{4r}\|a\|^2. \tag{2.45}
\end{aligned}
$$

Given $a = X(0)$, $b = Y(0)$ and $c = Z(0)$, using arguments as above and the formula for the triple product in (2.41), we see that the triple product $\{\cdot, \cdot, \cdot\}$ is continuous:

$$
\|\{a, b, c\}\| \le C\|a\|\|b\|\|c\|, \tag{2.46}
$$

where the constant $C > 0$ is independent of a, b and c.

Lemma 2.5.18 *Let D be the open unit ball of V such that D is a symmetric domain. For each $a \in V$ with $\|a\| = 1$, the operator $I - a\,\square\, a : V \longrightarrow V$ is not invertible, where I is the identity operator on V.*

Proof Assume that $I - a \square a$ is invertible. We deduce a contradiction.

For each $v \in V$, there is a unique $X_v \in \mathfrak{p}$ such that $X_v(0) = v$. Let

$$X^v = -\frac{1}{2}[X_v, X_a] + X_v \in \mathfrak{k} \oplus \mathfrak{p} = \text{aut } D.$$

By Remark 2.5.14 and (2.43), we have

$$X^v(a) = (-\frac{1}{2}[X_v, X_a] + X_v)(a) = (I - a \square a)(v).$$

Let $F : V \longrightarrow V$ be the holomorphic map

$$F(v) = \exp X^v(a).$$

Then the differential dF_0 is the evaluation map $dF_0(v) = X^v(a) = (I - a \square a)(v)$, which, by assumption, is a linear isomorphism. Hence, by the inverse function theorem, F maps an open neighbourhood of 0 homeomorphically onto an open subset of the orbit $\{\exp X^v(a) : v \in V\}$. This would contradict the fact that $\| \exp X^v(a) \| = 1$ for all $v \in V$. To see the latter, we observe that $\exp X^v : D \longrightarrow D$ is a biholomorphic map which is an isometry with respect to the Carathéodory metric δ_D. Let $\overline{D}$ be the closed unit ball in V. By continuity, we have $\exp X^v(\overline{D}) \subset \overline{D}$. If $\exp X^v(a) \in D$, then we have $\exp X^v(a) = \exp X^v(z)$ for some $z \in D$. Since δ_D is equivalent to the metric on D defined by the norm, we have

$$0 = \delta_D(\exp X^v(a), \exp X^v(z)) = \lim_{n \to \infty} \delta_D(\exp X^v(a - a/n), \exp X^v(z))$$
$$= \lim_{n \to \infty} \delta_D(a - a/n, z) = \delta_D(a, z),$$

which gives $a = z \in D$ and is impossible. Hence $\| \exp X^v(a) \| = 1$. This completes the contradiction and the proof. $\qquad \square$

We have already observed that the box operator $a \square a$ is hermitian and therefore its spectrum $\sigma(a \square a)$ must lie in $\mathbb{R}$. We will write $\sigma(a \square a) < t \in \mathbb{R}$ to mean that $\lambda < t$ for all $\lambda \in \sigma(a \square a)$.

Lemma 2.5.19 *Let D be the open unit ball of V such that D is a symmetric domain. Then we have*

$$\{a \in V : \sigma(a \square a) < 1\} \subset D.$$

Consequently $\sigma(a \square a) \le 0$ implies $a = 0$.

Proof Let $\sigma(a \square a) < 1$. If $\|a\| > 1$, then we have

$$\sigma \left(\frac{a}{\|a\|} \square \frac{a}{\|a\|} \right) = \frac{1}{\|a\|^2} \sigma(a \square a) < 1,$$

which implies that the operator $I - \frac{a}{\|a\|} \Box \frac{a}{\|a\|}$ is invertible, contradicting Lemma 2.5.18. Therefore we must have $\|a\| \leq 1$.

If $\sigma(a \Box a) \leq 0$, then $ta \in D$ for all $t \in \mathbb{R}$, which implies $a = 0$. □

We shall strengthen Lemma 2.5.17 by showing that the box operator $a \Box a$ has non-negative spectrum. Let $V(a)$ be the closed subtriple in V generated by a, which is the closed complex linear span of odd powers of a. We first compare the two spectra $\sigma(a \Box a)$ and $\sigma(a \Box a|_{V(a)})$.

Lemma 2.5.20 *Let $b \in V(a)$. Then the spectrum $\sigma(b \Box b|_{V(a)})$ of the operator $b \Box b|_{V(a)} : V(a) \longrightarrow V(a)$ is contained in the spectrum $\sigma(b \Box b)$ of $b \Box b$ in $L(V)$.*

Proof In $L(V)$, consider the closed subalgebra

$$\mathcal{B} = \{T \in L(V) : T(V(a)) \subset V(a)\}$$

which contains the identity and the operator $b \Box b$. Since $\sigma(b \Box b) \subset \mathbb{R}$, the complement $\mathbb{C} \backslash \sigma(b \Box b)$ is connected. It follows that $\sigma_{\mathcal{B}}(b \Box b) = \sigma(b \Box b)$ (see, for example, [99, p. 239]). Observe that the restriction map $T \in \mathcal{B} \mapsto T|_{V(a)} \in L(V(a))$ is an algebra homomorphism. In particular, we have $\sigma(b \Box b|_{V(a)}) \subset \sigma_{\mathcal{B}}(b \Box b)$ and the result follows. □

It follows from Lemma 2.5.17 that the operator $b \Box b|_{V(a)}$ is hermitian in the Banach algebra $L(V(a))$ of bounded complex linear operators on $V(a)$. Let $\overline{V(a)}_0$ be the closed linear span of $V(a) \Box V(a)|_{V(a)}$ in $L(V(a))$. Since $V(a)$ is an abelian Jordan triple system by power associativity, $\overline{V(a)}_0$ is an abelian closed subalgebra of $L(V(a))$. Moreover, polarisation gives

$$4x \Box y = (x + y) \Box (x + y) - (x - y) \Box (x - y)$$
$$+ i((x + iy) \Box (x + iy) - (x - iy) \Box (x - iy)) \qquad (x, y \in V(a)).$$

Hence $\overline{V(a)}_0$ is the complexification of the hermitian elements in $\overline{V(a)}_0$, and by the Vidav–Palmer theorem [10, p. 65], $\overline{V(a)}_0$ is an abelian C*-algebra with involution

$$(h + ik)^* = h - ik,$$

where h and k are hermitian elements in $\overline{V(a)}_0$.

Lemma 2.5.21 *Let V be a complex Banach space which is a Hermitian Jordan triple with continuous triple product. Let $V(a)$ be the closed subtriple generated by an element $a \in V$. Then we have*

$$\sigma(a \Box a) \subset \frac{1}{2}(S + S),$$

where $S = \sigma(a \square a|_{V(a)}) \cup \{0\}$. Also, the spectrum $\sigma(B(a, a))$ of the Bergmann operator $B(a, a) : V \longrightarrow V$ is contained in $(1 - S)(1 - S)$.

Proof Let V^a be the a-homotope of V and denote its Jordan algebra product by juxtaposition $xy = \{x, a, y\}$ (cf. Definition 1.2.9). By continuity of the triple product, we have

$$\|xy\| = \|\{x, a, y\}\| \leq C\|x\|\|y\| \qquad (x, y \in V)$$

for some constant $C > 0$. Replacing the norm $\| \cdot \|$ by $C\| \cdot \|$, we may assume $C = 1$.

Let $V_1^a = V^a \oplus \mathbb{C}$ be the unit extension of V^a. Since $V(a)$ is abelian, it is a closed associative subalgebra of V^a and $V(a)_1 := V(a) \oplus \mathbb{C}$ is a unital commutative Banach algebra.

Let $\sigma(a)$ be the spectrum of a in $V(a)_1$. It is identical with the spectrum $\sigma(L_a)$ of the left multiplication $L_a : V(a)_1 \longrightarrow V(a)_1$ given by

$$L_a(x) = ax = \{a, a, x\} = a \square a(x) \qquad (x \in V(a)).$$

Hence we have

$$\sigma(a) = \sigma(L_a) = \sigma(a \square a|_{V(a)}) \cup \{0\}.$$

The box operator $a \square a : V \longrightarrow V$ is the left multiplication by a on the Jordan algebra V^a, which extends naturally to the left multiplication $\widetilde{L}_a : V_1^a \longrightarrow V_1^a$ by a.

The set $L = \{\widetilde{L}_b : b \in V(a)_1\}$ of left multiplications on V_1^a mutually commute by power associativity (cf. Theorem 1.1.1). Let $\mathcal{A}$ be a maximal commutative subalgebra of $L(V_1^a)$ containing L.

Let $\lambda \in \sigma(a \square a) \subset \sigma(\widetilde{L}_a)$. Then there is a nonzero complex homomorphism χ of $\mathcal{A}$ such that $\chi(\widetilde{L}_a) = \lambda$. For each $b \in V(a)_1$, let $Q_b = 2\widetilde{L}_b^2 - \widetilde{L}_{b^2} : V_1^a \longrightarrow V_1^a$ be the quadratic operator defined in (1.1). We have $Q_b \in \mathcal{A}$ and, in particular, we can find $\lambda_1, \lambda_2 \in \mathbb{C}$ satisfying

$$\lambda_1 + \lambda_2 = 2\lambda \quad \text{and} \quad \lambda_1\lambda_2 = \chi(Q_a).$$

Let $a_j = a - \lambda_j \in V_1^a$ for $j = 1, 2$. Then $Q_{a_j} = Q_a - 2\lambda_j\widetilde{L}_a + \lambda_j^2$ and hence $\chi(Q_{a_j}) = \lambda_1\lambda_2 - 2\lambda_j\lambda + \lambda_j^2 = 0$. Therefore Q_{a_j} is not invertible in $\mathcal{A}$, and also not invertible in $L(V_1^a)$, by maximal commutativity of $\mathcal{A}$. It follows that a_j is not invertible in the Jordan algebra V_1^a, and also not invertible in $V(a)_1$. In other words, $\lambda_j \in \sigma(a)$. This gives

$$\lambda = \frac{1}{2}(\lambda_1 + \lambda_2) \in \frac{1}{2}(S + S).$$

The preceding argument also shows that the spectrum of the quadratic operator $Q_a : V_1^a \longrightarrow V_1^a$ is contained in SS, and that the Bergmann operator

$$B(a, a) = Q_a - 2a \square a + id_v$$

on V is the quadratic operator $Q_{1-a} : V^a \longrightarrow V^a$, where $\mathbf{1}$ is the adjoined identity in V_1^a. It follows that $\sigma(B(a, a)) \subset (1 - S)(1 - S)$.   □

Proposition 2.5.22 *Let V be a complex Banach space with a symmetric open unit ball and the induced Hermtian Jordan structure. Then for each $a \in V$, the spectrum $\sigma(a \square a)$ of the box operator $a \square a$ is contained in $[0, \infty)$.*

Proof Let $V(a)$ be the closed subtriple in V generated by a. By Lemma 2.5.21, it suffices to show that $\sigma(a \square a|_{V(a)}) \subset [0, \infty)$. As remarked before, the closed linear span $\overline{V(a)}_0$ of $V(a) \square V(a)|_{V(a)}$ is an abelien C*-algebra. The C*-sublagebra $\mathcal{A}(a \square a)$ generated by the hermitian element $a \square a|_{V(a)} \in \overline{V(a)}_0$ is isometrically isomorphic to the C*-algebra $C_0(\sigma(a \square a|_{V(a)}))$ of complex continuous functions on $\sigma(a \square a|_{V(a)})$, vanishing at infinity, and $a \square a|_{V(a)}$ identifies with the identity function on the spectrum $\sigma(a \square a|_{V(a)})$.

If $\sigma(a \square a|_{V(a)})$ contains any negative real number, then there is a nonzero function $f \in C_0(\sigma(a \square a|_{V(a)}))$ vanishing on $\sigma(a \square a|_{V(a)}) \cap [0, \infty)$. Identify f as an element in $\mathcal{A}(a \square a) \subset L(V(a))$, which is the limit of a sequence of polynomials p_n in $a \square a|_{V(a)}$. By power associativity, we have $p_n(a) \square p_n(a)|_{V(a)} = (p_n^2)(a \square a)|_{V(a)}$. It follows that $f(a) \square f(a)|_{V(a)} = (f^2)(a \square a)|_{V(a)}$. As an element in $C_0(\sigma(a \square a|_{V(a)}))$, the function $(f^2)(a \square a)|_{V(a)}$ has non-positive spectrum. Hence $\sigma(f(a) \square f(a)|_{V(a)}) \leq 0$ and also $\sigma(f(a) \square f(a)) \leq 0$. By Lemma 2.5.19, we have $f(a) = 0$, contradicting $(f^2)(a \square a)|_{V(a)} \neq 0$.

Hence $\sigma(a \square a|_{V(a)}) \subset [0, \infty)$.   □

Remark 2.5.23 Considering $a \square a$ as a hermitian element in a C*-algebra in the above proof, we have $\|(a \square a)^3\| = \|a \square a\|^3$.

Corollary 2.5.24 *Let V be a complex Banach space with a symmetric open unit ball and the induced Hermtian Jordan structure. Then we have*

$$\{a \in V : \sigma(a \square a|_{V(a)}) < 1\} = \{a \in V : \sigma(a \square a) < 1\},$$

where $V(a)$ denotes the closed subtriple generated by a, and

$$\|a\|_\infty := \|a \square a\|^{1/2} \qquad (a \in V)$$

defines a norm on V.

Proof The first assertion follows from Lemma 2.5.20 and Lemma 2.5.21.

Since $a \square a$ is an hermitian operator on V, the norm $\|a \square a\|$ equals the spectral radius of $a \square a$. If $\|a \square a\| = 0$, then Lemma 2.5.19 implies $a = 0$. It remains to show that $\| \cdot \|_\infty$ satisfies the triangle inequality.

Let $\|a\|_\infty, \|b\|_\infty \leq 1$. We show $\|a + b\|_\infty \leq 2$. Observe that

$$4 - (a + b) \square (a + b) = 2(I - a \square a) + 2(I - b \square b) + (a - b) \square (a - b),$$

where I is the identity operator on V and the operator on the right-hand side has non-negative spectrum. Hence $\sigma((a + b) \square (a + b)) \subset [0, 4]$ concludes the proof. $\square$

By Lemma 2.5.19, the ball $D_\infty := \{a \in V : \|a\|_\infty < 1\}$ is contained in the open unit ball D of V, equivalently, $\| \cdot \| \leq \| \cdot \|_\infty$. Hence these two norms are equivalent by (2.45). We shall see that they are actually identical.

The preceding lengthy exposition exhibits a number of important properties of the Banach space $(V, \| \cdot \|_\infty)$ and motivates the following definition.

Definition 2.5.25 A complex Banach space $(V, \| \cdot \|)$ is called a *JB*-triple* if it is a Hermitian Jordan triple with a continuous triple product and the box operator $a \square a$ of each element $a \in V$ satisfies the following conditions:

(i) $a \square a$ is a hermitian operator on V, that is, $\| \exp it(a \square a)\| = 1$ for all $t \in \mathbb{R}$;
(ii) $a \square a$ has non-negative spectrum;
(iii) $\|a \square a\| = \|a\|^2$.

To summarise, we have shown that $(V, \| \cdot \|_\infty)$ is a JB*-triple. Although our demonstration was based on the assumption that the given symmetric domain is the open unit ball of the complex Banach space V, we have nevertheless shown the central tenets of analogous arguments for arbitrary bounded symmetric domains D in V. Indeed, via a suitable local chart at a point $p \in D$, one can use similar arguments to show that $\|a\|_\infty := \|a \square a\|^{1/2}$ defines a complete norm on V, turning it into a JB*-triple. Further, it follows from Kaup [70, proposition 4.7] that the ball $D_\infty = \{a \in V : \|a\|_\infty < 1\}$ can be considered as a universal covering of the given domain D and the covering map is actually biholomorphic. This gives an infinite-dimensional version of the Riemann mapping theorem, as follows.

Theorem 2.5.26 *Let D be a bounded symmetric domain in a complex Banach space V. Then there is a norm $\| \cdot \|_\infty$ on V such that $(V, \| \cdot \|_\infty)$ is a JB*-triple and D is biholomorphic to the open unit ball D_∞ of $(V, \| \cdot \|_\infty)$.*

As a special case of this theorem, we show in the next theorem that $D = D_\infty$ if D is the original open unit ball of V. We first look into the symmetry of D_∞. Let $a \in D_\infty$ and let I be the identity operator on V, which is a JB*-triple in the norm $\| \cdot \|_\infty$. By Lemma 2.5.21, the Bergmann operator $B(a, a)$ has positive spectrum. For each $x \in D_\infty$, the linear operator $I - x \,\square\, a : V \longrightarrow V$ is invertible, since $\sigma(x \,\square\, a) < 1$. The latter follows from the observation that the polarization

$$x \,\square\, a = \sum_{\varepsilon^4 = 1} \varepsilon \left(\frac{x + \varepsilon a}{2} \right) \,\square\, \left(\frac{x + \varepsilon a}{2} \right)$$

implies that the numerical range $N(x \,\square\, a)$ of $x \,\square\, a$ is contained in the square $(-1, 1)^2 \subset \mathbb{C}$ by Lemma 2.5.21. Replacing x by λx in the above argument for all complex numbers λ of unit modulus, we see that

$$N(x \,\square\, a) \subset \bigcap_{|\lambda|=1} \lambda(-1, 1)^2 = \{z \in \mathbb{C} : |z| < 1\}. \tag{2.47}$$

Let X_a be the vector field $X_a(z) = a - \{z, a, z\}$ on D_∞, and X^a the constant vector field taking value $a \in D_\infty$. By Kaup [70, proposition 4.6], the map $g_a : D_\infty \longrightarrow D_\infty$ defined by

$$g_a = (\exp X^a) \circ B(a, a)^{1/2} \circ \exp(X_a - X^a)$$

is a biholomorphic map with inverse g_{-a} such that $g_a(0) = a$ and $g'_a(0) = B(a, a)^{1/2}$. This gives an explicit description of homogeneity of D_∞. In fact, $\exp X^a(z) = a + z$ is a translation by a and $\exp(X_a - X^a)(z) = (I + z \,\square\, a)^{-1}(z)$ for $z \in D_\infty$. Hence the map g_a, called the *Möbius transformation* induced by a, has the form

$$g_a(z) = a + B(a, a)^{1/2}(I + z \,\square\, a)^{-1}(z) \qquad (z \in D_\infty) \tag{2.48}$$

in terms of the Jordan structures of V.

We can now complete the special case of Theorem 2.5.26 and conclude this section with some examples.

Theorem 2.5.27 *A complex Banach space V is a JB*-triple if and only if its open unit ball D is a symmetric domain.*

Proof Given a JB*-triple V with open unit ball $D_\infty = \{a \in V : \sigma(a \,\square\, a) < 1\}$, the preceding argument shows that D_∞ is homogeneous and hence symmetric.

Conversely, let V be a complex Banach space in which the open unit ball D is symmetric. Then we have already shown that V is a JB*-triple in the equivalent norm $\|a\|_\infty = \|a \,\square\, a\|^{1/2}$ and $D_\infty = \{a \in V : \|a\|_\infty < 1\} \subset D$. We complete the proof by showing that $D_\infty = D$ and hence $\| \cdot \|_\infty$ is the original norm of V.

If there exists $\|v\| < 1$ and $\|v \square v\| = 1$, then $\sigma(tv \square tv) < 1$, that is, $tv \in D_\infty$, for $0 < t < 1$. Let $\gamma : (0, 1) \longrightarrow D_\infty$ be the curve $\gamma(t) = tv$. We show that γ has infinite length in D_∞, giving a contradiction and completing the proof.

At each point $tv \in D_\infty$, the tangent space $T_{tv} D_\infty = V$ has the Carathéodory norm

$$\|u\|_{tv} = \sup\{|f'(tv)(u)| : f \in H(D_\infty, U), f(tv) = 0\} \qquad (u \in V).$$

The Möbius transformation $g_{tv} : D_\infty \longrightarrow D_\infty$ defined in (2.48) satisfies $g_{tv}(0) = tv$ and $g'_{tv}(0) = B(tv, tv)^{1/2}$.

When restricting to the closed subtriple $V(v)$ generated by v in V, the Bergmann operator $B(tv, tv)$ has the form

$$B(tv, tv) = id_v - 2t^2 v \square v + t^4 (v \square v)^2$$

by power associativity, where id_v is the identity map on $V(v)$. Therefore $B(tv, tv)^{1/2} v = (id_v - t^2(v \square v))(v)$. Let $f = g_{tv}^{-1}$ so that $f(tv) = 0$ and

$$\begin{aligned} f'(tv)(v) = g'_{tv}(0)^{-1}(v) &= (id_v - t^2(v \square v))^{-1}(v) \\ &= v + t^2(v \square v)(v) + t^4(v \square v)^2(v) + \cdots . \end{aligned} \qquad (2.49)$$

Let $R(v)$ be the closed real linear span of odd powers of v in V and let $\mathcal{R} \subset L(V)$ be the closed real linear span of the box operators $R(v) \square R(v)$. Since $R(v)$ is flat, each box operator $a \square b = \frac{1}{2}(a \square b + b \square a)$ is hermitian for $a, b \in R(v)$, by Lemma 2.5.17. Hence we have $\|a \square b\| \leq \|a\|_\infty \|b\|_\infty$ since, as shown previously, $\|a\|_\infty, \|b\|_\infty < 1$ implies $\sigma(a \square b) < 1$. Let

$$\mathcal{A} = R(v) \oplus_{\ell_1} \mathcal{R}|_{V(v)}$$

be the ℓ_1-sum of real Banach spaces, where $R(v)$ is equipped with the norm $\| \cdot \|_\infty$, and define a product in $\mathcal{A}$ by

$$(a \oplus h)(b \oplus k) = (h(b) + k(a)) \oplus (a \square b|_{V(v)} + hk).$$

Then $\mathcal{A}$ is a real abelian Banach algebra. Since $\|v \square v|_{V(v)}\| = 1$, there exists a real linear character $\chi : \mathcal{A} \longrightarrow \mathbb{C}$ such that $\chi(v \square v|_{V(v)}) = 1$, which implies $\chi(v) \neq 0$.

From (2.49), we have

$$\|f'(tv)(v)\| \geq |\chi(f'(tv)(v))| = |\chi(v) + t^2 \chi(v) + t^4 \chi(v) + \cdots| = \frac{|\chi(v)|}{1 - t^2}.$$

It follows that

$$\int_0^s \|\gamma'(t)\|_{tv} dt = \int_0^s \|v\|_{tv} dt \geq \int_0^s \frac{|\chi(v)|}{1-t^2} = \frac{|\chi(v)|}{2} \log \frac{1+s}{1-s}$$

for $s < 1$. Letting $s \to 1$, we get the contradiction that γ has infinite length. $\square$

Remark 2.5.28 The above theorem and the remark before it imply immediately that, for all a, b in a JB*-triple, $\|a \,\square\, b\| \leq \|a\| \|b\|$ if $a \,\square\, b$ is hermitian. Since $a \,\square\, b + b \,\square\, a$ is hermitian, (2.47) also implies $\|a \,\square\, b + b \,\square\, a\| \leq 2\|a\| \|b\|$ and $\|a \,\square\, b - b \,\square\, a\| = \|ia \,\square\, b + b \,\square\, ia\| \leq 2\|a\| \|b\|$. It follows that

$$\|a \,\square\, b\| \leq 2\|a\| \|b\|. \tag{2.50}$$

By Remark 2.5.23, we also have

$$\|\{a, a, a\}\|^2 = \|\{a, a, a\} \,\square\, \{a, a, a\}\| = \|(a \,\square\, a)^3\| = \|a \,\square\, a\|^3 = \|a\|^6.$$

In particular, $\|a\| = 1$ if a is a nonzero tripotent.

Corollary 2.5.29 *Let W be a closed subtriple of a JB*-triple. Then W, with the inherited Jordan structure, is itself a JB*-triple.*

Proof This is because the open unit ball $D_W = D \cap W$ of W is homogeneous for if $a \in D_W$, then the Möbius transformation $g_a \in \text{Aut } D$ in (2.48) satisfies $g_a(0) = a$ and $g_a(D_W) \subset D \cap W$. $\square$

Definition 2.5.30 A JB*-triple V is called a *JBW^*-triple* if, as a Banach space, it has a predual.

By Theorem 2.5.27, the spaces of matrices in Theorem 2.5.9 are all finite dimensional JBW*-triples. Infinite-dimensional examples are given next.

Example 2.5.31 Let H and K be complex Hilbert spaces, and let $J : H \longrightarrow H$ be a conjugation; that is, J is a conjugate linear isometry and J^2 is the identity map on H.

The Hermitian Jordan triple $L(H, K)$ of bounded linear operators between H and K, defined in Example 1.2.6, has the trace-class operators as its predual. Consider the Hermitian Jordan triple $L(H) = L(H, H)$, which is a von Neumann algebra. For each $a \in L(H)$, the box operator $a \,\square\, a$ has the form

$$a \,\square\, a(x) = \frac{1}{2}(aa^*x + xa^*a)$$

and it follows that

$$\|a\|^2 = \|aa^* + a^*a\| = \|a \,\square\, a(\mathbf{1})\| \leq \|a \,\square\, a\| \leq \|a\|^2.$$

It is not difficult to show that $a \square a$ has a non-negative numerical range (cf. Lemma 3.1.3). Hence $L(H)$ is a JBW*-triple.

The Hermitian Jordan triple $L(H, K)$ can be embedded naturally as a closed subtriple of $L(H \oplus K)$ and is therefore a JBW*-triple. It is an infinite-dimensional generalization of the first space $M_{mn}(\mathbb{C})$ in Cartan's classification in Theorem 2.5.9. Hence it is called a *type* 1 or *rectangular type Cartan factor*.

The closed subtriples $\{a \in L(H) : a = -Ja^*J\}$ and $\{a \in L(H) : a = Ja^*J\}$ of $L(H)$ generalize the second and third spaces in Cartan's classification in Theorem 2.5.9, hence called respectively a *type* 2 (or *symplectic type*) and a *type* 3 (or *Hermitian type*) *Cartan factor*. They are JBW*-triples since they are weak* closed in $L(H)$.

The triple spin factor $Sp_n(H)$, defined in Theorem 2.5.9, also called a *type* 4 *Cartan factor*, is a JBW*-triple. In fact, it is a reflexive Banach space. The finite-dimensional JB*-triples $M_{1,2}(\mathcal{O})$ and $H_3(\mathcal{O})$ are called the *type* 5 and *type* 6 *Cartan factors*, respectively. They are also called the *exceptional Cartan factors*.

Example 2.5.32 Given a complex Hilbert space $(H, \langle \cdot, \cdot \rangle)$, the linear isometry $x \in L(\mathbb{C}, H) \mapsto x(1) \in H$ identifies the two spaces and induces a JB*-triple structure on H. The adjoint x^* of $x \in L(\mathbb{C}, H)$ is given by $x^*(h) = \langle h, x(1) \rangle$ for $h \in H$ and hence the triple product in H can be expressed as

$$\{x, y, z\} = \frac{1}{2}(\langle x, y \rangle z + \langle z, y \rangle x).$$

In all the examples of JBW*-triples given above, it is evident that the triple product is separately weak* continuous.

Example 2.5.33 Every C*-algebra is a closed subtriple of $L(H)$ for some Hilbert space H and hence carries the structure of a JB*-triple in which the tripotents are exactly the partial isometries. In particular, the abelian C*-algebra $C_0(\Omega)$ of complex continuous functions on a locally compact Hausdorff space Ω, vanishing at infinity, is a JB*-triple with the Jordan triple product

$$\{f, g, h\} = f\overline{g}h \qquad (f, g, h \in C_0(\Omega)),$$

where $\overline{g}$ denotes the complex conjugate of g.

A von Neumann algebra $\mathcal{A}$ is a JBW*-triple with unique predual and the triple product in $\mathcal{A}$ is separately weak* continuous, since involution x^* and multiplication xy in $\mathcal{A}$ are separately weak* continuous.

Definition 2.5.34 A closed subtriple of $L(H)$ for some Hilbert space H is called a *JC*-triple*. If V is a JC*-triple as well as a JBW*-triple, it is also called a *JW*-triple*.

JC*-triples are called J*-algebras in Harris [49]. We shall see that JB*-triples share many common features with C*-algebras and can be viewed as a generalization of C*-algebras. Nevertheless, it cannot be overemphasized that the predominant role of JB*-triples is in complex geometry. We see from the previous discussion that JB*-triples are a study of quadratic vector fields on bounded symmetric domains.

Notes

There is a substantial literature on infinite-dimensional manifolds, Lie groups and Lie algebras. The basic material presented in this chapter can be found in Bourbaki [12], Chevalley [20], Dieudonné [33], Helgason [51], Klingenberg [75], Lang [81], Omori [94], Upmeier [111] and Varadarajan [112].

The two key words in the chapter are "Jordan structures" and "symmetric manifolds." Our object is to expose their close relationship. An important link for this is the automorphism group Aut M of a particular symmetric manifold M, which carries the structure of a Lie group and induces a TKK Lie algebra. From this, the related Jordan structure is derived. For a bounded domain M in $\mathbb{C}^n$, it was first shown by H. Cartan [18] that Aut M is a Lie group. If M is a connected complex Banach manifold with a compatible metric, it has been shown by Upmeier [110] that Aut M is a Lie group in the topology of locally uniform convergence. For the special case of a bounded domain M in a complex Banach space of any dimension, this result has also been proved by Vigué [114]. All these results have been proved carefully and completely in Upmeier [111]. If M is a finite-dimensional connected Riemannian manifold, it is a well-known result of Myers and Steenrod [91] that the isometry group $G(M)$ of M is a Lie group. The proof of Theorem 2.2.17 that the biholomorphic isometry group $G_a(M)$ is a Lie group, where the Riemannian manifold M can be infinite-dimensional, follows the ideas and exposition of Upmeier [111]. The metrizable topology defined on $G_a(M)$ is the topology of locally uniform convergence mentioned previously.

The geometry of finite-dimensional symmetric cones and related Jordan structures, as well as analysis on these cones, has been treated comprehensively in Faraut and Koranyi [38]. Satake's book [103] also discusses connections between Jordan algebraic structures, finite-dimensional symmetric cones and bounded symmetric domains. Proposition 2.3.24 is an infinite-dimensional extension of the result in Hirzebruch [55, satz 2.1]. Theorem 2.3.32 is taken from Chu [21]. Manifolds of finite rank tripotents in JB*-triples have been studied in Chu and Isidro [26].

The extension of the concept of an *orthogonal involutive Lie algebra* to infinite dimension was introduced in Chu [22]. Its motivation comes from the correspondence between Riemannian symmetric spaces and orthogonal involutive Lie algebras, a well-known fact in finite dimensions [11, 63]. This fact in the infinite-dimensional case has been shown in Chu [22]. In Definition 2.4.4, orthogonality is defined in terms of a *completely* positive definite quadratic form instead of a positive definite quadratic form in Chu [22]. The definition of a JH-triple given in Definition 2.4.19 is more general than the one given in Chu [22]; the latter assumes non-degeneracy and continuity of the triple product. Hence the correspondence between JH-triples and a class of Riemannian symmetric spaces, shown in Theorem 2.4.31, covers a wider class of Jordan Hilbert triples than the one given in Chu [22].

Cartan domains are called *classical domains* in Hua's book [59]. In Friedman's book [39], bounded symmetric domains are called *homogeneous balls* and have been identified as a mathematical model for several areas of physics. The concept of a JB*-triple and the correspondence between bounded symmetric domains and JB*-triples are due to Kaup [69, 70]. Prior to this, Loos established the correspondence between finite-dimensional bounded symmetric domains and Jordan pairs by showing that, in finite dimensions, the circled bounded symmetric domains are exactly the open unit balls of Jordan triples with the spectral norm [85]. The key to the construction of the Jordan triple product from a symmetric Banach manifold D is the result that the complete holomorphic vector fields aut D are quadratic polynomial vector fields. This result is also crucial in deriving further properties of the triple product. An alternative construction of the triple product via a TKK Lie algebra is given in Proposition 2.5.12, which relates the triple product to the curvature tensor directly. Apart from this, the development in the last section leading to Theorem 2.5.26 and Theorem 2.5.27 can be found in the seminal papers [69] and [70].

3

Jordan structures in analysis

3.1 Banach spaces

The importance of Jordan structures in geometry has been amply demonstrated in the previous chapter. We go on to show their useful role in analysis. In this section, we discuss Jordan structures in Banach spaces. To begin, we observe that every complex Banach space V admits a subspace with a Hermitian Jordan structure in the following way. Let D be the open unit ball in V and let

$$\text{aut } D = \mathfrak{k} \oplus \mathfrak{p}$$

be the eigenspace decomposition induced by the symmetry $s_0(z) = -z$ at the origin $0 \in D$. Although D need not be homogeneous in general, and hence the evaluation map $X \in \text{aut } D \mapsto X(0) \in V$ need not be surjective, the vector fields $X \in \mathfrak{p}$ are still polynomials of degree 2, without a linear term, as derived in (2.42). Let

$$V_s = \{X(0) : X \in \text{aut } D\} = \{X(0) : X \in \mathfrak{p}\}.$$

The restriction of each $X \in \text{aut } D$ to the open unit ball $D_s = D \cap V_s$ of V_s is a complete vector field on D_s. By (2.44), V_s is a closed subspace of V and admits a Hermitian Jordan triple structure via $\mathfrak{p}$, as in Proposition 2.5.12. In fact, since the evaluation map $X \in \text{aut } D_s \mapsto X(0) \in V_s$ is surjective, the open unit ball of V_s is homogeneous by Lemma 2.5.10 and hence V_s is a JB*-triple. We call V_s the *symmetric part* of V; it is the largest subspace of V with a symmetric open unit ball under the action of Aut D. Of course, $V_s = V$ if, and only if, V itself is a JB*-triple.

The fact that every complex Banach space V contains a subspace V_s which is a JB*-triple is a remarkable phenomenon. This suggests that V_s and its relationship with V should be an interesting object of study. For instance, it can be used to characterize Banach algebras satisfying the von Neumann inequality. A Banach algebra $\mathcal{A}$ with identity $\mathbf{1}$ is said to satisfy the *von Neumann inequality* if for each $a \in \mathcal{A}$ with $\|a\| \leq 1$,

$$\|p(a)\| \leq \sup\{|p(\alpha)| : |\alpha| = 1\}$$

for every polynomial p with complex coefficients. It has been shown in Arazy [5] and, Chu and Mellon [29] that $\mathcal{A}$ satisfies the von Neumann inequality if, and only if, $\mathbf{1} \in \mathcal{A}_s$.

The study of Hermitian symmetric spaces and related Jordan structures motivates the following definition.

Definition 3.1.1 A complex Banach space V is called a *Hermitian Jordan Banach triple* or *J*-triple* if it is a Hermitian Jordan triple with a continuous triple product such that the box operator $a \,\square\, a : V \longrightarrow V$ is hermitian for all $a \in V$.

Remark 3.1.2 To avoid confusion, it should be noted that a Banach space which is a Hermitian Jordan triple need not be a *Hermitian Jordan Banach triple*! Indeed, any Banach *-algebra $\mathcal{A}$ is a Hermitian Jordan triple in the triple product

$$\{a, b, c\} = (a \circ b^*) \circ c + a \circ (b^* \circ c) - b^* \circ (a \circ c),$$

where $\circ$ is the special Jordan product in $\mathcal{A}$.

Both JH*-triples and JB*-triples are Hermitian Jordan Banach triples. Although the structures of JH*-triples and JB*-triples are now well understood, the larger class of Hermitian Jordan Banach triples, which correspond to symmetric Banach manifolds, are less so. This is an important class of Banach spaces from the geometric perspective and deserves further investigation. A Hermitian Jordan Banach triple is called *positive* if the spectrum of each box operator $a \,\square\, a$ lies in $[0, \infty)$ (cf. Definition 1.2.29). The *negative* Hermitian Jordan Banach triples are the ones in which the spectrum of each $a \,\square\, a$ lies in $(-\infty, 0]$. The sign of these Jordan triples relates to the *opposite* sign of the curvature of the corresponding symmetric manifolds. JB*-triples are positive.

Lemma 3.1.3 *A positive Hermitian Jordan Banach triple V is a JB*-triple if, and only if, $\|\{a, a, a\}\| = \|a\|^3$ for all $a \in V$. In particular, JB*-triples are anisotropic.*

Proof The necessity follows from Remark 2.5.28. Conversely, given that V is a positive Hermitian Jordan Banach triple, the proof of Corollary 2.5.24

reveals that

$$\|a\|_\infty := \|a \,\square\, a\|^{1/2} \qquad (a \in V)$$

defines a norm on V. The domain $D_\infty := \{a \in V : \|a\|_\infty < 1\}$ is contained in the open unit ball D of V, since

$$\|a\|_\infty^2 = \|a \,\square\, a\| \geq \left\|(a \,\square\, a)\left(\frac{a}{\|a\|}\right)\right\| = \|a\|^2 \qquad (a \in V).$$

Continuity of the triple product implies that the norm $\|\cdot\|_\infty$ is equivalent to the original norm $\|\cdot\|$. The bounded domain D_∞ has a symmetry $s_0(z) = -z$ at 0 and is homogeneous via the Möbius transformations in (2.48). Hence $(V, \|\cdot\|_\infty)$ is a JB*-triple by Theorem 2.5.27, and in particular, we have

$$\|\{a, a, a\}\|_\infty = \|a\|_\infty^3$$

by Remark 2.5.28. We complete the proof by showing that $\|\cdot\| = \|\cdot\|_\infty$. Otherwise, we would have $\|v\| < \|v\|_\infty = 1$ for some $v \in V$, which entails the contradiction

$$\lim_{n \to \infty} v^{3^n} = 0 \quad \text{and} \quad \lim_{n \to \infty} \|v^{3^n}\|_\infty = 1. \qquad \square$$

In this section, we derive the basic analytic properties of JB*-triples, but we present some examples first.

Example 3.1.4 Let $\{V_\alpha\}_{\alpha \in A}$ be a family of JB*-triples. Then the ℓ_∞-sum

$$\bigoplus_{\alpha \in A}^{\ell_\infty} V_\alpha$$

is a JB*-triple in the triple product

$$\{(x_\alpha), (y_\alpha), (z_\alpha)\} := (\{x_\alpha, y_\alpha, z_\alpha\}),$$

which is well defined by Remark 2.5.28 and satisfies $\|\{(x_\alpha), (x_\alpha), (x_\alpha)\}\| = \sup_\alpha \|x_\alpha\|^3$.

Definition 3.1.5 A complex Jordan algebra $\mathcal{B}$ with a conjugate linear algebra involution $*$ is called a JB^*-algebra or *Jordan C*-algebra* if it is also a Banach space in which the norm satisfies

$$\|ab\| \leq \|a\|\|b\|, \qquad \|a^*\| = \|a\|, \qquad \|\{a, a, a\}\| = \|a\|^3$$

for all $a, b \in \mathcal{B}$, where $\{\cdot, \cdot, \cdot\}$ denotes the canonical Hermitian Jordan triple product defined in (1.12). A JB*-algebra is called a *JBW*-algebra* if it has a predual.

Given a JB*-algebra $\mathcal{B}$, its *self-adjoint part*

$$\mathcal{B}_{sa} = \{a \in \mathcal{B} : a^* = a\}$$

forms a JB-algebra in the inherent Jordan product [47, 3.8.2] and we have

$$\mathcal{B} = \mathcal{B}_{sa} + i\mathcal{B}_{sa}.$$

Conversely, it has been shown in Wright [117] that a JB-algebra A can be complexified to a JB*-algebra $\mathcal{A}$ so that A identifies with the self-adjoint part $\mathcal{A}_{sa}$ of $\mathcal{A}$.

We see from Example 2.3.9 that the exceptional Jordan algebra $H_3(\mathcal{O})$ is a JB*-algebra. A C*-algebra $\mathcal{A}$ is a JB*-algebra in the canonical triple product

$$\{a, b, c\} = \frac{1}{2}(ab^*c + cb^*c).$$

Lemma 3.1.6 *A JB*-algebra $\mathcal{B}$, with the canonical Hermitian Jordan triple product*

$$\{a, b, c\} = (ab^*)c + a(b^*c) - b^*(ac),$$

is a JB-triple. Hence a JBW*-algebra is a JBW*-triple.*

Proof In the canonical triple product, $\mathcal{B}$ is a Hermitian Jordan triple with continuous triple product. In view of Lemma 3.1.3, we only need to show that $\mathcal{B}$ is a positive Hermitian Jordan Banach triple.

For each $z \in \mathcal{B}$, it has been shown in Youngson [119, theorem 6] that $z = z^*$ if, and only if, the left multiplication $L_z : \mathcal{B} \longrightarrow \mathcal{B}$ is hermitian.

Let $a = a^*$ in $\mathcal{B}$. Then $a \square a = L_{a^2}$ is hermitian. The closed subalgebra $\mathcal{A}$ generated by a in $\mathcal{B}$ is associative and is therefore an abelian C*-algebra, since each $x \in \mathcal{A}$ satisfies

$$\|x\|^3 = \|(xx^*)x\| \le \|xx^*\|\|x\| \le \|x\|^2\|x^*\| = \|x\|^3.$$

When $\mathcal{A}$ is identified as the algebra $C_0(\Omega)$ of complex continuous functions vanishing at infinity on some locally compact Hausdorff space Ω, it is readily seen that the operator

$$a \square a|_{\mathcal{A}} : \mathcal{A} \longrightarrow \mathcal{A}$$

has non-negative spectrum, since it is just the left multiplication by $|a|^2$. The closed subtriple $V(a)$ generated by a is contained in $\mathcal{A}$ and hence $a \square a|_{V(a)}$ also has non-negative spectrum, as in Lemma 2.5.20. By Lemma 2.5.21, we have $\sigma(a \square a) \subset [0, \infty)$.

Given $x \in \mathcal{B}$, we can write $x = a + ib$ with $a = a^*$ and $b = b^*$. Since

$$x \square x = a \square a + b \square b + 2i(L_bL_a - L_aL_b),$$

where the left multiplications L_a and L_b are hermitian, $x \,\square\, x$ is hermitian (cf. [10, p. 47]).

Observe that

$$x^* \,\square\, x^* = * \circ (x \,\square\, x) \circ *,$$

where $* : \mathcal{B} \longrightarrow \mathcal{B}$ is the involution of $\mathcal{B}$. Hence $\sigma(x^* \,\square\, x^*) = \sigma(x \,\square\, x)$. Now

$$x \,\square\, x + x^* \,\square\, x^* = 2a \,\square\, a + 2b \,\square\, b$$

implies $\sigma(x \,\square\, x) \subset [0, \infty)$. $\qquad\square$

An important feature of JB*-triples is that their norm and triple product determine each other completely. We first show that a complex Banach space can admit at most one JB*-triple structure.

Theorem 3.1.7 *Let $\varphi : V \longrightarrow W$ be a surjective linear isometry between two JB*-triples V and W. Then φ is a triple isomorphism.*

Proof Consider V and W as the tangent space at 0 of the open unit balls D_V and D_W, respectively, which are symmetric domains. As in (2.40), let aut $D_W = \mathfrak{k} \oplus \mathfrak{p}$ be the eigenspace decomposition of complete holomorphic vector fields on D_W, induced by the symmetry $s_0(w) = -w$ at $0 \in D_W$.

Let $a \in V$. Then there is a unique vector field $Y_{\varphi(a)} \in \mathfrak{p}$ such that $Y_{\varphi(a)}(0) = \varphi(a)$. Likewise, let X_a be the unique complete holomorphic vector field on D_V such that $X_a(0) = a$. Since $\varphi : D_V \longrightarrow D_W$ is a surjective linear isometry, $\varphi X_a \varphi^{-1}$ is a complete holomorphic vector filed on D_W satisfying

$$(\varphi X_a \varphi^{-1})(0) = \varphi(a).$$

Hence we have $\varphi X_a \varphi^{-1} = Y_{\varphi(a)}$. This gives, by Lemma 2.5.15,

$$\varphi(a) - \{\varphi(z), \varphi(a), \varphi(z)\}$$
$$= Y_{\varphi(a)}(\varphi(z)) = (\varphi X_a \varphi^{-1})(\varphi(z)) = \varphi(a) - \varphi\{z, a, z\} \qquad (z \in D_V)$$

and hence

$$\varphi\{z, a, z\} = \{\varphi(z), \varphi(a), \varphi(z)\} \qquad (z \in D_V),$$

which proves that φ is a triple isomorphism by polarisation. $\qquad\square$

We will prove the converse of the above theorem later. In the special case of C*-algebras, this theorem subsumes, but more importantly, provides a geometric perspective of a well-known result on linear isometries between them [65].

Given a complex Banach space V, we define its *conjugate* $\overline{V}$ to be the Banach space V, but with scalar multiplication $(\lambda, x) \in \mathbb{C} \times V \mapsto \overline{\lambda}x \in V$. The conjugate $\overline{V}$ of a JB*-triple V is also a JB*-triple in the same triple product.

In analogy to the fact that JB-algebras are real forms of JB*-algebras, we introduce JB-triples as real forms of JB*-triples. First, we define, as usual, a *conjugation* on a complex Banach space V to be a conjugate linear isometry $\tau : V \longrightarrow V$ such that τ^2 is the identity map. A conjugation τ on a JB*-triple V can be viewed as a complex linear isometry $\tau : V \longrightarrow \overline{V}$. In particular, it preserves the triple product, by Theorem 3.1.7.

Definition 3.1.8 A *real form* of a JB*-triple V is a closed real subtriple of the form

$$V^\tau = \{x \in V : \tau(x) = x\},$$

where τ is a conjugation on V. A real Banach space E which is a real Jordan triple is called a *JB-triple* or a *real JB*-triple* if its hermitification $(E_c, \{\cdot, \cdot, \cdot\}_h)$, defined in (1.24), can be normed to become a JB*-triple.

Lemma 3.1.9 *The JB-triples are exactly the real forms of JB*-triples and identify with the class of real closed subtriples of JB*-triples.*

Proof Let E be a JB-triple. Then its hermitification $E_c = E \oplus iE$ is a JB*-triple and $\tau(a + ib) = a - ib$ is a conjugation on E_c. Hence $E = \{x \in E_c : \tau(x) = x\}$ is a real form of E_c.

Given a real form V^τ of a JB*-triple V, the hermitification

$$V_c^\tau = V^\tau \oplus iV^\tau = V$$

is a JB*-triple.

Let E be a real closed subtriple of a JB*-triple V. Then the ℓ_∞-sum $V \oplus \overline{V}$ is a JB*-triple and E can be identified as the closed real subtriple $\{(x, x) \in V \oplus \overline{V} : x \in E\}$ of $V \oplus \overline{V}$. The hermitification $E \oplus iE$ of E identifies as a closed subtriple of $V \oplus \overline{V}$ via the embedding

$$x \oplus iy \in E \oplus iE \mapsto (x, x) + i(y, y) = (x + iy, x - iy) \in V \oplus \overline{V}$$

and is a JB*-triple by Corollary 2.5.29.   □

In the sequel, the (real) dual space of a real Banach space V, for instance, a JB-triple, will be denoted by V'.

Example 3.1.10 Real C*-algebras and JB-algebras are JB-triples. Of course, JB*-triples can be viewed as JB-triples when restricted to real scalar multiplication. Although Theorem 3.1.7 still holds for real C*-algebras [23] and JB-algebras [118], it is no longer true for arbitrary JB-triples. A surjective linear isometry φ between JB-triples need not preserve the triple product [30] but only satisfies the condition

$$\varphi(a^3) = \varphi(a)^3 \qquad (a \in V).$$

We discuss isometries further in Section 3.4.

It is evident that the geometry of symmetric domains hinges on the spectrum $\sigma(a \,\square\, a)$ of the box operators. We now develop further the spectral theory of JB*-triples. Let V be a JB*-triple and let $a \in V \backslash \{0\}$. If $\dim V < \infty$, then the element a admits a spectral decomposition, by Theorem 1.2.34, and, in Definition 1.2.35, one defines the triple spectrum of a. We now extend the concept of a triple spectrum to infinite dimension and derive useful spectral properties of JB*-triples.

As before, let $V(a)$ be the closed subtriple of V generated by a, and let $R(a)$ be the real closed subtriple of V generated by a. The latter is the closed real linear span of odd powers of a and

$$V(a) = R(a) + i R(a).$$

Let $\mathcal{R} \subset L(V)$ be the closed real linear span of $R(a) \,\square\, R(a)$. By power associativity, $V(a)$ is abelian and $R(a)$ is flat. Hence each operator $u \,\square\, v$ is hermitian for $u, v \in R(a)$ and $\|u \,\square\, v\| \leq \|u\| \|v\|$ by Remark 2.5.28. Also, $\mathcal{R}|_{V(a)}$ is a real abelian Banach subalgebra of $L(V(a))$. Let

$$\mathcal{A} = R(a) \oplus_{\ell_1} \mathcal{R}|_{V(a)}$$

be the ℓ_1-sum of Banach spaces and define a product in $\mathcal{A}$ by

$$(u \oplus h)(v \oplus k) = (h(v) + k(u)) \oplus (u \,\square\, v|_{V(a)} + hk).$$

Then $\mathcal{A}$ is a real abelian Banach algebra and $R(a)$ can be identified with $R(a) \oplus 0 \subset \mathcal{A}$. Let Ω be the spectrum of $\mathcal{A}$, consisting of all characters of $\mathcal{A}$, where a character of $\mathcal{A}$ is a *nonzero* real linear multiplicative functional $\chi : \mathcal{A} \longrightarrow \mathbb{C}$ which is necessarily continuous and is actually the restriction of a character on the complexification $\mathcal{A} \oplus i\mathcal{A}$ of $\mathcal{A}$. The spectrum Ω is locally compact in the topology of pointwise convergence, and $\Omega \cup \{0\}$ is compact.

The Banach algebra quasi-spectrum $\sigma'_{\mathcal{A}}(b)$ of each $b \in \mathcal{A}$ is given by

$$\sigma'_{\mathcal{A}}(b) = \{0\} \cup \{\chi(b) : \chi \in \Omega\}.$$

We note that $\chi(a) \neq 0$ for each $\chi \in \Omega$, since χ does not vanish on $\mathcal{A}$. Hence $\chi(a^2) = \chi(a)^2 \neq 0$, where $a^2 = a \,\square\, a|_{V(a)}$.

For each $b \in R(a) \subset \mathcal{A}$, the quasi-spectrum $\sigma'_{\mathcal{A}}(b^2)$ is the same as the quasi-spectrum of $b \,\square\, b|_{V(a)}$ in $\mathcal{R}|_{V(a)}$, which is a Banach subalgebra of $L(V(a))$.

Lemma 2.5.20 implies that the quasi-spectrum of $b \,\square\, b|_{V(a)}$ in $L(V(a))$ is contained in $[0, \infty)$ and is therefore the same as the quasi-spectrum $\sigma'(b \,\square\, b|_{V(a)})$ of the element $b \,\square\, b|_{V(a)}$ in the subalgebra $\mathcal{R}|_{V(a)} \subset L(V(a))$, by the same reason as given in the proof of Lemma 2.5.20. It follows that $\sigma'_{\mathcal{A}}(b^2) = \sigma'(b \,\square\, b|_{V(a)}) \subset [0, \infty)$ and $\chi(b^2) \in [0, \infty)$ for each $\chi \in \Omega$. This implies $\chi(b) \in \mathbb{R}$ for each $\chi \in \Omega$.

The spectrum Ω is the disjoint union of two open subsets,

$$S = \{\chi \in \Omega : \chi(a) > 0\} \quad \text{and} \quad T = \{\chi \in \Omega : \chi(a) < 0\}.$$

Let $C_0(S, \mathbb{R})$ be the real JB*-triple of real continuous functions on S, vanishing at infinity. The Gelfand transform

$$\widehat{} : R(a) \longrightarrow C_0(S, \mathbb{R})$$

defined by

$$\widehat{b}(\chi) = \chi(b) \qquad (b \in R(a), \chi \in S)$$

is a real triple homomorphism and extends naturally to a complex triple homomorphism $\widehat{} : V(a) \longrightarrow C_0(S, \mathbb{C})$. These two triple homomorphisms are actually isomorphisms.

Observe that for each $b \in V(a)$, the element $b^2 = b \,\square\, b|_{V(a)}$ is hermitian in the complex abelian Banach algebra $\mathcal{A} \oplus i\mathcal{A}$ since

$$\| \exp(it b \,\square\, b|_{V(a)}) \| \leq \| \exp(it b \,\square\, b) \| \leq 1 \qquad (t \in \mathbb{R}).$$

Hence the norm $\|b \,\square\, b|_{V(a)}\|$ coincides with the spectral radius of $b \,\square\, b|_{V(a)}$ [10, p. 54]; that is,

$$\|b \,\square\, b|_{V(a)}\| = \sup\{|\chi(b \,\square\, b|_{V(a)})| : \chi \in \Omega_c\}$$
$$= \sup\{|\chi(b \,\square\, b|_{V(a)})| : \chi \in \Omega\},$$

where each character $\chi \in \Omega$ is the restriction of a character in the spectrum Ω_c of $\mathcal{A} \oplus i\mathcal{A}$ and we use the same symbol for both characters.

We first show that the map $\widehat{}: R(a) \longrightarrow C_0(S, \mathbb{R})$ is isometric. Indeed, for $b \in R(a)$, we have

$$
\begin{aligned}
\|\widehat{b}\|^2 &= \sup\{|\widehat{b}(\chi)|^2 : \chi \in S\} \\
&= \sup\{\chi(b)^2 : \chi \in S\} \\
&= \sup\{\chi(b)^2 : \chi \in S \cup T\} \\
&= \sup\{\chi(b \square b|_{V(a)}) : \chi \in \Omega\} \\
&= \|b \square b|_{V(a)}\| \leq \|b \square b\| = \|b\|^2.
\end{aligned}
$$

On the other hand,

$$
\|b\|^2 = \|(b \square b)(\|b\|^{-1}b)\| \leq \|b \square b|_{V(a)}\| = \|\widehat{b}\|^2.
$$

Next, each $z \in V(a)$ is an element in the complexification $\mathcal{A} \oplus i\mathcal{A}$ of $\mathcal{A}$ and by the definition of the algebraic product, $z^2 = z \square z|_{V(a)}$. Let $z = b + ic \in R(a) + iR(a)$. Then we have

$$
z \square z = b \square b + c \square c + i(c \square b - b \square c)
$$

and it follows that

$$
\chi(z^2) = \chi(z)\overline{\chi(z)} \qquad (\chi \in \Omega).
$$

Hence, as before, we have

$$
\begin{aligned}
\|\widehat{z}\|^2 &= \sup\{|\widehat{z}(\chi)|^2 : \chi \in S\} \\
&= \sup\{\chi(z)\overline{\chi(z)} : \chi \in S\} \\
&= \sup\{|\chi(z \square z|_{V(a)})| : \chi \in \Omega\} \\
&= \|z \square z|_{V(a)}\|.
\end{aligned}
$$

Again, this leads to $\|z\| = \|\widehat{z}\|$.

Finally, we show that the map $\widehat{}: R(a) \longrightarrow C_0(S, \mathbb{R})$ is surjective. We first observe that the map

$$
\widehat{a}: \chi \in S \mapsto \widehat{a}(\chi) = \chi(a) \in \widehat{a}(S) = |\sigma'(a)\backslash\{0\}| \subset (0, \infty)
$$

is a homeomorphism, since the algebra $\mathcal{A}$ is generated by a. Therefore the map

$$
f \in C_0(\widehat{a}(S), \mathbb{R}) \mapsto f \circ \widehat{a} \in C_0(S, \mathbb{R})
$$

is an isometric isomorphism.

Let $p \in C_0(\widehat{a}(S), \mathbb{R})$ be an odd polynomial. We have

$$
p \circ \widehat{a}(\chi) = p(\widehat{a}(\chi)) = p(\chi(a)) = \chi(p(a)) = \widehat{p(a)}(\chi) \qquad (\chi \in S),
$$

where $p(a)$ is an odd polynomial in a and belongs to $R(a)$.

Let $F \in C_0(S, \mathbb{R})$. Then $F = f \circ \hat{a}$ for some $f \in C_0(\hat{a}(S), \mathbb{R})$, and f extends to an odd function $\tilde{f} \in C_0(\hat{a}(S) \cup -\hat{a}(S), \mathbb{R})$ satisfying

$$\tilde{f}(-\hat{a}(\chi)) = -f(\hat{a}(\chi)) \qquad (\chi \in S).$$

Hence $\tilde{f}$ is the uniform limit of a sequence (p_n) of odd polynomials in $C_0(\hat{a}(S) \cup -\hat{a}(S), \mathbb{R})$. Each p_n restricts to an odd polynomial $p_n|_{\hat{a}(S)}$ in $C_0(\hat{a}(S), \mathbb{R})$, and it follows that $p_n|_{\hat{a}(S)} \circ \hat{a} = \widehat{p_n(a)}$ with $p_n(a) \in R(a)$. This gives

$$F = f \circ \hat{a} = \tilde{f}|_{\hat{a}(S)} \circ \hat{a} = \lim_n p_n|_{\hat{a}(S)} \circ \hat{a} = \lim_n \widehat{p_n(a)} \in \widehat{R(a)},$$

proving the surjectivity of $\widehat{\ } : R(a) \longrightarrow C_0(S, \mathbb{R})$ as well as that of its extension to $V(a)$.

We note that the nonzero real triple homomorphisms $\psi : C_0(S, \mathbb{R}) \longrightarrow \mathbb{R}$ of $C_0(S, \mathbb{R})$ are exactly the point evaluations $\{\pm\delta_\chi : \chi \in S\}$. Identifying $R(a)$ with $C_0(S, \mathbb{R})$, we see that the nonzero real triple homomorphisms of $R(a)$ are exactly the restrictions $\{\chi|_{R(a)} : \chi \in \Omega\}$, and the nonzero complex triple homomorphisms of $V(a)$ are exactly the restrictions $\Omega|_{V(a)}$.

The above discussion leads to a natural notion of spectrum in JB*-triples.

Definition 3.1.11 Let V be a JB*-triple and let $a \in V \setminus \{0\}$. Let $V(a)$ be the closed subtriple generated by a. We denote by $\Sigma(a)$ the spectrum of the operator $a \,\square\, a|_{V(a)} : V(a) \longrightarrow V(a)$, where $\Sigma(a)$ is a compact subset of $[0, \infty)$. The *symmetrized triple spectrum* of a is defined to be the compact set

$$\mathrm{Sp}\,(a) = \{t \in \mathbb{R} : t^2 \in \Sigma(a)\},$$

where $\mathrm{Sp}\,(a) = -\mathrm{Sp}\,(a)$ is the union of two compact subsets

$$\mathrm{Sp}\,(a)^+ = \{t \in [0, \infty) : t^2 \in \Sigma(a)\} \quad \text{and}$$
$$\mathrm{Sp}\,(a)^- = \{t \in (-\infty, 0] : t^2 \in \Sigma(a)\}.$$

We define the *triple spectrum* of a to be the set

$$s(a) = \mathrm{Sp}\,(a)^+ \setminus \{0\}.$$

We see from the preceding construction of the Gelfand map $\widehat{\ }$ that

$$\Sigma(a) \setminus \{0\} = \sigma'_{\mathcal{A}}(a^2) \setminus \{0\} = \{\chi(a^2) : \chi \in \Omega\}, \quad \mathrm{Sp}\,(a) \setminus \{0\} = \{\chi(a) : \chi \in \Omega\}$$

and also

$$\mathrm{Sp}\,(a)^+ \setminus \{0\} = \{\chi(a) : \chi \in S\} = \hat{a}(S).$$

We note that $s(a) \cup \{0\}$ is compact by compactness of $S \cup \{0\}$. Moreover, we have

$$\mathrm{Sp}\,(a) \cup \{0\} = \{\chi(a) : \chi \text{ is a (real) triple homomorphism of } R(a)\}$$
$$= \{\chi(a) : \chi \text{ is a (complex) triple homomorphism of } V(a)\}$$

and

$$\|a\| = \sup s(a) = \sup\{|\chi(a)| : \chi \text{ is a (real) triple homomorphism of } R(a)\}.$$

Summarising, we draw the following conclusion.

Theorem 3.1.12 *Let V be a JB*-triple and $a \in V \setminus \{0\}$. Then there is an isometric triple isomorphism Φ from the closed subtriple $V(a)$ generated by a to the C*-algebra $C_0(s(a), \mathbb{C})$ of complex continuous functions on the triple spectrum $s(a)$, vanishing at infinity, such that*

$$\Phi(a)(s) = s \qquad (s \in s(a)).$$

The restriction of Φ to the real closed subtriple $R(a)$ generated by a is an isometric triple isomorphism to the real C*-algebra $C_0(s(a), \mathbb{R})$ of real continuous functions on $s(a)$, vanishing at infinity.

Let V^τ be a JB-triple which is a real closed subtriple of a JB*-triple V. Given $a \in V^\tau$, it is evident that the real closed subtriple $R(a)$ of V^τ generated by a is also the smallest real closed subtriple of V containing a. Hence $R(a)$ identifies with $C_0(s(a), \mathbb{R})$ as before.

Remark 3.1.13 We note that $C_0(s(a), \mathbb{C})$ is isometrically triple isomorphic to

$$C_{\mathrm{odd}}(\mathrm{Sp}\,(a), \mathbb{C}) = \{f \in C(\mathrm{Sp}\,(a), \mathbb{C}) : f(-s) = -f(s), \forall s \in \mathrm{Sp}\,(a)\}.$$

Example 3.1.14 If e is a tripotent in a JB*-triple, then $R(e) = \mathbb{R}e$ and e has triple spectrum $s(e) = \{1\}$. Let $V = M_2(\mathbb{C})$ be the C*-algebra of 2×2 complex matrices. Then the C*-algebra spectrum of the tripotent $\begin{pmatrix} 0 & 1 \\ 0 & 0 \end{pmatrix}$ is $\{0\}$.

Let $\{e_{ij}\}$ be the matrix units in $M_2(\mathbb{C})$ and let $a = \begin{pmatrix} 1 & 0 \\ 0 & -2 \end{pmatrix}$. Then the box operator $a \square a : M_2(\mathbb{C}) \longrightarrow M_2(\mathbb{C})$ has a 4×4 diagonal matrix representation with respect to the ordered basis $\{e_{11}, e_{12}, e_{21}, e_{22}\}$, with diagonal entries $\{1, 5/2, 5/2, 4\}$. Hence $a \square a$ has spectrum $\sigma(a \square a) = \{1, 5/2, 4\}$. We have $V(a) = \mathbb{C}e_{11} + \mathbb{C}e_{22}$, and the spectrum of the restriction operator $a \square a|_{V(a)} : V(a) \longrightarrow V(a)$ is $\{1, 4\}$. The triple spectrum of a is $s(a) = \{1, 2\}$ and the symmetrized triple spectrum $\mathrm{Sp}\,(a) = \{\pm 1, \pm 2\}$. The C*-algebra spectrum of $a \in M_2(\mathbb{C})$ is $\{1, -2\}$. In fact, we have the following result.

Corollary 3.1.15 *Let a be a nonzero self-adjoint element in a C*-algebra $\mathcal{A}$. Let $\sigma'_{\mathcal{A}}(a)$ be the C*-algebra quasi-spectrum of a. Then the triple spectrum $s(a)$ is just the set $|\sigma'_{\mathcal{A}}(a)\backslash\{0\}| = \{|\alpha| : \alpha \in \sigma'(a)\backslash\{0\}\}$.*

Proof We have $V(a) \subset C^*(a)$, where $C^*(a)$ is the abelian C*-subalgebra generated by a and the C*-algebra quasi-spectrum of a is given by

$$\sigma'_{\mathcal{A}}(a) = \{0\} \cup \{\rho(a) : \rho \text{ is a character of } C^*(a)\}.$$

Since each character ρ of $C^*(a)$ restricts to a nonzero complex triple homomorphism of $V(a)$, we have $\rho(a) \in s(a)$ or $-\rho(a) \in s(a)$.

On the other hand, Lemma 2.5.20 implies that $\Sigma(a) = \sigma(a \square a|_V(a))$ is contained in the spectrum $\sigma(a \square a)$ of the box operator $a \square a : C^*(a) \longrightarrow C^*(a)$, which is the left multiplication on $C^*(a)$ by a^2. Identifying $C^*(a)$ with the C*-algebra $C_0(\sigma'_{\mathcal{A}}(a)\backslash\{0\})$, and the element a with the identity function on $\sigma'_{\mathcal{A}}(a)\backslash\{0\}$, we see that

$$\sigma(a \square a) \subset \{t^2 : t \in \sigma'_{\mathcal{A}}(a)\},$$

which implies $s(a) \subset |\sigma'_{\mathcal{A}}(a)\backslash\{0\}|$. $\square$

Corollary 3.1.16 *A Hermitian Jordan Banach triple V is a JB*-triple if, and only if, the closed subtriple $V(a)$ generated by each $a \in V$ is isometrically triple isomorphic to an abelian C*-algebra.*

Proof If V is a Hermitian Jordan Banach triple and if $V(a)$ is isometrically triple isomorphic to an abelian C*-algebra $C_0(s(a), \mathbb{C})$, then we have $\|\{a, a, a\}\| = \|a\|^3$ and $a \square a|_{V(a)}$ has non-negative spectrum. By Lemma 2.5.21, $\sigma(a \square a) \subset [0, \infty)$. The sufficiency now follows from Lemma 3.1.3. $\square$

Let J be a closed *triple* ideal of an abelian C*-algebra $\mathcal{A}$. It is readily seen that J is an algebra ideal of $\mathcal{A}$ and hence the quotient $\mathcal{A}/J$ is an abelian C*-algebra. If $\pi : \mathcal{A} \longrightarrow V$ is a triple homomorphism onto a Banach space V which is a Hermitian Jordan triple, then $\mathcal{A}/\pi^{-1}(0)$ is an abelian C*-algebra and triple isomorphic to V. Since π induces an abelian Banach algebraic structure on V, it follows that $\mathcal{A}/\pi^{-1}(0)$ is isometric to V.

Remark 3.1.17 In fact, the closed triple ideals in *any* C*-algebra are exactly the closed two-sided algebra ideals. We refer to Barton and Timoney [7] for a proof. It can also be shown that a closed subspace J of a JB*-triple V is a triple ideal if, and only if, $\{V, J, J\} \subset J$ [15].

Corollary 3.1.18 *Let V be a JB*-triple and J a closed triple ideal of V. Then the quotient space V/J is a JB*-triple.*

Proof The quotient map $\pi : V \longrightarrow V/J$ is a surjective triple homomorphism and V/J is a Hermitian Jordan Banach triple. For each $a \in V$, let $V(\pi(a))$ be the closed subtriple generated by $\pi(a)$ in V/J. Then $V(\pi(a)) = \pi(V(a))$, where $V(a)$ is isometrically triple isomorphic to an abelian C*-algebra $\mathcal{A}$. By the above observation, a continuous triple homomorphic image of $\mathcal{A}$, and hence $V(\pi(a))$, is isometrically isomorphic to an abelian C*-algebra. $\qquad\square$

Corollary 3.1.19 *Let* $\varphi : V \longrightarrow W$ *be a triple homomorphism between two JB*-triples* V *and* W*. Then* φ *is contractive. The same conclusion holds if* V *and* W *are JB-triples.*

Proof Let $a \in V \backslash \{0\}$ and let $V(a)$ be the JB*-subtriple generated by a. Then $\varphi(V(a))$ is the JB*-subtriple $W(\varphi(a))$ generated by $\varphi(a)$ in W. Further, for each complex triple homomorphism χ of $W(\varphi(a))$, the composite $\chi \circ \varphi$ is a complex triple homomorphism of $V(a)$. It follows that

$$\|\varphi(a)\| = \sup\{|\chi(\varphi(a))| : \chi \text{ is a triple homomorphism of } W(\varphi(a))\} \leq \|a\|.$$

The arguments using $R(a)$ for JB-triples do not change. $\qquad\square$

The following result is the converse of Theorem 3.1.7 and shows uniqueness of the norm in a JB*-triple.

Theorem 3.1.20 *Let* $\varphi : V \longrightarrow W$ *be a triple isomorphism between two JB*-triples* V *and* W*. Then* φ *is a linear isometry. The same conclusion holds if* V *and* W *are JB-triples.*

Proof Apply Corollary 3.1.19 to φ and its inverse φ^{-1}. $\qquad\square$

Corollary 3.1.21 *Let* V *be a JB*-triple which is the vector space direct sum* $V = V_1 \oplus V_2$ *of two closed subtriples* V_1 *and* V_2*. If* $V_1 \,\square\, V_2 = \{0\}$*, then* $\|v_1 + v_2\| = \max\{\|v_1\|, \|v_2\|\}$ *for* $v_j \in V_j$ *and* $j = 1, 2$*.*

Proof The ℓ_∞-sum $V_1 \oplus^{\ell_\infty} V_2$ is a JB*-triple which is triple isomorphic, and hence isometric, to V. $\qquad\square$

3.2 Holomorphic mappings

From the perspective of the Riemann mapping theorem, the open unit ball of a JB*-triple can be regarded as an infinite-dimensional generalization of the open unit disc in $\mathbb{C}$, since every bounded symmetric domain in a complex Banach space is biholomorphic to such a ball. Cartan domains, as well as the two exceptional domains, are open unit balls of JB*-triples. It is therefore natural to study function theory on the open unit ball of a JB*-triple and expect some fruitful applications of Jordan structures. Of course, analysis on *finite-dimensional* bounded symmetric domains is already a vast and rich enterprise, and literature abounds. The novelty here is the application of Jordan methods in the infinite-dimensional setting. We will show several infinite-dimensional results on holomorphic maps to highlight the use of Jordan structures.

Let D be the open unit ball of a JB*-triple V. A natural question comes to mind is the boundary structure of D. Answering this question would be a lengthy task and we refer to Loos [85], Sauter [102] and Wolf [116] for details. It involves analyzing the manifold of tripotents in V and its components. A simple task, however, is to identify the extreme points $\operatorname{ext} \overline{D}$ of the closed unit ball $\overline{D}$. We show they are complete tripotents of V. As before, let

$$\operatorname{aut} D = \mathfrak{k} \oplus \mathfrak{p}$$

be the eigenspace decomposition induced by the symmetry s_0 at $0 \in D$. Integration of vector fields in $\operatorname{aut} D$ has been computed explicitly in Kaup [70].

A point $u \in \overline{D}$ is called a *complex extreme point* if whenever $v \in V$ satisfies $u + \lambda v \in \overline{D}$ for all complex numbers λ of modulus at most 1, we have $v = 0$. Evidently, an extreme point of $\overline{D}$ is also a complex extreme point.

A point $u \in \overline{D}$ is called a *holomorphic extreme point* if for every open neighbourhood $\Omega \subset \mathbb{C}$ of 0 and holomorphic map $f : \Omega \longrightarrow \overline{D} \subset V$ satisfying $f(0) = u$, we have $f'(0) = 0$. Plainly, a holomorphic extreme point is also a complex extreme point.

Lemma 3.2.1 *Let u be a tripotent in a JB*-triple V. Then the Peirce projections $P_k(u)$ ($k = 0, 1, 2$) induced by u are all contractive and we have*

$$\exp it(u \,\square\, u) = P_0(u) + e^{it/2} P_1(u) + e^{it} P_2(u).$$

Proof Since the box operator $u \,\square\, u$ has eigenvalues 0, $1/2$ and 1, with corresponding eigenspaces $P_0(u)V$, $P_1(u)V$ and $P_2(u)V$, the spectral mapping theorem implies that the linear isometry $\exp it(u \,\square\, u)$ has eigenvalues 0, $e^{it/2}$ and e^{it} as well as

$$\exp it(u \,\square\, u) = P_0(u) + e^{it/2} P_1(u) + e^{it} P_2(u) \qquad (t \in \mathbb{R}).$$

For each $v \in V$, we have

$$\|v\| = \|\exp it(u \,\square\, u)v\| = \|P_0(u)v + e^{it/2} P_1(u)v + e^{it} P_2(u)v\| \quad \text{for all } t \in \mathbb{R}.$$

It follows that

$$\|P_k(u)v\| \le \|v\| \qquad (k = 0, 1, 2).$$

$\square$

For $t = 2\pi$ in Lemma 3.2.1, we have

$$\exp 2\pi i(u \,\Box\, u) = P_0(u) - P_1(u) + P_2(u) = B(u, 2u),$$

where $B(u, 2u) : V \longrightarrow V$ is the Bergmann operator and the last equality follows from (1.33). The isometry $\exp 2\pi i(u \,\Box\, u)$ is involutive and fixes u, but is not necessarily a symmetry at u in the manifold of tripotents of V (cf. [26, p. 747]). However, it is called the *Peirce symmetry* at u. Since

$$P_0(u) + P_2(u) = \frac{1}{2}(1 + B(u, 2u)),$$

it is also a contractive projection on V.

We observe that, if a JB*-triple V admits a nonzero tripotent u, then the u-homotope of the Peirce 2-space $P_2(u)V$ is a JB*-algebra in the inherited norm, with involution defined by

$$a^* := \{u, a, u\},$$

and u becomes the identity of the algebra. This often enables us to make use of results in JB*-algebras and reduce a problem concerning JB*-triples to one of JB*-algebras. For simplicity, we use the same notation $P_2(u)V$ for its u-homotope and write the product by juxtaposition:

$$ab = \{a, u, b\}.$$

Remark 3.2.2 Since we have not yet removed the constant 2 from the inequality in (2.50), one can show $\|ab\| \leq \|a\|\|b\|$ in $P_2(u)V$ in the following way. For a self-adjoint element $a = a^* = \{u, a, u\}$, the triple identity implies $\{u, a, b\} = \{a, u, b\}$ for $b \in P_2(u)V$, and hence the box operator $a \,\Box\, u|_{P_2(u)V}$ is hermitian on the JB*-triple $P_2(u)V$. Therefore

$$\|ab\| = \|a \,\Box\, u(b)\| \leq \|a\|\|u\|\|b\| = \|a\|\|b\|,$$

by Remark 2.5.28. Also, the triple product $[\cdot, \cdot, \cdot]$ of the u-homotope with involution $*$ coincides with the original triple product on $P_2(u)V$:

$$[x, y, x] = \{x, y^*, x\}_u = \{x, \{u, y^*, u\}, x\} = \{x, y, x\}.$$

It follows that the self-adjoint part $\{a \in P_2(u)V : a = a^*\}$ of $P_2(u)V$ is a JB-algebra and its complexification is $P_2(u)V$ itself which, as noted before, can be equipped with a norm $\|\cdot\|$ to become a JB*-algebra. Since a JB*-algebra is also a JB*-triple and the norm in a JB*-triple is unique, $\|\cdot\|$ is the original norm of $P_2(u)V$.

We recall that a tripotent u in a Jordan triple system V is called *complete* if the Peirce space $V_0(u)$ equals $\{0\}$.

Theorem 3.2.3 *Let D be the open unit ball of a JB*-triple V and let $u \in V$. The following conditions are equivalent:*

  (i) *u is an extreme point of $\overline{D}$.*
  (ii) *u is a complex extreme point of $\overline{D}$.*
 (iii) *u is a complete tripotent.*
 (iv) *u is a holomorphic extreme point of $\overline{D}$.*

Proof (ii) $\Rightarrow$ (iii). The closed subtriple $V(u)$ generated by u is triple isomorphic to an abelian C*-algebra $\mathcal{A}$ in which the complex extreme points of the closed unit ball are partial isometries. It follows that u is a tripotent. One needs to show that $P_0(u)V = \{0\}$. We first observe that $\{z, u, z\} = 0$ for $z \in P_0(u)V$, by the Peirce multiplication rule.

Let $X(\cdot) = u - \{\cdot, u, \cdot\}$ be the unique vector field in $\mathfrak{p}$ such that $X(0) = u$. The integral curve $\exp t X(z)$ of X can be described explicitly by [70, theorem 2.25]. In our case, however, one can verify easily that, for $z \in P_0(u)V$, the flow

$$g_t(z) = (\tanh t)u + z$$

solves the differential equation

$$\frac{dg_t(z)}{dt} = X(g_t(z)).$$

Hence

$$D \supset \exp t X(P_0(u)V \cap D) = (\tanh t)u + (P_0(u)V \cap D).$$

Letting $t \to \infty$, we see that $u + (P_0(u)V \cap D) \subset \overline{D}$, which implies that $P_0(u)V = \{0\}$.

(iii) $\Rightarrow$ (iv). For $v \in V$, we write

$$v = v_1 + v_2 \in V_1(u) \oplus V_2(u)$$

for its Peirce decomposition. Let Ω be an open neighbourhood of $0 \in \mathbb{C}$ and $f : \Omega \longrightarrow \overline{D}$ a holomorphic map satisfying $f(0) = u$.

Let X be the unique vector field in $\mathfrak{p}$ such that $X(0) = u$. For each $t \in \mathbb{R}$, write $g_t = \exp(-tX)$. We have

$$g_t(u) = u \quad \text{and} \quad g_t'(u)(v_1 + v_2) = e^t v_1 + e^{2t} v_2.$$

Observe that $\{g_t \circ f : t \in \mathbb{R}\}$ is a family of holomorphic maps from Ω to V, uniformly bounded, and $g_t \circ f(0) = u$. By the Cauchy inequality (2.1), $\{g_t'(u)(f'(0)) : t \in \mathbb{R}\}$ is bounded in V and hence $f'(0) = 0$.

(iv) $\Rightarrow$ (i). By (iv), u is a complex extreme point of $\overline{D}$ and hence a complete tripotent by (ii) $\Rightarrow$ (iii). Let $u + \lambda v \in \overline{D}$ for all $\lambda \in [-1, 1]$. We show $v = 0$ to complete the proof.

Let $v = v_1 + v_2 \in P_1(u)V \oplus P_2(u)V$ be the Peirce decomposition of v. By Lemma 3.2.1, we have

$$u + \lambda v_2 \in \overline{D} \quad \text{and} \quad u + \lambda e^{it} v_1 + \lambda v_2 \in \overline{D} \qquad (t \in \mathbb{R}).$$

Now $P_2(u)V$ is a JB*-algebra as noted before and u is a complex extreme point of its closed unit ball. By the result [14, lemma 4.1] for JB*-algebras, u is an extreme point of the ball and hence $v_2 = 0$. It follows that $v_1 = 0$ also. $\qquad\square$

An immediate consequence of the above result is that JBW*-triples contain many (complete) tripotents by the Krein–Milman theorem, although the JB*-triple $C_0[0, 1]$ of complex continuous functions on $[0, 1]$, vanishing at 0, contains none except 0. In fact, we can say much more.

Corollary 3.2.4 *Let V be a JBW*-triple. Then V is the norm-closed linear span of the extreme points of its closed unit ball $\overline{D}$.*

Proof Pick any nonzero a in the open unit ball D. Evidently, the Möbius transformation

$$g_a(z) = a + B(a, a)^{1/2}(I + z \square a)^{-1}(z)$$

can be extended to a biholomorphic map $\widetilde{g}_a$ on a neighbourhood of D. Given any extreme point $u \in \overline{D}$, we have $\widetilde{g}_a(\alpha u) \in \overline{D}$ for $|\alpha| = 1$ and it is easily verified that $\widetilde{g}_a(\alpha u)$ is a holomorphic extreme point. We can define a holomorphic map $f : \Omega \longrightarrow V$ on a neighbourhood Ω of the open unit disc in $\mathbb{C}$ by

$$f(\alpha) = \widetilde{g}_a(\alpha u) \qquad (\alpha \in \Omega)$$

so that $f(0) = a$. By the mean value property of f, we have

$$a = \frac{1}{2\pi} \int_0^{2\pi} \widetilde{g}_a(e^{i\theta} u) d\theta,$$

which is in the norm-closed linear span of the extreme points of $\overline{D}$. $\qquad\square$

The facial structure of the closed unit ball of a JBW*-triple and its predual has been studied in detail in Edwards and Rüttiman [35]. We will not pursue this linear aspect of JB*-triples. Instead, we are concerned with two holomorphic maps which are fundamental in the geometry of the open unit ball D of a JB*-triple V, namely, the Bergmann operator on V and the Möbius transformation on D. The significance of the Bergmann operator can be seen from the fact that, in finite dimensions, the Bergmann metric h on D, which is a Hermitian metric, is given by

$$h_p(z, w) = h_0(B(p, p)^{-1}z, w) \qquad (p \in D),$$

where the metric at 0 is given by

$$h_0(z, w) = \text{Trace}\,(z \,\square\, w)$$

(cf. [85, theorem 2.10]). In Corollary 3.2.4, we have already made use of the Möbius transformation g_a at $a \in D$, introduced in (2.48):

$$g_a(z) = a + B(a, a)^{1/2}(I + z \,\square\, a)^{-1}(z) \qquad (z \in D),$$

which is related to the Bergman operator and $g'_a(0) = B(a, a)^{1/2}$.

Example 3.2.5 Let D be the open unit ball of a Hilbert space $(V, \langle \cdot, \cdot \rangle)$. Then

$$\text{ext}\,\overline{D} = \{v \in V : \|v\| = 1\}.$$

For each $a \in D$, the Bergmann operator $B(a, a)$ is given by

$$\begin{aligned} B(a, a)(v) &= v - 2\{a, a, v\} + \{a, \{a, v, a\}, a\} \\ &= (1 - \|a\|^2)(v - \langle v, a \rangle a), \end{aligned}$$

which is a self-adjoint operator on V.

Let $u \in \text{ext}\,\overline{D}$ be such that $a = \|a\|u$. Let $V = V_1(u) \oplus V_2(u)$ be the Peirce decomposition induced by u. Then we have

$$B(a, a)(v) = \begin{cases} (1 - \|a\|^2)v & (v \in V_1(u)) \\ (1 - \|a\|^2)^2 v & (v \in V_2(u)). \end{cases}$$

The inverse of $B(a, a)$ can be easily computed, and

$$B(a, a)^{-1}(z) = \frac{1}{1 - \|a\|^2}\left(z + \frac{\langle z, a \rangle a}{1 - \|a\|^2}\right) \qquad (z \in V).$$

Also, one can verify that

$$B(a, a)^{1/2}(z) = \sqrt{1 - \|a\|^2}\left(z + (\sqrt{1 - \|a\|^2} - 1)\langle z, a \rangle \frac{a}{\|a\|^2}\right).$$

It follows that $B(a, a)^{1/2}(a) = (1 - \|a\|^2)a = a - \{a, a, a\}$ and

$$\begin{aligned} B(a, a)^{-1/2}(z) &= B(a, a)^{-1} B(a, a)^{1/2}(z) \\ &= B(a, a)^{-1}\left(\sqrt{1 - \|a\|^2}\left(z + (\sqrt{1 - \|a\|^2} - 1)\langle z, a \rangle \frac{a}{\|a\|^2}\right)\right) \\ &= \frac{1}{\sqrt{1 - \|a\|^2}}\left(z + \frac{1 - \sqrt{1 - \|a\|^2}}{\|a\|^2\sqrt{1 - \|a\|^2}}\langle z, a \rangle a\right). \end{aligned}$$

In particular,

$$B(a, a)^{-1/2}(a) = \frac{a}{1 - \|a\|^2}.$$

Direct computation gives

$$(I + z \square a)^{-1} z = z - z \square a(z) + (z \square a)^2(z) - (z \square a)^3(z) + \cdots$$

$$= z - \langle z, a \rangle z + \langle z, a \rangle^2 z - \langle z, a \rangle^3 z + \cdots$$

$$= \frac{z}{1 + \langle z, a \rangle} \qquad (z \in D).$$

It follows that the Möbius transformation g_a is given by

$$g_a(z) = a + B(a, a)^{1/2}(I + z \square a)^{-1}(z)$$

$$= a + \frac{\sqrt{1 - \|a\|^2}}{1 + \langle z, a \rangle} \left(z + (\sqrt{1 - \|a\|^2} - 1)\langle z, a \rangle \frac{a}{\|a\|^2} \right)$$

$$= \frac{a + E_a(z) + \sqrt{1 - \|a\|^2}(I - E_a)(z)}{1 + \langle z, a \rangle} \qquad (z \in D),$$

where E_a is the projection from V onto the subspace $\mathbb{C}a$.

Now let $V = \mathbb{C}^n$. The Bergmann metric on D is given by

$$h_0(z, w) = \frac{n + 1}{2} \langle z, w \rangle$$

$$h_a(z, w) = \frac{n + 1}{2} \langle B(a, a)^{-1} z, w \rangle \qquad (a \in D, z, w \in \mathbb{C}^n)$$

$$= \frac{n + 1}{2(1 - \|a\|^2)^2}((1 - \|a\|^2)\langle z, w \rangle + \langle z, a \rangle \langle a, w \rangle).$$

We have already discussed the case $n = 1$ in Example 2.5.5.

The Möbius transformations g_a $(a \in D)$ on the open unit ball D in a JB*-triple V determine its automorphism group $\operatorname{Aut} D$. We first note that, using the Hahn–Banach theorem, the Schwarz lemma for the open unit disc in $\mathbb{C}$ can be readily extended to D; that is, given a homomorphic map $h : D \longrightarrow D$ satisfying $h(0) = 0$, we must have $\|h(z)\| \le \|z\|$ for all $z \in D$.

Proposition 3.2.6 *Let D be the open unit ball of a JB*-triple V. Then*

$$\operatorname{Aut} D = \{\varphi \circ g_a : a \in D, \ \varphi \text{ is a linear isometry on } V\}.$$

Proof Let $g \in \operatorname{Aut} D$ and $a = g^{-1}(0)$. Let $\varphi = g_{-a}g^{-1}$. Then $\varphi \in \operatorname{Aut} D$ and $\varphi(0) = 0$. Since for all $t \in \mathbb{R}$, the automorphisms $\varphi(e^{it} \cdot)$ and $e^{it}\varphi(\cdot)$ have the same value and the same derivative at 0, they are equal, by Cartan's uniqueness theorem. It follows that φ is a linear isometry and we have $g = \varphi^{-1}g_{-a}$. $\square$

Example 3.2.7 The bidisc $U \times U$ in $\mathbb{C}^2$ is not biholomorphic to the open ball

$$B = \{(z, w) \in \mathbb{C}^2 : |z|^2 + |w|^2 < 1\}.$$

We have

$$\text{Aut}\,(U \times U) = \{g : g(z_1, z_2) = (g_1(z_{\sigma_1}), g_2(z_{\sigma_2})),\ g_1, g_2 \in \text{Aut}\,U\},$$

where σ is a permutation of $\{1, 2\}$. On the other hand,

$$\text{Aut}\,B = \{\varphi \circ g_a : \varphi \in U(2),\ a \in B\}.$$

The isotropy group

$$(\text{Aut}\,B)_0 = \{\varphi \in \text{Aut}\,B : \varphi(0) = 0\} = U(2)$$

is not abelian, whereas, by Example 2.5.5, we have the isotropy group

$$\text{Aut}\,(U \times U)_0 = \{(e^{i\alpha}, e^{i\beta}) : \alpha, \beta \in \mathbb{R}\},$$

which is abelian.

In analysis, one often needs to estimate the norm of the Bergmann operator. In the case of Hilbert spaces and abelian C*-algebras, the norm can be computed exactly from the explicit expression of the operator.

Lemma 3.2.8 *Let D be the open unit ball of a Hilbert space V and let $a \in D$. Then we have*

$$\|B(a, a)^{1/2}\|^2 = \|B(a, a)\| = \begin{cases} (1 - \|a\|^2)^2 & \text{if } \dim V = 1 \\ 1 - \|a\|^2 & \text{if } \dim V \geq 2. \end{cases}$$

Proof The first equality holds, since $B(a, a)$ is self-adjoint. From Example 3.2.5, we have

$$B(a, a)(z) = (1 - \|a\|^2)(z - \langle z, a\rangle a) \qquad (z \in V),$$

where

$$\begin{aligned} \|z - \langle z, a\rangle a\|^2 &= \langle z - \langle z, a\rangle a,\ z - \langle z, a\rangle a\rangle \\ &= \|z\|^2 - 2|\langle z, a\rangle|^2 + |\langle z, a\rangle|^2 \|a\|^2 \\ &< \|z\|^2 - |\langle z, a\rangle|^2 \leq \|z\|^2. \end{aligned}$$

Hence $\|B(a, a)(z)\| \leq (1 - \|a\|^2)\|z\|$.

If $V = \mathbb{C}$, we have actually the equality $|B(a, a)(z)| = (1 - |a|^2)^2 |z|$.

If $\dim V \geq 2$, we can pick a unit vector $z_0 \in V$ orthogonal to a, and therefore

$$\|B(a, a)\| \geq \|B(a, a)z_0\| = \|(1 - \|a\|^2)z_0\| = 1 - \|a\|^2,$$

which proves $\|B(a, a)\| = 1 - \|a\|^2$. $\qquad\square$

If V is a JC*-triple contained in some $L(H)$, then the Bergmann operator has the form

$$B(a, b)(x) = x - 2\{a, b, x\} + \{a, \{b, x, b\}, a\}$$
$$= x - ab^*x - xb^*a + ab^*xb^*a$$
$$= (\mathbf{1} - ab^*)x(\mathbf{1} - b^*a),$$

where $\mathbf{1}$ denotes the identity operator in $L(H)$. In particular, we have

$$B(a, a)(x) = (\mathbf{1} - aa^*)x(\mathbf{1} - a^*a)$$

and

$$B(a, a)^{1/2}(x) = (\mathbf{1} - aa^*)^{1/2}x(\mathbf{1} - a^*a)^{1/2} \qquad (x \in V)$$

This readily gives the estimate

$$\|B(a, a)^{1/2}\| \le \|(\mathbf{1} - aa^*)^{1/2}\|\|(\mathbf{1} - a^*a)^{1/2}\|$$
$$= \|\mathbf{1} - aa^*\|^{1/2}\|\mathbf{1} - a^*a\|^{1/2} \le 1. \tag{3.1}$$

One can compute the norm $\|\mathbf{1} - aa^*\|$ via the Hilbert space H. For each $a \in V \subset L(H)$, we define

$$\alpha(a) = \inf\{\|a\xi\| : \xi \in H, \|\xi\| = 1\} \quad \text{and}$$
$$\beta(a) = \inf\{\|a^*\xi\| : \xi \in H, \|\xi\| = 1\}.$$

Lemma 3.2.9 *Let D be the open unit ball of a JC*-triple $V \subset L(H)$ and let $a \in D$. We have $\|B(a, a)^{1/2}\|^2 \le (1 - \alpha(a)^2)(1 - \beta(a)^2)$.*

Proof This follows from (3.1) and $\|(\mathbf{1} - aa^*)^{1/2}\|^2 = \sup_{\|\xi\|=1} \langle(\mathbf{1} - aa^*)\xi, \xi\rangle = 1 - \beta(a)^2$ as well as $\|(\mathbf{1} - a^*a)^{1/2}\|^2 = 1 - \alpha(a)^2$. $\qquad\square$

For JB*-algebras, one can obtain an estimate of $\|B(a, a)\|$ independent of Hilbert spaces. For a self-adjoint element b in a unital JB*-algebra $\mathcal{A}$, the closed subalgebra generated by b and the identity $\mathbf{1}$ is an unital abelian C*-algebra. The spectrum $\sigma(b)$ of b in the latter algebra is called the *C*-spectrum* of b.

Lemma 3.2.10 *Let D be the open unit ball of a unital JB*-algebra $\mathcal{A}$ and let $a \in D$. Then we have*

$$\|B(a, a)^{1/2}\|^2 \le (1 - \inf \sigma(aa^*))(1 - \inf \sigma(a^*a)),$$

where $\sigma(aa^*)$ and $\sigma(a^*a)$ denote respectively the C-spectra of aa^* and a^*a.*

Proof This can be seen easily from functional calculus, since the closed subalgebra of $\mathcal{A}$ generated by aa^* and the identity $\mathbf{1}$ identifies with the algebra

$C(\sigma(aa^*))$ of complex continuous functions on the spectrum $\sigma(aa^*) \subset [0, \infty)$, where aa^* identifies with the identity function $id : \sigma(aa^*) \to \sigma(aa^*)$. Hence $\|\mathbf{1} - aa^*\| = 1 - \inf \sigma(aa^*)$ and likewise $\|\mathbf{1} - a^*a\| = 1 - \inf \sigma(a^*a)$. □

Example 3.2.11 On the open unit ball D of a JC*-triple V, the Möbius transformation g_a induced by $a \in D$ has the form

$$
\begin{aligned}
g_a(x) &= a + B(a, a)^{1/2}(1 + x \square a)^{-1}(x) \\
&= a + (1 - aa^*)^{1/2}(1 + x \square a)^{-1}(x)(1 - a^*a)^{1/2} \\
&= a + (1 - aa^*)^{1/2}(1 - x \square a + (x \square a)^2 \\
&\quad - (x \square a)^3 + \cdots)(x)(1 - a^*a)^{1/2} \\
&= a + (1 - aa^*)^{1/2}(x - xa^*x + xa^*xa^*x \\
&\quad - xa^*xa^*xa^*x + \cdots)(1 - a^*a)^{1/2} \\
&= a + (1 - aa^*)^{1/2}(1 - xa^* + (xa^*)^2 - (xa^*)^3 + \cdots)x(1 - a^*a)^{1/2} \\
&= a + (1 - aa^*)^{1/2}(1 - xa^*)^{-1}x(1 - a^*a)^{1/2}.
\end{aligned}
$$

Example 3.2.12 Let D be the open unit ball of an abelian C*-algebra, represented as a continuous function space $C_0(\Omega)$ on some locally compact Hausdorff space Ω. Let $a \in D$. The Bergmann operator $B(a, a)$ has the form

$$
\begin{aligned}
B(a, a)(f) &= f - 2\{a, a, f\} + \{a, \{a, f, a\}, a\} \\
&= f - 2|a|^2 f + |a|^2 f |a|^2 \\
&= (1 - |a|^2)^2 f \qquad (f \in C_0(\Omega)),
\end{aligned}
$$

where the last factorisation can be regarded as a convenient shorthand, since $C_0(\Omega)$ may not contain an identity. It follows immediately that

$$
B(a, a)^{1/2}(f) = (1 - |a|^2)f
$$

and

$$
\|B(a, a)^{1/2}\| = \|1 - |a|^2\| = 1 - \inf |a|^2.
$$

We also have

$$
\begin{aligned}
(I + z \square a)^{-1}z &= z - z \square a(z) + (z \square a)^2(z) - (z \square a)^3(z) + \cdots \\
&= z - \bar{a}z^2 + \bar{a}^2 z^3 - \bar{a}^3 z^4 + \cdots \\
&= \frac{z}{1 + \bar{a}z} \qquad (z \in D).
\end{aligned}
$$

Hence the Möbius transformation g_a has the familiar form

$$g_a(z) = a + B(a, a)^{1/2}(\mathbf{1} + z \square a)^{-1}(z)$$
$$= a + \frac{(1 - |a|^2)z}{1 + \bar{a}z}$$
$$= \frac{a + z}{1 + \bar{a}z} \qquad (z \in D).$$

We note that the linear fractional transformation $z \in D \mapsto F_a(z) = (I + z \square a)^{-1}z \in C_0(\Omega)$ has the derivative

$$F'_a(z)(v) = \frac{v}{(1 + z\bar{a})^2} = B(z, -a)^{-1}(v) \qquad (z \in D, v \in C_0(\Omega)),$$

and it follows that

$$g'_a(z) = B(a, a)^{1/2}B(z, -a)^{-1}, \qquad (3.2)$$

which is in fact true for *all* JB*-triples (cf. [70, (2.18)]).

Returning to an arbitrary JB*-triple V, let D be the open unit ball and let $a \in D$. The Bergmann operator $B(a, a)$ is invertible since it has positive spectrum. Although $B(a, a)$ need not be a hermitian operator on V [73, example 4.5], its square root can be expressed as the exponential of a hermitian operator on V. To see this, we make use of the formula $g'_a(0) = B(a, a)^{1/2}$ and write the Möbius transformation g_a as the exponential of a vector field $X : D \longrightarrow V$.

We first identify, via Theorem 3.1.12, the real closed subtriple $R(a)$ generated by a with the JB-triple $C_0(s(a), \mathbb{R})$ of continuous functions on the triple spectrum $s(a)$. Denote by $f(a)$ the element in $R(a)$ corresponding to the function $f \in C_0(s(a), \mathbb{R})$. Let $\alpha = \tanh^{-1}(a) \in R(a)$. Then $R(\alpha) = R(a)$. Let $X(\cdot) = \alpha - \{\cdot, \alpha, \cdot\} \in \mathfrak{p}$ be the unique vector field such that $X(0) = \alpha$. By [70, Proposition 4.6], $g_a = \exp X$ and the one-parameter group $\exp tX : D \longrightarrow D$ satisfies $\exp tX(0) = \tanh(t\alpha) \in R(a)$. Since

$$\frac{d}{dt}(\exp tX(z)) = X(\exp tX(z)) = \alpha - \{\exp tX(z), \alpha, \exp tX(z)\} \qquad (z \in D),$$

differentiation gives

$$\frac{d}{dt}(\exp tX)'(0) = -2\{\exp tX(0), \alpha, (\exp tX)'(0)\}$$
$$= -2(\exp tX(0) \square \alpha)(\exp tX(0)),$$

where $(\exp tX)'(0) : V \longrightarrow V$ is the identity map at $t = 0$. For this initial condition, the solution of the equation is given by

$$(\exp tX)'(0) = \exp\left(-2\int_0^t (\exp sX(0) \square \alpha)ds\right)$$

and in particular,

$$B(a, a)^{1/2} = g'_a(0) = (\exp X)'(0) = \exp\left(-2\int_0^1 (\exp s X(0) \,\square\, \alpha)ds\right).$$

$$(3.3)$$

As noted in the proof of Theorem 2.5.27, the box operator $u \,\square\, v : V \longrightarrow V$ is hermitian for $u, v \in R(a)$ and hence the integral is a hermitian linear operator on V, since the hermitian operators in $L(V)$ form a real closed subspace [10, p. 47].

It follows from Bonsall and Duncan [10, p. 54] that the norm of $B(a, a)^{1/2}$ coincides with its spectral radius, and the same applies to the norm $\|B(a, a)^{-1/2}\|$.

By Lemma 2.5.21, the spectrum of $B(a, a)$ is contained in $(1 - S)(1 - S)$, where $\sup S = \|a \,\square\, a\| = \|a\|^2$. Therefore $\|B(a, a)\| \leq \|B(a, a)^{1/2}\|^2 \leq 1$ and

$$\|B(a, a)^{-1/2}\| \leq \frac{1}{1 - \|a\|^2}. \qquad (3.4)$$

If $V \subset L(H)$ is a JC*-triple, this inequality can be deduced directly from the formula

$$B(a, a)^{-1/2}(x) = (1 - aa^*)^{-1/2}x(1 - a^*a)^{-1/2} \qquad (x \in V).$$

In fact, (3.4) can be strengthened to an equality.

Proposition 3.2.13 *For every a in the open unit ball D of a JB*-triple V, we have*

$$\|B(a, a)^{-1/2}\| = \frac{1}{1 - \|a\|^2}.$$

Proof Let g_a be the Möbius transformation induced by a. We recall that $g_{-a}(a) = 0$ and $g'_{-a}(a) = B(a, a)^{-1/2}$.

Pick a functional $h \in V^*$ satisfying

$$h(a) = \|a\| \quad \text{and} \quad \|h\| = 1.$$

Define a holomorphic function $\varphi : D \longrightarrow U = \{\lambda \in \mathbb{C} : |\lambda| < 1\}$ by

$$\varphi(z) = \frac{h(z) - \|a\|}{1 - \|a\|h(z)} \qquad (z \in D).$$

Then we have

$$\varphi'(z)(v) = \frac{h(v)(1 - \|a\|^2)}{(1 - \|a\|h(z))^2}.$$

As noted before, the Carathéodory norm

$$\|z\|_0 = \sup\{|f'(0)(z)| : f \in H(D, U), f(0) = 0\}$$

coincides with the norm $\|z\|$. Since $\varphi \circ g_{-a}^{-1}(0) = 0$, it follows that, for each $z \in D$,

$$\begin{aligned}
\|B(a, a)^{-1/2}(z)\| &= \|g'_{-a}(a)(z)\|_0 \\
&\geq |(\varphi \circ g_{-a}^{-1})'(0)(g'_{-a}(a)(z))| \\
&= |\varphi'(a) \circ g'_{-a}(a)^{-1} \circ g'_{-a}(a)(z)| \\
&= |\varphi'(a)(z)| = \frac{|h(z)|(1 - \|a\|^2)}{(1 - \|a\|h(a))^2} \\
&= \frac{|h(z)|}{1 - \|a\|^2}.
\end{aligned}$$

Hence

$$\|B(a, a)^{-1/2}\| = \sup\{\|B(a, a)^{-1/2}(z)\| : z \in D\} \geq \frac{1}{1 - \|a\|^2}$$

and by (3.4), the last inequality becomes an equality. □

Corollary 3.2.14 *Let D be the open unit ball of a JB*-triple V. Then we have*

$$\|g'(0)^{-1}\| = \frac{1}{1 - \|a\|^2}$$

for each $g \in \operatorname{Aut} D$ satisfying $g(0) = a$.

Proof By Proposition 3.2.6, every automorphism of D is the composite of a Möbius transformation and a linear isometry. If $g(0) = a$, then $g^{-1} = \ell \circ g_{-a}$ for some linear isometry ℓ on D. Hence

$$\|g'(0)^{-1}\| = \|(g^{-1})'(a)\| = \|g'_{-a}(a)\| = \|B(a, a)^{-1/2}\|.$$

□

As examples of applications of Jordan methods in infinite-dimensional holomorphy, we now use the Bergmann operator and the Möbius transformation to deduce some properties of holomorphic maps on the open unit balls of JB*-triples.

First, the Schwarz–Pick lemma can easily be deduced for JB*-triples.

Lemma 3.2.15 *Let f be a holomorphic self-map on the open unit ball D of a JB*-triple. Then we have*

$$\|g_{-f(w)}(f(z))\| \leq \|g_{-w}(z)\| \qquad (z, w \in D),$$

where g_a denotes the Möbius transformation induced by $a \in D$.

Proof Since $g_{-f(w)} \circ f \circ g_w(0) = 0$, the Schwarz lemma implies

$$\|g_{-f(w)} \circ f \circ g_w(g_{-w}(z))\| \leq \|g_{-w}(z)\| \qquad (z, w \in D),$$

which gives

$$\|g_{-f(w)}(f(z))\| \leq \|g_{-w}(z)\|.$$

$\square$

Theorem 3.2.16 *Let D be the open unit ball of a JB*-triple V and let f : $D \longrightarrow V$ be a biholomorphic map onto $f(D)$, which is convex. Given $f(0) = 0$ and that $f'(0)$ is the identity map, we have, for $a \in D$,*

(i) $\dfrac{1}{(1 + \|a\|)^2} \leq \|f'(a)\| \leq \dfrac{1}{(1 - \|a\|)^2}.$

(ii) $\dfrac{(1 - \|a\|)\|z\|}{(1 + \|a\|)\|B(a, a)^{1/2}\|} \leq \|f'(a)(z)\| \leq \dfrac{\|z\|}{(1 - \|a\|)^2} \qquad (z \in V).$

Proof We make use of the following distortion result in [46]:

$$\frac{1 - \|a\|}{1 + \|a\|}\|z\|_a \leq \|f'(a)(z)\| \leq \frac{1 + \|a\|}{1 - \|a\|}\|z\|_a,$$

where $\| \cdot \|_a$ is the Carathéodory norm which is invariant under the automorphisms of D by (2.39).

Let g_{-a} be the Möbius transformation induced by $-a$. We deduce from Proposition 3.2.13 that

$$\begin{aligned}
\|f'(a)(z)\| &\leq \frac{1 + \|a\|}{1 - \|a\|}\|z\|_a = \frac{1 + \|a\|}{1 - \|a\|}\|g'_{-a}(a)(z)\|_{g_{-a}(a)} \\
&= \frac{1 + \|a\|}{1 - \|a\|}\|B(a, a)^{-1/2}(z)\|_0 \leq \frac{1 + \|a\|}{1 - \|a\|}\left(\frac{\|z\|}{1 - \|a\|^2}\right) \\
&= \frac{\|z\|}{(1 - \|a\|)^2}.
\end{aligned}$$

Likewise we have

$$\|f'(a)(z)\| \geq \frac{1 - \|a\|}{1 + \|a\|}\|B(a, a)^{-1/2}(z)\| \qquad (z \in V),$$

which gives the lower bound in (i).

The lower bound in (ii) follows from

$$\begin{aligned}
\|f'(a)(z)\| &\geq \frac{1 - \|a\|}{1 + \|a\|}\|z\|_a \\
&= \frac{1 - \|a\|}{1 + \|a\|}\|g'_{-a}(a)(z)\|,
\end{aligned}$$

where $\|z\| = \|B(a,a)^{1/2}g'_{-a}(a)(z)\| \leq \|B(a,a)^{1/2}\|\|g'_{-a}(a)(z)\|$ gives

$$\|f'(a)(z)\| \geq \frac{(1-\|a\|)\|z\|}{(1+\|a\|)\|B(a,a)^{1/2}\|}.$$

$\square$

We conclude this section with a brief discussion of the dynamics of a holomorphic map, namely, the behavior of the iterates f^n of a holomorphic self-map f on the open unit ball D of a JB*-triple V. We retain the same notation in the remainder of the section. The objective is to seek invariant domains of f. This can be achieved by means of the Bergmann operator.

Lemma 3.2.17 *For $a, b \in D$, the norm $\|g_{-b}(a)\|$ for the Möbius transformation g_{-b} is related to the Bergmann operator $B(a,b)$ by*

$$\frac{1}{1-\|g_{-b}(a)\|^2} = \|B(a,a)^{-1/2}B(a,b)B(b,b)^{-1/2}\|.$$

Proof Since $g_{g_{-b}(a)}(0) = g_{-b} \circ g_a(0)$, the Cartan uniqueness theorem implies that

$$g_{g_{-b}(a)} = \varphi \circ g_{-b} \circ g_a$$

for some linear isometry φ on D. It follows that

$$\begin{aligned}
B(g_{-b}(a), g_{-b}(a))^{1/2} &= g_{g_{-b}}(a)'(0) \\
&= \varphi \circ g'_{-b}(a) \circ g'_a(0) \\
&= \varphi \circ g'_{-b}(a) \circ B(a,a)^{1/2}
\end{aligned}$$

and hence

$$\frac{1}{1-\|g_{-b}(a)\|^2} = \|B(g_{-b}(a), g_{-b}(a))^{-1/2}\| = \|B(a,a)^{-1/2}g'_{-b}(a)^{-1}\|,$$

where

$$g'_{-b}(a)^{-1} = B(a,b)B(b,b)^{-1/2}$$

by (3.2), which completes the proof. $\square$

Theorem 3.2.18 *Let D be the open unit ball of a JB*-triple V and let $f : D \longrightarrow D$ be a holomorphic map without fixed point. Let f be a compact map; that is, $f(D)$ is relatively compact in V. Then there exists a sequence (z_k) in D converging to a boundary point $\xi \in \partial B$ with $\lim_k f(z_k) = \xi$. Further, if the sequence of operators*

$$(1-\|z_k\|^2)B(z_k, z_k)^{-1/2} : V \longrightarrow V$$

converges uniformly to an operator $T \in L(V)$, then for any $\lambda > 0$, the set

$$D(\xi, \lambda) := \{x \in D : \|B(x, x)^{-1/2} B(x, \xi) T\| < \lambda\}$$

is f-invariant; that is, $f(D(\xi, \lambda)) \subset D(\xi, \lambda)$.

Proof Choose an increasing sequence (α_k) in $(0, 1)$ with limit 1. Let $f_k = \alpha_k f$. Then f_k maps D strictly inside D and by Earle and Hamilton [34], f_k has a fixed point $z_k \in D$. Since $f(D)$ is relatively compact, by choosing a subsequence, we may assume (z_k) converges to some $\xi \in \overline{D}$. If $\xi \in D$, then $f(\xi) = \xi$, which is impossible. Hence $\xi \in \partial D$ and $\lim_k f(z_k) = \xi$.

Let

$$T = \lim_k (1 - \|z_k\|^2) B(z_k, z_k)^{-1/2}.$$

Then $\|T\| = 1$, by Proposition 3.2.13. Let $x \in D$. By Lemma 3.2.15 and Lemma 3.2.17, we have

$$\|B(f_k(x), f_k(x))^{-1/2} B(f_k(x), z_k) B(z_k, z_k)^{-1/2}\|$$
$$\leq \|B(x, x)^{-1/2} B(x, z_k) B(z_k, z_k)^{-1/2}\|$$

and hence

$$\|B(f_k(x), f_k(x))^{-1/2} B(f_k(x), z_k)(1 - \|z_k\|^2) B(z_k, z_k)^{-1/2}\|$$
$$\leq \|B(x, x)^{-1/2} B(x, z_k)(1 - \|z_k\|^2) B(z_k, z_k)^{-1/2}\|$$

for all k. Letting $k \to \infty$, we get

$$\|B(f(x), f(x))^{-1/2} B(f(x), \xi) T\| \leq \|B(x, x)^{-1/2} B(x, \xi) T\|,$$

from which it follows immediately that $f(D(\xi, \lambda)) \subset D(\xi, \lambda)$. $\square$

In the above theorem, we always have

$$\|B(x, x)^{-1/2} B(x, \xi) T\| > 0, \tag{3.5}$$

since $\|T\| = 1$ and $B(x, x)^{-1/2} B(x, \xi)$ is invertible by the remark after (3.2). Also, for $0 < \alpha < 1$, we have

$$\|B(\alpha\xi, \alpha\xi)^{-1/2} B(\alpha\xi, \xi) T\| \leq \frac{(1 - \alpha)^2 \|B(\xi, \xi)\| \|T\|}{1 - \alpha^2} \leq \frac{1 - \alpha}{1 + \alpha},$$

which implies $\alpha\xi \in D(\xi, \lambda)$ for $\lambda > \frac{1-\alpha}{1+\alpha}$. In fact, it can be shown that $D(\xi, \lambda)$ is a convex domain. The invariant domain $D(\xi, \lambda)$ is a key to the study of the iterates of f. We illustrate this with a Hilbert ball, but refer to Mellon [89] for details in the general case as well as relevant literature on iterations of holomorphic maps.

Let D be the open unit ball of a Hilbert space V in Theorem 3.2.18. Then

$$T(y) = \lim_k (1 - \|z_k\|^2) B(z_k, z_k)^{-1/2}(y)$$

$$= \lim_k \left(\sqrt{1 - \|z_k\|^2}\, y + \frac{(1 - \sqrt{1 - \|z_k\|^2})\langle y, z_k \rangle z_k}{\|z_k\|^2} \right) = \langle y, \xi \rangle \xi.$$

Hence

$$B(x, \xi) T(y) = \langle y, \xi \rangle B(x, \xi)(\xi) = \langle y, \xi \rangle (1 - \langle x, \xi \rangle)(\xi - x).$$

It follows that

$$\|B(x, x)^{-1/2} B(x, \xi) T\| = |1 - \langle x, \xi \rangle|\, \|B(x, x)^{-1/2}(\xi - x)\|$$

$$= \frac{|1 - \langle x, \xi \rangle|}{1 - \|x\|^2} \|(1 - \|x\|^2) B(x, x)^{-1/2}(\xi) - x\|.$$

Let (x_k) be a sequence in D converging to some $\zeta \in \partial B$. Then

$$\lim_k (1 - \|x_k\|^2) B(x_k, x_k)^{-1/2}(\xi) = \langle \xi, \zeta \rangle \zeta.$$

Hence

$$\lim_k \|(1 - \|x_k\|^2) B(x_k, x_k)^{-1/2}(\xi) - x_k\| = |1 - \langle \xi, \zeta \rangle| = \lim_k |1 - \langle x_k, \xi \rangle|.$$

It follows that, from some k onwards,

$$2\|B(x_k, x_k)^{-1/2} B(x_k, \xi) T\|$$

$$= \frac{2|1 - \langle x_k, \xi \rangle|}{1 - \|x_k\|^2} \|(1 - \|x_k\|^2) B(x_k, x_k)^{-1/2}(\xi) - x_k\|$$

$$\geq \frac{|1 - \langle x_k, \xi \rangle|^2}{1 - \|x_k\|^2}$$

$$\geq \frac{1}{2}\|B(x_k, x_k)^{-1/2} B(x_k, \xi) T\|.$$

For $\beta > 0$, we define

$$E(\xi, \beta) = \left\{ x \in D : \frac{|1 - \langle x, \xi \rangle|^2}{1 - \|x\|^2} < \beta \right\},$$

which can be viewed as the *ellipsoid*

$$E(\xi, \beta) = \left\{ x \in D : \frac{|\langle x, \xi \rangle - (1 - c)|^2}{c^2} + \frac{\|x - \langle x, \xi \rangle \xi\|^2}{c} < 1 \right\},$$

where $c = \frac{\beta}{1+\beta} < 1$. This implies

$$\overline{E(\xi, \beta)} \cap \partial D = \{\xi\}.$$

The above inequalities give

$$\overline{E(\xi, \lambda/2)} \subset \overline{D(\xi, \lambda)} \subset \overline{E(\xi, 2\lambda)}$$

and in particular

$$\overline{D(\xi, \lambda)} \cap \partial D = \{\xi\}.$$

We have the following extension of the Denjoy–Wolff theorem for $\mathbb{C}$ to Hilbert spaces.

Theorem 3.2.19 *Let $f : D \longrightarrow D$ be a fixed-point-free compact holomorphic map on the open unit ball D of a Hilbert space. Then there exists a unique boundary point $\xi \in \partial D$, called the Wolff point, such that*

$$\lim_{n \to \infty} f^n(x) = \xi \qquad (x \in D),$$

where (f^n) is the sequence of iterates of f. Moreover, this convergence is locally uniform in the sense that the convergence is uniform on each open ball $B \subset D$ satisfying $d(B, \partial D) > 0$.

Proof Uniqueness of the Wolff point is obvious. We show its existence. Let $\xi \in \partial D$ be the boundary point in Theorem 3.2.18. Since f is a compact map, one can show that every subsequence of (f^n) admits a subsequence converging to a function $h : D \longrightarrow \overline{D}$ locally uniformly. This is shown in the next lemma.

To complete the proof, it suffices to show that every subsequential limit h of (f^n) is the constant map $h(\cdot) = \xi$. For this, one first shows that h is a constant map. To highlight the role of the Bergmann operator, we suppress the arguments for this, which have been given in Chu and Mellon [28]. We now show $h(\cdot) = \xi$.

Given $y \in D$ with

$$h(y) = \lim_{k \to \infty} f^{n_k}(y),$$

we must have $f^{n_k}(y) \in \overline{D(\xi, \lambda_y)}$, since $D(\xi, \lambda_y)$ is f-invariant, where

$$\lambda_y = \| B(y, y)^{-1/2} B(y, \xi) T \| > 0.$$

Since f has no fixed point in D, we must have $h(y) \in \partial D$. It follows that $h(y) \in \overline{D(\xi, \lambda_y)} \cap \partial D = \{\xi\}$, which completes the proof.   □

Remark 3.2.20 The example in Stachura [107] shows that the preceding theorem is false without the compactness assumption on f, even if f is biholomorphic.

Lemma 3.2.21 *Let $f : D \longrightarrow D$ be a compact holomorphic map on the open unit ball D of a Hilbert space. Then the sequence (f^n) of iterates has a subsequence converging locally uniformly to a function on D. The same result holds for each subsequence of (f^n).*

Proof Choose a sequence (r_n) in $(0, 1)$ such that $r_n \uparrow 1$ and $f(D) \cap r_1 D \neq \emptyset$. We have

$$f(D) = \bigcup_{n=1}^{\infty} \big(f(D) \cap r_n D \big).$$

We first find a subsequence of (f^n) converging uniformly on $f(D) \cap r_1 D$. By compactness of $\overline{f(D) \cap r_1 D} \subset \overline{f(D)}$, there is a countable set $\{z_n\}$ in $f(D) \cap r_1 D$, which is dense in $\overline{f(D) \cap r_1 D}$.

Since f is compact, (f^n) has a subsequence $(f^{(n,1)})$ such that $\big(f^{(n,1)}(z_1)\big)$ converges. Likewise, $(f^{(n,1)})$ has a subsequence $(f^{(n,2)})$ such that $\big(f^{(n,2)}(z_2)\big)$ converges. Proceed to find subsequences $\big(f^{(n,k)}\big)_n$ which converge at $z_1, \ldots, z_k$. We show that the diagonal sequence $\big(f^{(k,k)}\big)$ converges uniformly on $f(D) \cap r_1 D$. It suffices to show that it is uniformly Cauchy on $f(D) \cap r_1 D$. Let $\varepsilon > 0$. Since $d(r_1 D, \partial D) = 1 - r_1 > 0$, the Cauchy inequality and the mean value theorem give

$$\|h(z) - h(w)\| \leq \frac{\|z - w\|}{1 - r_1} \qquad (z, w \in r_1 D) \tag{3.6}$$

for each holomorphic map $h : D \longrightarrow D$. By compactness, there exist $z_{n_1}, \ldots, z_{n_l}$ in $\{z_n\}$ such that

$$\overline{f(D) \cap r_1 D} \subset \bigcup_{i=1}^{l} D\big(z_{n_i}, \frac{\varepsilon}{3}(1 - r_1)\big),$$

where $D(x, r)$ denotes the open ball centred at x, of radius r. There exists N such that $j, k > N$ implies

$$\big\| f^{(j,j)}(z_{n_i}) - f^{(k,k)}(z_{n_i}) \big\| < \frac{\varepsilon}{3}$$

for $i = 1, \ldots, l$. Hence, for any $z \in f(D) \cap r_1 D$, we have $z \in D(z_{n_i}, \frac{\varepsilon}{3}(1 - r_1))$ for some i, and

$$\begin{aligned}
\big\| f^{(j,j)}(z) - f^{(k,k)}(z) \big\| &\leq \big\| f^{(j,j)}(z) - f^{(j,j)}(z_{n_i}) \big\| + \big\| f^{(j,j)}(z_{n_i}) - f^{(k,k)}(z_{n_i}) \big\| \\
&\quad + \big\| f^{(k,k)}(z_{n_i}) - f^{(k,k)}(z) \big\| \\
&< \frac{\varepsilon(1 - r_1)}{3(1 - r_1)} + \frac{\varepsilon}{3} + \frac{\varepsilon(1 - r_1)}{3(1 - r_1)} = \varepsilon
\end{aligned}$$

whenever $j, k > N$. This shows that $\left(f^{(k,k)}\right)$ is uniformly convergent on $f(D) \cap r_1 D$.

We repeat the diagonal process as follows. Choose a subsequence (f^{n_1}) of (f^n) converging uniformly on $f(D) \cap r_1 D$. Then choose a subsequence (f^{n_2}) of (f^{n_1}) converging uniformly on $f(D) \cap r_2 D$, and so on. The diagonal sequence (f^{n_n}) then converges uniformly on $f(D) \cap r_k D$ for $k = 1, 2, \ldots$

Finally, we show that (f^{n_n+1}) converges locally uniformly on D. Pick $x \in D$ and choose $r, R > 0$ such that $r + R = 1 - \|x\|$ and $\frac{r}{R} < 1 - \|f(x)\|$. Then $D(x, r)$ and $D(f(x), r/R)$ are contained in D. Using (3.6), we see that $d(D(x, r), \partial D) \geq R > 0$ implies

$$f\big(D(x, r)\big) \subset f(D) \cap D\big(f(x), r/R\big) \subset f(D) \cap r_k D$$

for some k. It follows that (f^{n_n}) converges uniformly on $f\big(D(x, r)\big)$ and hence (f^{n_n+1}) converges uniformly on $D(x, r)$.

The last assertion can be proved by the same arguments. $\qquad\square$

3.3 Contractive projections on JB*-triples

Contractive projections on Banach spaces are important objects of study. The Peirce projections induced by a tripotent in a JB*-triple are contractive. In this section, we present a fundamental result which states that the category of JB*-triples is stable under contractive projections. The proof, given in Kaup [72], is an elegant combination of Jordan theory, geometry and complex analysis. This result has many applications and subsumes several functional-analytic results concerning contractive projections on C*-algebras.

Let V be a JB*-triple with open unit ball B and boundary

$$\partial B = \{z \in V : \|z\| = 1\}.$$

As before, *we identify a holomorphic vector field X on B as a holomorphic map $X : B \longrightarrow V$.* Let X be a holomorphic vector field on B such that X can be extended holomorphically to a neighbourhood containing the closure $\overline{B}$. It can be seen that X is complete if and only if, for every boundary point $z_0 \in \partial B$ and every continuous real linear map $\psi : V \longrightarrow \mathbb{R}$ satisfying $\psi(B) > \psi(z_0)$, we have $\psi(X(z_0)) \geq 0$.

Theorem 3.3.1 *Let V be a JB*-triple with triple product $\{\cdot, \cdot, \cdot\}$ and let $P : V \longrightarrow V$ be a contractive projection; that is, $\|P\| \leq 1$ and $P^2 = P$. Then the range $P(V)$ is a JB*-triple in the inherited norm from V and the triple product*

$$\{Px, Py, Pz\}_v := P\{x, y, z\} \qquad (x, y, z \in V).$$

Proof First, $P(V)$ is a Banach space in the inherited norm, since P has closed range. To show that $W = P(V)$ has the structure of a JB*-triple, we need only show that its open unit ball $D = P(B) = B \cap W$ is homogeneous, by Lemma 2.5.11.

Let $z \in V$ and consider the complete analytic vector field $X : B \longrightarrow V$ given by

$$X(u) = z - \{u, z, u\} \qquad (u \in B),$$

where $z = X(0)$. Define an analytic vector field $Y_z : D \longrightarrow V$ by

$$Y_z(v) = P(z - \{v, z, v\}) \qquad (v \in D).$$

Let $v_0 \in \partial D$ and let $\psi : V \longrightarrow \mathbb{R}$ be a continuous real linear map satisfying $\psi(D) > \psi(v_0)$. Then $v_0 \in \partial B$ and $\psi \circ P(B) > \psi(v_0) = \psi \circ P(v_0)$. By completeness of X, we have $\psi(Y_z(v_0)) = \psi \circ P(X(v_0)) \geq 0$. Hence Y_z is a complete vector field on D, by the above remark. The same argument implies that the holomorphic vector field

$$Y_P(v) := Pz - P\{v, Pz, v\} \qquad (v \in D)$$

on D is also complete.

Since $Y_z(0) = Pz = Y_P(0)$ and $Y_z'(0) = Y_P'(0)$, where $Y_z'(v) = -2P\{v, z, \cdot\}$ and $Y_P'(v) = -2P\{v, Pz, \cdot\}$, we must have $Y_z = Y_P$, by Cartan's uniqueness theorem. It follows that

$$P\{Px, z, Py\} = P\{Px, Pz, Py\} \qquad (x, y, z \in V). \tag{3.7}$$

Let aut D be the Lie algebra of complete holomorphic vector fields on D. Then the evaluation map $X \in \text{aut } D \mapsto X(0) \in W$ is surjective, since each $Pz \in W$ is the image of the complete vector field Y_z. It follows from Lemma 2.5.10 that the open unit ball D is homogeneous and W is a JB*-triple, with the triple product $P\{x, y, z\}$. □

Corollary 3.3.2 *Let $\mathcal{A}$ be a C*-algebra with identity $\mathbf{1}$ and let $P : \mathcal{A} \longrightarrow \mathcal{A}$ be a contractive projection such that $P(\mathbf{1}) = \mathbf{1}$. Then the range $P(\mathcal{A})$ is a unital JB*-algebra with Jordan product*

$$x \circ y = \frac{1}{2} P(xy + yx)$$

and involution

$$x^* = P(\{\mathbf{1}, x, \mathbf{1}\}) \qquad (x \in P(\mathcal{A})).$$

If $\mathcal{A}$ is abelian, then $(P(\mathcal{A}), \circ, *)$ is an abelian C*-algebra.

Proof By Definition 1.2.9, the JB*-triple $P(\mathcal{A})$ is a unital Jordan algebra as the a-homotope $P(\mathcal{A})^{(a)}$ of $a = P(1) = 1$, with Jordan product

$$x \circ_a y = \{x, a, y\}_{P(\mathcal{A})} = P(\{x, 1, y\}) = \frac{1}{2} P(xy + yx).$$

For $x \in P(\mathcal{A})$, we have, from (3.7),

$$x^{**} = P\{1, x^*, 1\} = P\{1, P\{1, x, 1\}, 1\} = P\{1, \{1, x, 1\}, 1\} = P(x) = x$$

and

$$\|x^*\| \le \|x\| = \|x^{**}\| \le \|x^*\|.$$

Hence $P(\mathcal{A})$ is a JB*-algebra.

Finally, if $\mathcal{A}$ is abelian, then the product $x \circ y = P(xy)$ is associative and we have

$$\|x\|^3 = \|\{x, x, x\}\| = \|(x \circ x^*) \circ x\| \le \|x \circ x^*\| \|x\| \le \|x\|^2 \|x\|,$$

which implies that $P(\mathcal{A})$ is a C*-algebra. $\qquad\square$

Remark 3.3.3 One can replace the contractive assumption in the above corollary with a positive condition, since a positive linear map P on a unital C*-algebra is continuous with norm $\|P\| = \|P(1)\|$.

One can also use Theorem 3.3.1 to show that the second dual V^{**} of a JB*-triple V is JBW*-triple, since V^{**} is the range of a contractive projection on an ultraproduct of V. We recall that a *filter* on a nonempty set S is a family $\mathcal{F}$ of nonempty subsets of S such that

(i) $A \in \mathcal{F}$ and $A \subset B \Rightarrow B \in \mathcal{F}$;
(ii) $A_1, \ldots, A_n \in \mathcal{F} \Rightarrow A_1 \cap \ldots \cap A_n \in \mathcal{F}$.

A filter is called an *ultrafilter* if it is not properly contained in any other filter. Let $(V_\alpha)_{\alpha \in S}$ be a family of Banach spaces and let $\mathcal{U}$ be an ultrafilter on S. Then for any family $(x_\alpha)_{\alpha \in S}$ of elements in a compact Hausdorff space Ω, the limit

$$\lim_{\mathcal{U}} x_\alpha = x$$

exists in Ω, which means that, for every neighbourhood O of x, the set $\{\alpha \in S : x_\alpha \in O\}$ belongs to $\mathcal{U}$. In particular, for any (v_α) in the ℓ_∞-sum $\bigoplus_\alpha V_\alpha$, the limit $\lim_{\mathcal{U}} \|v_\alpha\|$ exists. Let

$$N = \{(v_\alpha) \in \bigoplus_\alpha V_\alpha : \lim_{\mathcal{U}} \|v_\alpha\| = 0\}.$$

The quotient space

$$(V_\alpha)_\mathcal{U} := \bigoplus_\alpha V_\alpha / N$$

is a Banach space in the quotient norm and is called the *ultraproduct* of the Banach spaces $(V_\alpha)_\alpha$. We note that

$$\|(v_\alpha)_\mathcal{U}\| = \lim_\mathcal{U} \|v_\alpha\|$$

for $(v_\alpha)_\mathcal{U} \in (V_\alpha)_\mathcal{U}$. If each $V_\alpha = V$, then the ultraproduct $(V)_\mathcal{U}$ is called an *ultrapower* of V. We refer to Heinrich [50, proposition 6.7] for the proof of the following fundamental result.

Lemma 3.3.4 *Given any Banach space V, there are an ultrafilter $\mathcal{U}$ and an isometric embedding $J : V^{**} \longrightarrow (V)_\mathcal{U}$ such that $J(V^{**})$ is the range of a contractive projection $P : (V)_\mathcal{U} \longrightarrow V^{**}$ defined by*

$$P(v_\alpha)_\mathcal{U} = \lim_\mathcal{U} v_\alpha,$$

where the limit is taken in the weak* topological space V^{**} and $Jv = (v_\alpha)_\mathcal{U}$ with $v_\alpha = v$ for $v \in V$.

Corollary 3.3.5 *The second dual V^{**} of a JB*-triple V is a JBW*-triple in the triple product*

$$\{u, v, w\} = \lim_\mathcal{U} \{u_\alpha, v_\alpha, w_\alpha\}$$

for $Ju = (u_\alpha)_\mathcal{U}$, $Jv = (v_\alpha)_\mathcal{U}$ and $Jw = (w_\alpha)_\mathcal{U}$, where J is the embedding in Lemma 3.3.4 and the limit is taken in the weak* topological space V^{**}. The triple product of V^{**} restricts to the original triple product of V.

Proof By Example 3.1.4, an ℓ_∞-sum $\bigoplus_\alpha V_\alpha$ of JB*-triples is a JB*-triple. Since $\|\{u, v, w\}\| \leq 2\|u\|\|v\|\|w\|$ by Remark 2.5.28, the closed subspace

$$N = \{(v_\alpha) \in \bigoplus_\alpha V_\alpha : \lim_\mathcal{U} \|v_\alpha\| = 0\}$$

is a subtriple of $\bigoplus_\alpha V_\alpha$. Hence, by Lemma 3.1.18, the ultrapower $(V)_\mathcal{U}$ in Lemma 3.3.4 is a JB*-triple with triple product

$$\{(u_\alpha)_\mathcal{U}, (v_\alpha)_\mathcal{U}, (w_\alpha)_\mathcal{U}\} = (\{u_\alpha, v_\alpha, w_\alpha\})_\mathcal{U}.$$

By Theorem 3.3.1, $J(V^{**})$ is a JB*-triple in the triple product

$$P\{(u_\alpha)_\mathcal{U}, (v_\alpha)_\mathcal{U}, (w_\alpha)_\mathcal{U}\} = P(\{u_\alpha, v_\alpha, w_\alpha\})_\mathcal{U} = \lim_\mathcal{U} \{u_\alpha, v_\alpha, w_\alpha\}$$

for $(u_\alpha)_\mathcal{U}, (v_\alpha)_\mathcal{U}, (w_\alpha)_\mathcal{U} \in J(V^{**})$.

The last assertion is immediate from the description of J in Lemma 3.3.4. $\square$

If $V \subset L(H)$ is a JC*-triple, then the embedding $V^{**} \subset L(H)^{**}$ shows directly that V^{**} is a JW*-triple and that the triple product in V^{**} is separately weak* continuous. In the case of a JB*-triple V, the triple product in V^{**} is also separately weak*-continuous. This fact, proved in Barton and Timoney [7] by a refinement of the ultrafilter $\mathcal{U}$, has the consequence that a JBW*-triple W has a unique predual $W_* \subset W^*$; that is, W_* is the only closed subspace of W^* which is a predual of W in the canonical duality. We provide some details for the latter.

Let W be a JBW*-triple with a predual N such that the triple product $\{\cdot, \cdot, \cdot\}$ is separately continuous on W in the weak topology $\sigma(W, N)$, called the *w*-topology* for convenience in the following discussion. Every w*-closed subtriple of W has a predual and is a JBW*-triple by Corollary 2.5.29. Consider N as a subspace of W^* via the natural embedding. Then N consists of functionals $f \in W^*$ which are w*-continuous on W. The Peirce projections $P_j(e)$ induced by a tripotent $e \in W$ are w*-continuous. In particular, the e-homotope $W_2(e) = P_2(e)W$ is w*-closed and has a predual $W_2(e)_* = N/W_2(e)^0$, where

$$W_2(e)^0 = \{f \in N : f(W_2(e)) = \{0\}\} = \{f \in N : f \circ P_2(e) = 0\}.$$

Therefore we have the identification

$$W_2(e)_* = \{f \circ P_2(e) : f \in N\} = \{f \in N : f = f \circ P_2(e)\}. \quad (3.8)$$

The e-homotope $W_2(e)$ is a JBW*-algebra on which the involution $a^* = \{e, a, e\}$ is w*-continuous. Hence the self-adjoint part

$$W_2(e)_{sa} = \{a \in W_2(e) : a^* = a\}$$

is w*-closed and has a predual. It follows that $W_2(e)_{sa}$ is a JBW-algebra and has a *unique* predual $(W_2(e)_{sa})_*$ by Hanche-Olsen and Størmer [47, 4.4.16]. This enables us to apply properties of JBW-algebras to $W_2(e)$ via

$$W_2(e) = W_2(e)_{sa} + i W_2(e)_{sa}.$$

Denote by $W_2(e)'_{sa}$ the (real) dual space of $W_2(e)_{sa}$.

Given $f \in W_2(e)^*$, define

$$f^*(a) = \overline{f(a^*)} \qquad (a \in W_2(e)).$$

Then $f + f^*$ and $i(f - f^*)$ are real-valued on $W_2(e)_{sa}$ and

$$f = \frac{f + f^*}{2} + i \frac{f - f^*}{2i}$$

with $\frac{f+f^*}{2}|_{W_2(e)_{sa}} \in W_2(e)'_{sa}$ and $\frac{f-f^*}{2i}|_{W_2(e)_{sa}} \in W_2(e)'_{sa}$. Conversely, given $g \in W_2(e)'_{sa}$, we define its complexification $g_c \in W_2(e)^*$ by

$$g_c(a + ib) = g(a) + i g(b) \qquad (a, b \in W_2(e)_{sa})$$

so that $g_c|_{W_2(e)_{sa}} = g$. The mapping $g \in W_2(e)'_{sa} \mapsto g_c \in W_2(e)^*$ identifies $W_2(e)'_{sa}$ as a real subspace of $W_2(e)^*$ and we have the canonical identifications

$$W_2(e)^* = W_2(e)'_{sa} + i W_2(e)'_{sa}$$

and $(W_2(e)_{sa})_* \subset W_2(e)'_{sa}$.

Let $(W_2(e)_*)_h = \{f \in W_2(e)_* : f = f^*\}$ which is a (norm) closed real subspace of $W_2(e)_*$. Let $(W_2(e)_*)'_h$ be the (real) dual space of $(W_2(e)_*)_h$. For each $f \in (W_2(e))_*$, we have $f^* \in (W_2(e))_*$ by (3.8) and hence $f + f^* \in (W_2(e)_*)_h$. It follows that

$$W_2(e)_* = (W_2(e)_*)_h + i(W_2(e)_*)_h \quad \text{and} \quad W_2(e) = (W_2(e)_*)'_h + i(W_2(e)_*)'_h.$$

In the latter identification, $W_2(e)_{sa}$ identifies as a subspace of $(W_2(e)_*)'_h$ via the map $a \in W_2(e)_{sa} \mapsto \tilde{a} \in (W_2(e)_*)'_h$, where $\tilde{a}(f) = f(a)$ for $f \in (W_2(e)_*)_h$.

Given $b \in (W_2(e)_*)'_h \subset W_2(e)$ and $b = b_1 + ib_2$ with $b_1, b_2 \in W_2(e)_{sa}$, and given any $f \in (W_2(e)_*)_h$, we have $f = f^*$ and $f(b) \in \mathbb{R}$. This implies $f(ib_2) = f(b) - f(b_1) \in \mathbb{R}$ and

$$f(ib_2) = f^*(ib_2) = \overline{f(-ib_2)} = -\overline{f(ib_2)} = -f(ib_2),$$

which gives $f(ib_2) = 0$. Hence $g(ib_2) = 0$ for all $g \in W_2(e)_*$ and $b_2 = 0$. Therefore $b = b_1 \in W_2(e)_{sa}$ and we have shown $W_2(e)_{sa} = (W_2(e)_*)'_h$. By the uniqueness of predual, we have $(W_2(e)_*)_h = (W_2(e)_{sa})_*$ and

$$W_2(e)_* = (W_2(e)_{sa})_* + i(W_2(e)_{sa})_*. \tag{3.9}$$

Remark 3.3.6 If $e' \in W$ is a tripotent and if $\varphi : W_2(e) \longrightarrow W_2(e')$ is a surjective linear isometry such that $\varphi(e) = e'$, then φ is a triple isomorphism by Theorem 3.1.7, and the restriction $\varphi|_{W_2(e)_{sa}} : W_2(e)_{sa} \longrightarrow W_2(e')_{sa}$ is a Jordan algebra isomorphism between JBW-algebras. By Hanche-Olsen and Størmer [47, 4.5.6], $\varphi|_{W_2(e)_{sa}}$ is continuous in the weak* topologies of $W_2(e)_{sa}$ and $W_2(e')_{sa}$. Hence φ is w*-continuous on $W_2(e)$ by (3.9).

Let $z \in W$ and let $W(z)$ be the w*-closed subtriple generated by z in the JBW*-triple W. Then $W(z)$ is an abelian JBW*-triple and has a complete tripotent v. By Lemma 1.2.38, $W(z) = W(z)_2(v)$, the Peirce 2-space of $v \in W(z)$, which is an associative JBW*-algebra, and hence an abelian von Neumann algebra, in the homotope product $ab = \{a, v, b\}$ and involution $a^* = \{v, a, v\}$. In particular, z has a polar decomposition

$$z = u(z)|z|$$

where $u(z)$ is a partial isometry in $W(z)_2(u)$ and $u(z)^*u(z)$ is the range projection of $|z|$. In the von Neumann algebra $W(z)_2(v)$, we have

$$u(z)z^*u(z) = u(z)|z|u(z)^*u(z) = u(z)|z| = z.$$

Hence $u(z)$ is a tripotent in W and satisfies $\{u(z), z, u(z)\} = z$. We call $u(z)$ the *support tripotent of z*. Actually we can also find a complete tripotent $e \in W$ satisfying $\{e, z, e\} = z$.

Lemma 3.3.7 *Let W be a JBW*-triple with a predual N and separately $\sigma(W, N)$-continuous triple product. Given any $z \in W$, there is a complete tripotent $e \in W$ such that $\{e, z, e\} = z$.*

Proof If W is an abelian JBW*-triple, we can take any complete tripotent $e \in W$ since, by Lemma 1.2.38, we have $W = W_2(e)$, the Peirce 2-space, in this case.

Consider the case where W is not abelian. Let M be a maximal abelian w*-closed subtriple of W containing z. Then M is an abelian JBW*-triple. Let $e \in M$ be a complete tripotent of M. Then, as before, we have $\{e, z, e\} = z$ and $M \subset W_2(e)$. We show e is a complete tripotent in W, that is, $W_0(e) = \{0\}$. If there exists $x \in W_0(e)\backslash\{0\}$, then the w*-closed subtriple $W(x)$ generated by x is abelian and is contained in $W_0(e)$. The Peirce multiplication rules imply

$$M \,\square\, W(x) := \{a \,\square\, b : a \in M, b \in W(x)\} = \{0\}.$$

Hence M is properly contained in the abelian w*-closed subtriple $\overline{M + W(x)}$ generated by $M + W(x)$. By maximality of M, we have $W = \overline{M + W(x)}$ and W is abelian. This is impossible, and therefore $W_0(e) = \{0\}$. $\square$

Given a family $\{e_\alpha\}_{\alpha \in A}$ of mutually orthogonal tripotents in the JBW*-triple W, let $\mathcal{F}$ be the directed set of finite subsets of A, ordered by inclusion. For $F \in \mathcal{F}$, let $u_F = \sum_{\alpha \in F} e_\alpha$ which is a tripotent by orthogonality. The net $(u_F)_{F \in \mathcal{F}}$ is an increasing net; that is, $F \subset F_1$ implies $u_F \le u_{F_1}$ in the partial ordering $\le$ of tripotents introduced in Definition 1.2.42. By w*-compactness of the unit ball, there is a subnet $(v_\beta)_{\beta \in B}$ of (u_F) w*-converging to some $e \in W$. The subnet $(v_\beta)_{\beta \in B}$ is increasing and for each $F \in \mathcal{F}$, we have $u_F \le v_\beta$ for some $\beta \in B$. For each $\beta_0 \in B$, we have the w*-convergence

$$\{e, v_{\beta_0}, e\} = \lim_\beta \{v_\beta, v_{\beta_0}, e\} = \{v_{\beta_0}, v_{\beta_0}, e\}$$

$$= \lim_\beta \{v_{\beta_0}, v_{\beta_0}, v_\beta\} = \{v_{\beta_0}, v_{\beta_0}, v_{\beta_0}\} = v_{\beta_0},$$

since $v_{\beta_0} \le v_\beta$ eventually. Hence

$$\{e, e, e\} = \lim_\beta \{e, v_\beta, e\} = \lim_\beta v_\beta = e.$$

Therefore e is a tripotent and $v_\beta \in P_2(e)W = W_2(e)$ for all $\beta \in B$. It follows that $u_F \in W_2(e)$ for all $F \in \mathcal{F}$ and (u_F) is an increasing net of projections in the JBW*-algebra $W_2(e)$, or rather, in the JBW-algebra $W_2(e)_{sa}$. By (3.9), the weak* topology $\sigma(W_2(e)_{sa}, (W_2(e)_{sa})_*)$ on $W_2(e)_{sa}$ coincides with the relative w*-topology $\sigma(W_2(e), W_2(e)_*)$. By Hanche-Olsen and Størmer [47, 4.2.9], the net (u_F) weak* converges to a limit in $W_2(e)_{sa}$ and this limit must be e. We write

$$\sum_{\alpha \in A} e_\alpha = e = \lim_F u_F.$$

Lemma 3.3.8 *Let W be a JBW*-triple with a predual N and separately $\sigma(W, N)$-continuous triple product. Let $f \in W^*$. The following conditions are equivalent:*

(i) $f \in N$.

(ii) $f \circ P_2(e) \in W_2(e)_*$; *equivalently*, $f|_{W_2(e)} \in W_2(e)_*$, *for every tripotent $e \in W$.*

(iii) f *is completely additive; that is,* $f\left(\sum_\alpha e_\alpha\right) = \sum_\alpha f(e_\alpha)$ *for every family (e_α) of orthogonal tripotents in W.*

Proof (i) $\Rightarrow$ (ii). This follows from (3.8).

(ii) $\Rightarrow$ (iii). By the remark before this lemma, $e = \sum_\alpha e_\alpha$ is a tripotent and is the $\sigma(W_2(e)_{sa}, (W_2(e)_{sa})_*)$-limit of the increasing net $(\sum_{\alpha \in F} e_\alpha)_F$ of projections in the JBW-algebra $(P_2(e)W)_{sa}$. Since $f \circ P_2(e) \in W_2(e)_* = (W_2(e)_{sa})_* + i(W_2(e)_{sa})_*$, we have

$$f\left(\sum_\alpha e_\alpha\right) = f \circ P_2(e)\left(\sum_\alpha e_\alpha\right) = \lim_F \sum_{\alpha \in F} f \circ P_2(e)(e_\alpha) = \sum_\alpha f(e_\alpha).$$

(iii) $\Rightarrow$ (i). Let $e \in W$ be any tripotent and consider the JBW*-algebra $W_2(e) = P_2(e)W$. For any orthogonal family (p_α) of projections in $W_2(e)$, condition (iii) implies that $f\left(\sum_\alpha p_\alpha\right) = \sum_\alpha f(p_\alpha)$; that is, f is completely additive on the JBW*-algebra $W_2(e)$. Analogously to the case of von Neumann algebras and as noted in Horn [57, 3.18], it can be shown that $f|_{W_2(e)}$ is $\sigma(W_2(e), W_2(e)_*)$-continuous; that is, $f \circ P_2(e) \in W_2(e)_*$.

To show $f \in N$, it suffices to show f restricts to a w*-continuous function on the closed unit ball D of W, by the Krein–Smulyan theorem.

Let (z_β) be a net in D w*-converging to $z \in D$. We need to show $f(z) = \lim_\beta f(z_\beta)$. Let $\varepsilon > 0$. By the Bishop–Phelps theorem [9], there exist $g \in W^*$

and $a \in D$ such that $\|g\| = g(a)$ and $\|g - f\| < \varepsilon/4$. By Lemma 3.3.7, there is a complete tripotent $e \in W$ such that $a = \{e, a, e\}$. By w*-continuity of $f \circ P_2(e)$, we have

$$|f \circ P_2(e)(z_\beta - z)| < \varepsilon/2$$

from some β onwards. Since $\|g \circ P_2(e)\| = \|g\|$, we have $g \circ P_2(e) = g$ by Lemma 3.3.14 below. It follows that, from some β onwards, we have

$$
\begin{aligned}
|f(z_\beta - z)| &\le |f \circ P_2(e)(z_\beta - z)| + |f \circ P_1(e)(z_\beta - z)| \\
&\le \frac{\varepsilon}{2} + |(f - g) \circ P_1(e)(z_\beta - z)| + |g \circ P_1(e)(z_\beta - z)| \\
&\le \frac{\varepsilon}{2} + \frac{\varepsilon}{2} = \varepsilon,
\end{aligned}
$$

where $\|z_\beta - z\| \le 2$ and $g \circ P_1(e) = 0$. $\qquad\square$

Up to this point, the above lemma requires the condition of separate $\sigma(W, N)$-continuity of the triple product. However, using Lemma 3.3.8 and Remark 3.3.6, together with separate weak* continuity of the triple product on the second dual V^{**} of a JB*-triple V, one can now follow the arguments in Barton and Timoney [7] to show that every JBW*-triple W has a *unique* predual W_* and hence it is unambiguous to speak of the weak* topology $\sigma(W, W_*)$ on W. We show next that uniqueness of the predual of W actually implies that the triple product on W is always separately weak* continuous. Consequently, one can replace N in Lemma 3.3.8 by W_* and also omit the assumption of separate w*-continuity of the triple product.

Theorem 3.3.9 *Let W be a JBW*-triple. Then the triple product in W is separately weak* continuous.*

Proof The uniqueness of the predual of W enables us to apply the fact that surjective linear isometries on W are weak* continuous. Let $a \in W$. Then the isometries $\exp it(a \square a) : W \longrightarrow W$ are weak* continuous for $t \in \mathbb{R}$. Therefore the box operator

$$ia \square a = \left.\frac{d}{dt}\right|_{t=0} \exp it(a \square a) = \lim_{t \to 0} \frac{1}{t}(\exp it(a \square a) - I)$$

is weak* continuous, where the limit is in operator norm and I is the identity map on W. For any $b \in W$, weak*-continuity of the box operators $(a + b) \square (a + b)$ and $(a + ib) \square (a + ib)$ implies that of $a \square b$. In particular, the Peirce projections on W are weak* continuous. To complete the proof, it suffices to show weak* continuity of the quadratic operator Q_a for $a \in W$.

Let $f \in W_* \subset W^*$. Then weak* continuity implies that f achieves its norm at an extreme point e of the closed unit ball of W. By Theorem 3.2.3, e is a (complete) tripotent in W. By Lemma 3.3.14 below, we have $f = f \circ P_2(e)$. The quadratic map Q_e restricts to the involution on the JBW*-algebra $P_2(e)W$ and is weak* continuous on $P_2(e)W$. It follows that Q_e is weak* continuous on W, since it is the composite map $Q_e|_{P_2(e)W} \circ P_2(e)$.

Next, the composite map $Q_e \circ Q_a$ is weak* continuous, since

$$Q_e \circ Q_a = 2(e \,\square\, a)^2 - e \,\square\, \{a, e, a\}.$$

Hence $f \circ Q_a = f \circ P_2(e) \circ Q_a = f \circ Q_e \circ (Q_e \circ Q_a)$ is weak* continuous. Since $f \in W_*$ was arbitrary, we conclude that Q_a is weak* continuous. □

A linear functional f on a JBW-algebra A is called *normal* if it preserves bounded increasing nets in A. By [47, 4.5.6], the normal functionals on A are exactly the $\sigma(A, A_*)$-continuous functionals on A, which form the predual A_* of A. A weak* continuous linear functional on a JBW*-triple W is sometimes called a *normal* functional. In this context, normality is synonymous with complete additivity.

It is well known that the predual of a von Neumann algebra is weakly sequentially complete [101]. The proof of this fact in Akemann [1] carries over to preduals of JBW*-algebras.

Proposition 3.3.10 *Let W be a JBW*-triple. Then its predual W_* is weakly sequentially complete.*

Proof Let (f_n) be a weakly Cauchy sequence in $W_* \subset W^*$. Then the sequence is bounded and hence weak* converges to some $f \in W^*$. For each tripotent $e \in W$, the sequence $(f_n|_{P_2(e)W})$ is weakly Cauchy in the predual $(P_2(e)W)_*$ of the JBW*-algebra $P_2(e)W$ and therefore weakly converges to some $h \in (P_2(e)W)_*$. It follows that $f|_{P_2(e)W} = h$. Hence $f \in W_*$ by Lemma 3.3.8 and (f_n) converges to f weakly. □

Let W be a JBW*-triple. Given an element $a \in W$, we denote by $J(a)$ the smallest weak* closed triple ideal of W containing a. A tripotent $u \in W$ is called *abelian* if $J(u)$ is an abelian Jordan triple system as defined in Definition 1.2.12. We call u a *minimal tripotent* if $P_2(u)W = \mathbb{C}u$.

Definition 3.3.11 A JBW*-triple W is said to be of *type* I if it admits an abelian tripotent u such that $W = J(u)$.

A JBW*-triple W is called a *factor* if it does not contain any weak* closed triple ideal other than $\{0\}$ and W. We note that, however, a factor *can* contain a non-trivial norm-closed triple ideal. For example, the factor $L(H)$ of bounded operators on a Hilbert space H contains the norm-closed ideal $K(H)$ of compact operators on H. A minimal tripotent u in a JBW*-triple is clearly abelian. It follows that a factor must be of type I if it contains a minimal tripotent.

The structures of JBW*-triples can be described concretely. Type I JBW*-triples have been classified in Horn [56], where the following two fundamental structure theorems are proved.

Theorem 3.3.12 *A type I factor is triple isomorphic to one of the six types of Cartan factor in Example 2.5.31.*

Theorem 3.3.13 *A type I JBW*-triple is triple isomorphic to an ℓ_∞-sum*

$$\bigoplus_\alpha C(\Omega_\alpha, C_\alpha),$$

where Ω_α is a hyerstonean space, C_α is a Cartan factor and $C(\Omega_\alpha, C_\alpha)$ is the complex Banach space of C_α-valued continuous functions on Ω_α, which is a JBW*-triple in the pointwise triple product

$$\{f, g, h\}(\omega) = \{f(\omega), g(\omega), h(\omega)\} \qquad (f, g, h \in C(\Omega_\alpha, C_\alpha), \omega \in \Omega_\alpha).$$

The preceding theorems play a useful role in the theory of JB*-triples. We will show that a JB*-triple embeds as a norm-closed subtriple of a type I JBW*-triple. For this, we need some preliminary results.

Lemma 3.3.14 *Let u be a tripotent in a JB*-triple V with Peirce projections $P_0(u)$ and $P_2(u)$. Then for each $f \in V^*$, we have*

(i) $\|f \circ P_0(u) + f \circ P_2(u)\| = \|f \circ P_0(u)\| + \|f \circ P_2(u)\|$;
(ii) $\|f \circ P_2(u)\| = \|f\|$ *implies* $f \circ P_2(u) = f$.

Proof (i) Since the Peirce projections are contractive, we have

$$\|P_0(u)x\| = \|P_0(u)(P_0(u)x + P_2(u)x)\| \le \|P_0(u)x + P_2(u)x\|$$

and likewise $\|P_2(u)x\| \le \|P_0(u)x + P_2(u)x\|$ for all $x \in V$. Let $\varepsilon > 0$. Pick $x_0 \in P_0(u)V$ and $x_2 \in P_2(u)V$ such that $\|x_0\|, \|x_2\| \le 1$ and

$$f \circ P_0(u)(x_0) + \varepsilon \ge \|f \circ P_0(u)\|, \qquad f \circ P_2(u)(x_2) + \varepsilon \ge \|f \circ P_2(u)\|.$$

Then we have

$$(f \circ P_0(u) + f \circ P_2(u))(x_0 + x_2) = f \circ P_0(u)(x_0) + f \circ P_2(u)(x_2)$$
$$\ge \|f \circ P_0(u)\| + \|f \circ P_2(u)\| - 2\varepsilon.$$

By Corollary 3.1.21, $\|x_0 + x_2\| = \max\{\|x_0\|, \|x_2\|\} \leq 1$, and it follows that

$$\|f \circ P_0(u) + f \circ P_2(u)\| \geq \|f \circ P_0(u)\| + \|f \circ P_2(u)\| - 2\varepsilon,$$

which proves (i).

(ii) Since $P_0(u) + P_2(u)$ is contractive, we have $f \circ P_0(u) = 0$ by (i). To see that $f \circ P_1(u) = 0$, we show $f(x) = 0$ for all $x \in P_1(u)V$. We may assume $\|f\| = 1 \geq \|x\|$ and $f(x) \geq 0$. Let $1 > \varepsilon > 0$ and pick $z \in P_2(u)V$ such that $\|z\| = 1$ and $f(z) \geq 1 - \varepsilon$. We have

$$\|z + tx\| \geq f(z + tx) \geq 1 - \varepsilon + tf(x) \qquad (t \in \mathbb{R}).$$

Since $\{z, x, z\} = 0$ by the Peirce multiplication rule, we have

$$\|\{z + tx, z + tx, z + tx\}\| \leq \|\{z, z, z\}\| + 2|t|\|z \,\square\, z(x)\| + O(|t|^2)$$

and iterating gives the inequality

$$\|(z + tx)^{3^n}\| \leq \|z^{3^n}\| + 2|t|\|x\| + O(|t|^2)$$

for the odd powers. It follows that

$$|(1 - \varepsilon)^{3^n} + 3^n tf(x)(1 - \varepsilon)^{3^n - 1} + O(t^2)|$$
$$= |(1 - \varepsilon + tf(x))^{3^n}| \leq \|z + tx\|^{3^n}$$
$$= \|(z + tx)^{3^n}\| \leq 1 + 2^n t\|x\| + O(|t|^2).$$

Since $\varepsilon > 0$ was arbitrary, we have

$$f(x) + O(|t|) \leq \left(\frac{2}{3}\right)^n \|x\| + O(|t|)$$

for all n and $|t| > 0$. Hence $f(x) = 0$ and the proof is complete. $\qquad\square$

We continue to use the partial ordering $\leq$ introduced in Definition 1.2.42 for tripotents.

Lemma 3.3.15 *Let W be a JBW*-triple and let f be an extreme point of the closed unit ball of the predual of W. Then there is a unique minimal tripotent $u \in W$ such that $f(u) = 1$.*

Proof The Hahn–Banach theorem implies that the set $F = \{w \in W : \|w\| = 1 = f(w)\}$ is a non-empty weak* closed face of the closed unit ball of W. Hence F contains an extreme point w of the closed unit ball. By Theorem 3.2.3, w is a tripotent. The w-homotope $P_2(w)W$ is a JBW*-algebra and the restriction $f|_{P_2(w)W}$ is a normal state of $P_2(w)W$ since

$$\|f\| = 1 = f(w).$$

We show that $f|_{P_2(w)W}$ is an extreme point of the closed unit ball B of the predual of $P_2(w)W$.

Let $f|_{P_2(w)W} = \frac{1}{2}(g+h)$ for $g, h \in B$. Then

$$f = f \circ P_2(w) = \frac{1}{2}g \circ P_2(w) + \frac{1}{2}h \circ P_2(w)$$

since $\|f \circ P_2(w)\| = 1 = \|f\|$. Hence $f = g \circ P_2(w) = h \circ P_2(w)$. It follows that there is a unique minimal projection u in $P_2(w)W$ such that $f(u) = f|_{P_2(w)W}(u) = 1$. Plainly u is a tripotent in W. It is also minimal, since

$$P_2(u)W = P_2(u)P_2(w)W = \mathbb{C}u.$$

To see that u is unique, let $v \in W$ be a minimal tripotent satisfying $f(v) = 1$. We have $P_2(u)v = \alpha u$ and $1 = f(v) = f \circ P_2(u)(v) = \alpha$. Hence $P_2(u)v = u$ or $u \leq v$. Likewise $P_2(v)u = v$ and therefore $v = u$. $\qquad\square$

We call the above minimal tripotent u the *support tripotent* of f and note that

$$f(\cdot) = f\{u, \cdot, u\}, \quad \{u, \cdot, u\} = f(\cdot)u. \qquad (3.10)$$

Indeed, $\|f \circ P_2(u)\| = 1$ implies $f = f \circ P_2(u)$. Given $P_2(u)x = \alpha u$, we have $\{u, x, u\} = \{u^3, x, u^3\} = \{u, P_2(u)x, u\} = \alpha u$ and $f(x) = f(P_2(u)(x)) = \alpha f(u) = \alpha$.

Lemma 3.3.16 *Let J be a weak* closed triple ideal in a JBW*-triple W. Then there is a weak* closed triple ideal $J^\square$ in W such that $J^\square \square J = \{0\}$ and $W = J \oplus J^\square$.*

Proof Let u be a complete tripotent in the JBW*-triple J. Then u is a tripotent in W. Let $P_j(u) : W \longrightarrow W$ be the Peirce projections. For $x = x_1 + x_2 \in P_1(u)W + P_2(u)W$, we have $jx_j = 2\{u, u, x\} \in J$ and hence

$$J = P_1(u)W + P_2(u)W.$$

Let $J^\square = P_0(u)W$ which is a weak* closed subtriple of W. We have $J^\square \square P_2(u)W = \{0\}$. For $y \in J^\square$ and $z \in P_1(u)W$, we have $\{z, z, y\} \in P_0(u)W \cap J = \{0\}$. Hence $J^\square \square J = \{0\}$ and it follows that $\{W, J^\square, J^\square\} = \{J \oplus J^\square, J^\square, J^\square\} \subset J^\square$; that is, $J^\square$ is a triple ideal in W. $\qquad\square$

Remark 3.3.17 In the preceding lemma, it is easy to see that $J^\square = \{w \in W : w \square J = \{0\}\}$ since the latter set contains $J^\square$ and has intersection $\{0\}$ with J.

Lemma 3.3.18 *Let u be a minimal tripotent in a JBW*-triple W. Then the smallest weak* closed triple ideal $J(u)$ in W containing u is a type I factor.*

If $v \in W$ *is another minimal tripotent, then either* $J(u) = J(v)$ *or* $J(u) \square J(v) = \{0\}$.

Proof If J is a nonzero proper weak* closed triple ideal in $J(u)$, then Lemma 3.3.16 implies $J(u) = J \oplus J^\square$ for some nonzero proper weak* closed triple ideal $J^\square$ with $J^\square \square J = \{0\}$. Hence $u = u_1 + u_2 \in J \oplus J^\square$ and u_1, u_2 must be tripotents, contradicting minimality of u. Therefore $J(u)$ is a factor which is of type I. The second assertion is an immediate consequence. $\square$

Let W be a JBW*-triple. It follows from Lemma 3.3.16, Lemma 3.3.18 and Corollary 3.1.21 that W can be decomposed into an ℓ_∞-sum

$$W = \left(\bigoplus_u J(u) \right) \oplus \left(\bigoplus_u J(u) \right)^\square \tag{3.11}$$

of two weak* closed triple ideals, where the first summand sums over all minimal tripotents u in W and is called the *atomic part* of W. By Theorem 3.3.12, it is triple isomorphic to an ℓ_∞-sum of Cartan factors. The second summand does not contain any minimal tripotent.

Theorem 3.3.19 *A JB*-triple V is triple isomorphic to a closed subtriple of an ℓ_∞-sum of Cartan factors.*

Proof By (3.11), the second dual V^{**} is an ℓ_∞-sum

$$V^{**} = V_a \oplus V_a^\square$$

of two weak* closed triple ideals in which V_a is an ℓ_∞-sum of Cartan factors. Let $P : V^{**} \longrightarrow V_a$ be the canonical contractive projection and $\widehat{\ } : V \longrightarrow V^{**}$ the canonical embedding. Both maps are triple homomorphisms and therefore $P(\widehat{V})$ is a closed subtriple of V_a. If $P(\widehat{v}) = 0$, then $\widehat{v} \in V_a^\square$. For each extreme point f in the closed unit ball of V^* with support tripotent $u_f \in V_a$, we have

$$f(\widehat{v}) = f\{u_f, \widehat{v}, u_f\} = 0.$$

Hence $\widehat{v} = 0$ and $P \circ \widehat{\ } : V \longrightarrow P(\widehat{V})$ is a triple isomorphism. $\square$

Remark 3.3.20 The above proof reveals that $\widehat{V} \subset V_a$ and indeed, if V is a C*-algebra, then $\widehat{\ } : V \longrightarrow V_a$ is just the atomic representation of V. For each minimal tripotent $u \in V^{**}$, let $P_u : V^{**} \longrightarrow J(u)$ be the canonical projection. Then the map $P_u \circ \widehat{\ } : V \longrightarrow J(u)$ is called a *factor representation* of V which generalizes the notion of a factor representation of a C*-algebra.

An immediate consequence of Theorem 3.3.19 is that the inequality

$$\|\{x, y, z\}\| \leq \|x\| \|y\| \|z\|$$

holds for all elements x, y and z in a JB*-triple, since this is so in all Cartan factors.

Finally, to describe the structure of an arbitrary JBW*-triple, we need to recall some definitions first. A von Neumann algebra $\mathcal{A}$ is said to be *continuous* if it does not contain a nonzero projection p such that $p\mathcal{A}p$ is abelian. If $\beta : \mathcal{A} \longrightarrow \mathcal{A}$ is a linear *-antiautomorphism of period 2, we let

$$H(\mathcal{A}, \beta) = \{a \in \mathcal{A} : \beta(a) = a\},$$

which is a weak* closed subtriple of $\mathcal{A}$ since β is weak* continuous. We can now state the following structure theorem, which has been proved in Horn and Neher [58].

Theorem 3.3.21 . *A JBW*-triple is triple isomorphic to an ℓ_∞-sum*

$$\bigoplus_\alpha C(\Omega_\alpha, C_\alpha) \oplus \mathcal{R} \oplus H(\mathcal{A}, \beta)$$

where $\mathcal{R}$ is a weak closed right ideal of a continuous von Neumann algebra and $H(\mathcal{A}, \beta)$ is as defined earlier.*

Example 3.3.22 A complemented subspace E of a Banach space V is said to be 1-*complemented* if there is a contractive projection from V onto E. Weak* closed triple ideals in JBW*-triples are 1-complemented. Looking at each Cartan factor in Example 2.5.31, we observe that the predual of a Cartan factor is 1-complemented in the predual of a JBW*-algebra (see also Example 1.2.8). Using the preceding structure theorem, it is not difficult to see that the predual W_* of a JBW*-triple W is 1-complemented in the predual of a JBW*-algebra and W_* is also a complemented subspace of the predual $\mathcal{A}_*$ of a von Neumann algebra $\mathcal{A}$. Moreover, if W does not contain as a subtriple the exceptional Cartan factors $M_{1,2}(\mathcal{O})$ and $H_3(\mathcal{O})$, then W_* is 1-complemented in $\mathcal{A}_*$ for some von Neumann algebra $\mathcal{A}$. Consequently, a JBW*-triple W is isomorphic as a Banach space to a complemented subspace of a von Neumann algebra, and W is triple isomorphic to a 1-complemented subtriple of a von Neumann algebra if and only if W does not contain the exceptional Cartan factors. We refer to Chu and Iochum [25] for more details.

Example 3.3.23 A linear projection $P : W \longrightarrow W$ on a JBW*-triple is called a *structural projection* if it satisfies

$$P\{a, Pb, c\} = \{Pa, b, Pc\} \qquad (a, b, c \in W).$$

Such a projection is contractive and weak* continuous. In fact, it has been shown in Edwards *et al.* [36] that a subtriple of W is a weak* closed inner ideal if and only if it is the range of a unique structural projection on W, and if W is a von Neumann algebra, then the structural projections on W are of the form

$$P : w \in W \mapsto pwq \in W,$$

where $p, q \in W$ are projections with equal central support.

3.4 Isometries between JB-triples

Non-surjective linear isometries between Banach spaces often occur in applications. This section discusses the case of JB*-triples and the larger class of JB-triples. We begin by extending Theorem 3.1.20 to triple monomorphisms.

Theorem 3.4.1 *Let $\gamma : V \longrightarrow W$ be a Jordan triple monomorphism between two JB*-triples, V and W. Then γ is a linear isometry. The same conclusion holds if V and W are JB-triples.*

Proof Let $a \in V \backslash \{0\}$ and let $V(a)$ be the JB*-subtriple generated by a in V. Then $\varphi(V(a))$ is the closed subtriple $W(\gamma(a))$ generated by $\gamma(a)$ in W. The restriction $\gamma : V(a) \longrightarrow W(\gamma(a))$ is a triple isomorphism between JB*-triples and hence an isometry by Theorem 3.1.20.

If V and W are JB-triples, then by considering the real closed subtriple $R(a)$ generated a and the restriction of γ to $R(a)$, and applying Theorem 3.1.20, one also concludes that $\|\gamma(a)\| = \|a\|$. $\qquad\qquad\square$

We have seen from Theorem 3.1.7 that the converse of Theorem 3.4.1 holds if γ is surjective. Does a *non-surjective* linear isometry between JB*-triples still preserve the triple product? The answer is negative and in fact, it is easy to find a simple counterexample. Here is one.

Example 3.4.2 Let $\gamma : \mathbb{C} \longrightarrow M_2(\mathbb{C})$ be defined by

$$\gamma(a) = \begin{pmatrix} 0 & \frac{a}{2} \\ a & 0 \end{pmatrix}.$$

Then γ is a linear isometry and $\gamma(\mathbb{C})$ is not a subtriple of $M_2(\mathbb{C})$. Also, $\gamma(1)$ is not unitary in $M_2(\mathbb{C})$ and $\gamma(\mathbb{C})$ contains no nonzero positive matrix.

Nevertheless, in view of the fact that the norm and the triple structure determine each other in JB*-triples, it would be interesting to see how a non-surjective isometry is related to the underlying Jordan structure. This is the question concerning us. We are going to show that a linear isometry

$\varphi : V \longrightarrow W$ between JB-triples is, none the less, *locally* a triple monomorphism; that is, it preserves the triple product after a reduction by a tripotent in the second dual W''. This interesting phenomenon is in contrast to the fact that even a surjective linear isometry between JB-triples need not preserve the triple product *globally*, as noted in Example 3.1.10 (see also Example 3.4.10). We first prove some basic results which will be needed later.

A JB-triple V is always considered as a closed subtriple of its second dual V'' via the canonical embedding.

Lemma 3.4.3 *Let f be a continuous linear functional on a JB*-triple V such that $\|f\| = 1 = f(u)$ for some $u \in V^{**}$. Then we have the Cauchy–Schwarz inequality*

$$|f\{x, y, u\}| \leq f\{x, x, u\} f\{y, y, u\}$$

for all $x, y \in V$.

Proof Define a sesquilinear form $\langle \cdot, \cdot \rangle : V^2 \longrightarrow \mathbb{C}$ by

$$\langle x, y \rangle = f\{x, y, u\} \qquad (x, y \in V).$$

The inequality will follow once it is shown that the form is Hermitian and positive semidefinite.

For each $x \in V$, the box operator $x \,\square\, x$ is hermitian with non-negative spectrum $\sigma(x \,\square\, x)$. The numerical range $N(x \,\square\, x)$ is the convex hull of $\sigma(x \,\square\, x)$ and therefore resides in $[0, \infty)$. The functional $a \in L(V) \mapsto f(a(u)) \in \mathbb{C}$ has unit norm and value 1 at the identity $\mathbf{1} \in L(V)$. It follows that

$$\langle x, x \rangle = f(x \,\square\, x(u)) \geq 0$$

for all $x \in V$. Expanding the real number

$$f\{x + y, \, x + y, \, u\} - f\{x - y, \, x - y, \, u\},$$

we find that $\operatorname{Im} f\{x, y, u\} = -\operatorname{Im} f\{y, x, u\}$. Repeating this with $f\{x + iy, \, x + iy, \, u\} - f\{x - iy, \, x - iy, \, u\}$, we get $\operatorname{Re} f\{x, y, u\} = \operatorname{Re} f\{y, x, u\}$. Hence $f\{x, y, u\} = \overline{f\{y, x, u\}}$ and $\langle \cdot, \cdot \rangle$ is Hermitian. $\square$

Let W^τ be a real form of a JB*-triple W; that is, W^τ is a JB-triple as defined in Definition 3.1.8. We call W^τ a *JBW-triple* or a *real JBW*-triple* if W is a JBW*-triple. Using the conjugation τ, many properties of the JBW*-triple W can be carried over to W^τ. For instance, it can be shown that W^τ has a unique predual and that $\{u \in W : \tau(u) = u = \{u, u, u\}\}$ is the set of tripotents in W^τ [37]. In fact, $\tau : W \longrightarrow W$ is weak* continuous and its predual $\sigma : W_* \longrightarrow W_*$

is a conjugate linear isometry such that the predual W_*^τ of W^τ is given by

$$W_*^\tau = \{\varphi \in W_* : \sigma(\varphi) = \varphi\} \quad \text{and} \quad W_* = W_*^\tau \oplus i W_*^\tau.$$

Now let V^τ be a JB triple which is a real form of a JB*-triple V. Then its (real) dual space $(V^\tau)'$ identifies with the fixed point set

$$(V^\tau)' = \{\varphi \in V^* : \tau^*(\varphi) = \varphi\}$$

and the second dual $(V^\tau)'' = (V^{**})^{\tau^{**}}$ is a JBW-triple. Let $\varphi \in (V^\tau)'$ be an extreme point of the closed unit ball of $(V^\tau)'$. Since τ^* has period 2, it is easily seen that φ is also an extreme point of the closed unit ball of V^*. By Lemma 3.3.15, φ has a support u_φ which is the unique minimal tripotent in V^{**} satisfying $\varphi(u_\varphi) = 1$. It is evident that $\tau^{**}(u_\varphi)$ is also a minimal tripotent in V^{**}. Since

$$\varphi(\tau^{**}(u_\varphi)) = (\tau^*\varphi)(u_\varphi) = \varphi(u_\varphi) = 1,$$

we must have $\tau^{**}u_\varphi = u_\varphi$ by uniqueness; that is, u_φ belongs to the JBW-triple $(V^\tau)''$ and we also have the Cauchy–Schwarz inequality

$$|\varphi\{x, y, u_\varphi\}| \leq \varphi\{x, x, u_\varphi\}\varphi\{y, y, u_\varphi\}$$

for all $x, y \in V^\tau$.

Lemma 3.4.4 *Let V be a JB-triple and let φ be an extreme point of the closed unit ball of the dual V', with support tripotent $u \in V''$. Let*

$$N_\varphi = \{b \in V'' : \varphi\{b, b, u\} = 0\}.$$

Then $N_\varphi = P_0(u)V''$, which is the Peirce 0-space of the minimal tripotent u.

Proof Let $x \in P_0(u)V''$. By the Peirce multiplication rules in Theorem 1.2.44, we have $\{x, x, u\} = 0$ and hence $x \in N_\varphi$.

Conversely, let $b \in N_\varphi$. By the Cauchy–Schwarz inequality, we have $\varphi\{u, b, u\} = 0$ and hence minimality of u implies $P_2(u)(b) = 0$ by (3.10), and also $\{u, b, u\} = 0$. It remains to show $b_1 := P_1(u)(b) = 0$. Since

$$\varphi\{u, b_1, b\} = \varphi\{u, b_1 + P_0(u)(b), b\} = \varphi\{u, b, b\} = 0,$$

we have $\varphi\{u, b_1, b_1\} = 0$, again by the Cauchy–Schwarz inequality. The Peirce multiplication rule implies $\{u, b_1, b_1\} \in P_2(u)V''$ and therefore $\{u, b_1, b_1\} = 0$ by (3.10).

We now show that $(u \,\square\, b_1)^2 = 0$. Let $x \in V''$ with Peirce decomposition $x = x_0 + x_1 + x_2$. We have $u \,\square\, b_1(x_2) = 0$ by the Peirce multiplication rule. The Cauchy–Schwarz inequality gives $\varphi\{u, b_1, x_1\} = 0$ and it follows from (3.10) that $\{u, b_1, x_1\} = 0$, since $\{u, b_1, x_1\}$ is in the Peirce 2-space of u. Applying a similar argument to

$$\varphi\{u, b_1, \{u, b_1, x_0\}\} = \varphi\{\{u, b_1, u\}, b_1, x_0\} = 0,$$

we obtain $(u \,\Box\, b_1)^2(x_0) = 0$. It follows from the identity (1.18) that

$$0 = 2(u \,\Box\, b_1)^2 = Q_u(b_1) \,\Box\, b_1 + Q_u Q_{b_1} = Q_u Q_{b_1}.$$

Hence $Q_{\{b_1, u, b_1\}} = Q_{b_1} Q_u Q_{b_1} = 0$ and $\{b_1, u, b_1\} = 0$. Applying the identity (1.19),

$$2Q(b_1, u)(u \,\Box\, b_1) = Q(Q_{b_1}(u), u) + (u \,\Box\, u)Q_{b_1} = (u \,\Box\, u)Q_{b_1},$$

to b_1, we get

$$0 = 2\{b_1, \{u, b_1, b_1\}, u\} = \{u, u, \{b_1, b_1, b_1\}\} = \frac{1}{2}\{b_1, b_1, b_1\},$$

which yields $b_1 = 0$. □

Remark 3.4.5 The space N_φ is called the *left kernel* of φ.

Now we are ready to discuss isometries between JB-triples. In what follows, $C_0(S, \mathbb{R})$ denotes the JB-triple of real continuous functions on a locally compact Hausdorff space S, vanishing at infinity.

We denote by $\operatorname{ext} V_1$ the set of extreme points of the closed unit ball of a Banach space V.

Theorem 3.4.6 *Let W be a JB-triple with second dual V'' and let $\Gamma : C_0(S, \mathbb{R}) \longrightarrow W$ be a linear isometry, which need not be surjective. Then either Γ is a triple monomorphism or there is a tripotent $u \in W''$ such that*

$$\{u, \Gamma\{f, g, h\}, u\} = \{u, \{\Gamma f, \Gamma g, \Gamma h\}, u\}$$

for all $f, g, h \in C_0(S, \mathbb{R})$ and

$$\{u, \Gamma(\cdot), u\} : C_0(S, \mathbb{R}) \longrightarrow W''$$

is an isometry.

Proof Let $E = \Gamma(C_0(S, \mathbb{R}))$ be the range of Γ. Then the dual map $\Gamma' : E' \to C_0(S, \mathbb{R})'$ of $\Gamma : C_0(S, \mathbb{R}) \to E$ is a surjective linear isometry. We also denote by Γ' the dual map of $\Gamma : C_0(S, \mathbb{R}) \longrightarrow W$, since no confusion is likely. Let

$$Q = \{\varphi \in \operatorname{ext} W_1' : \varphi|_E \in \operatorname{ext} E_1'\}.$$

Then Q is non-empty, since each extreme point $\psi \in \operatorname{ext} E_1'$ extends to an extreme point $\varphi \in \operatorname{ext} W_1'$.

Let $\varphi \in Q$ with $\psi = \varphi|_E \in \operatorname{ext} E_1'$. Then $\Gamma'\varphi = \Gamma'\psi$ is an extreme point of the closed unit ball of $C_0(S, \mathbb{R})'$ and hence there exists $x_\varphi \in S$ such that $\Gamma'\psi = \alpha\delta_{x_\varphi}$ with $|\alpha| = 1$, where δ_{x_φ} is a point evaluation. Let $u_\varphi \in W''$ be the support tripotent of φ. By (3.10), we have

$$\{u_\varphi, b, u_\varphi\} = \varphi(b)u_\varphi \qquad (b \in W'').$$

From $\varphi \circ \Gamma(f) = (\Gamma'\varphi)(f) = (\Gamma'\psi)(f) = \alpha f(x_\varphi)$, we obtain, in W'',

$$\{u_\varphi, \Gamma(f), u_\varphi\} = \alpha f(x_\varphi)u_\varphi \qquad (f \in C_0(S, \mathbb{R}))$$

and $\{u_\varphi, \Gamma(\cdot), u_\varphi\}$ is a triple homomorphism. In particular,

$$\begin{aligned}
\alpha f^3(x_\varphi)u_\varphi = \{u_\varphi, \Gamma f, u_\varphi\}^3 &= \{u_\varphi, \{\Gamma f, P_2(u_\varphi)(\Gamma f), \Gamma f\}, u_\varphi\} \\
&= \alpha f(x_\varphi)\{u_\varphi, \{\Gamma f, u_\varphi, \Gamma f\}, u_\varphi\}
\end{aligned}$$

and hence $\varphi\{u_\varphi, \{\Gamma f, u_\varphi, \Gamma f\}, u_\varphi\} = (\alpha f(x_\varphi))^2$ or

$$\varphi\{\Gamma f, u_\varphi, \Gamma f\} = f(x_\varphi)^2. \tag{3.12}$$

We prove that

$$\{u_\varphi, \Gamma(f^3), u_\varphi\} = \{u_\varphi, (\Gamma f)^3, u_\varphi\} \qquad (f \in C_0(S, \mathbb{R})).$$

It suffices to show that

$$\varphi\{u_\varphi, (\Gamma f)^3, u_\varphi\} = \alpha f^3(x_\varphi).$$

We first show that

$$\{u_\varphi, u_\varphi, \Gamma h\} = u_\varphi$$

for $h \in C_0(S, \mathbb{R})$ satisfying $\|h\| = 1$ and $h(x_\varphi) = \alpha$. We have, by the Cauchy–Schwarz inequality,

$$\begin{aligned}
1 = |\varphi(\Gamma h)|^2 = |\varphi\{u_\varphi, \Gamma h, u_\varphi\}|^2 \\
\le \varphi\{u_\varphi, u_\varphi, u_\varphi\}\varphi\{\Gamma h, \Gamma h, u_\varphi\} \le \|\Gamma h\|^2 = \|h\|^2 = 1,
\end{aligned}$$

giving $\varphi\{\Gamma h, \Gamma h, u_\varphi\} = 1$. Let

$$N_\varphi = \{b \in W'' : \varphi\{b, b, u_\varphi\} = 0\}$$

be the left kernel of φ. Then we have

$$N_\varphi = P_0(u_\varphi)(W'') \tag{3.13}$$

by Lemma 3.4.4. We show $\Gamma h - u_\varphi \in N_\varphi$. Indeed, we have

$$\begin{aligned}
\varphi\{\Gamma h - u_\varphi, \Gamma h - u_\varphi, u_\varphi\} \\
= \varphi\{\Gamma h, \Gamma h, u_\varphi\} - \varphi\{u_\varphi, \Gamma h, u_\varphi\} + \varphi\{u_\varphi, u_\varphi, u_\varphi\} - \varphi\{\Gamma h, u_\varphi, u_\varphi\} = 0,
\end{aligned}$$

where $\varphi\{\Gamma h, u_\varphi, u_\varphi\} = \varphi\{u_\varphi, \Gamma h, u_\varphi\} = 1$. Hence, by (3.13), we have $\{u_\varphi, u_\varphi, \Gamma h - u_\varphi\} = 0$ and $\{u_\varphi, u_\varphi, \Gamma h\} = u_\varphi$.

We next show that $\varphi\{\Gamma g, \Gamma g, u_\varphi\} = 0$ whenever $g \in C_0(S, \mathbb{R})$ satisfies $g(x_\varphi) = 0$. We may assume, by Urysohn's lemma, that g vanishes on a neighbourhood of x_φ, in which case, we can choose $k \in C_0(S, \mathbb{R})$ such that $\|k\| = 1$, $k(x_\varphi) = \alpha$ and $kg = 0$. Then $\|k + g\| = 1$ and $(k + g)(x_\varphi) = \alpha$. Therefore, by the preceding argument, we have $\Gamma(k + g) + N_\varphi = u_\varphi + N_\varphi = \Gamma k + N_\varphi$, which yields $\Gamma g \in N_\varphi$; that is, $\varphi\{\Gamma g, \Gamma g, u_\varphi\} = 0$.

Now let $f \in C_0(S, \mathbb{R})$. Pick $h \in C_0(S, \mathbb{R})$ with $\|h\| = 1$ and $h(x_\varphi) = \alpha$. Then $(f - \alpha f(x_\varphi)h)(x_\varphi) = 0$ and therefore we have $\Gamma f - \alpha f(x_\varphi)\Gamma h \in N_\varphi$ and by (3.13) again,

$$\{u_\varphi, u_\varphi, \Gamma f - \alpha f(x_\varphi)\Gamma h\} = 0,$$

giving

$$\{u_\varphi, u_\varphi, \Gamma f\} = \alpha f(x_\varphi)\{u_\varphi, u_\varphi, \Gamma h\} = \alpha f(x_\varphi)u_\varphi.$$

Moreover, for any $g \in C_0(S, \mathbb{R})$, we have

$$\begin{aligned}
\alpha f(x_\varphi)\{u_\varphi, \Gamma g, u_\varphi\} &= \{u_\varphi, \Gamma g, \{u_\varphi, u_\varphi, \Gamma f\}\} \\
&= \{\{u_\varphi, \Gamma g, u_\varphi\}, u_\varphi, \Gamma f\} - \{u_\varphi, \{\Gamma g, u_\varphi, u_\varphi\}, \Gamma f\} + \{u_\varphi, u_\varphi, \{u_\varphi, \Gamma g, \Gamma f\}\} \\
&= \alpha g(x_\varphi)\{u_\varphi, u_\varphi, \Gamma f\} - \alpha g(x_\varphi)\{u_\varphi, u_\varphi, \Gamma f\} + \{u_\varphi, u_\varphi, \{u_\varphi, \Gamma g, \Gamma f\}\} \\
&= \{u_\varphi, u_\varphi, \{u_\varphi, \Gamma g, \Gamma f\}\}
\end{aligned} \tag{3.14}$$

and hence

$$\varphi\{u_\varphi, \Gamma g, \Gamma f\} = \varphi(\{u_\varphi, u_\varphi, \{u_\varphi, \Gamma g, \Gamma f\}\}) = \alpha f(x_\varphi)\varphi\{u_\varphi, \Gamma g, u_\varphi\}. \tag{3.15}$$

Therefore we have

$$\begin{aligned}
\varphi\{u_\varphi, (\Gamma f)^3, u_\varphi\} &= \varphi\{u_\varphi, u_\varphi, \{\Gamma f, \Gamma f, \Gamma f\}\} \\
&= \varphi(\{\{u_\varphi, u_\varphi, \Gamma f\}, \Gamma f, \Gamma f\} - \{\Gamma f, \{u_\varphi, u_\varphi, \Gamma f\}, \Gamma f\} \\
&\quad + \{\Gamma f, \Gamma f, \{u_\varphi, u_\varphi, \Gamma f\}\}) \\
&= 2\alpha f(x_\varphi)\varphi\{u_\varphi, \Gamma f, \Gamma f\} - \alpha f(x_\varphi)\varphi\{\Gamma f, u_\varphi, \Gamma f\} \\
&= \alpha f^3(x_\varphi)
\end{aligned}$$

using (3.12). It follows that

$$\{u_\varphi, \Gamma(f^3), u_\varphi\} = \alpha f^3(x_\varphi)u_\varphi = \{u_\varphi, (\Gamma f)^3, u_\varphi\}.$$

Although polarization in terms of cubes is unavailable in real Jordan triples, one can still extend the previous identity to

$$\{u_\varphi, \Gamma\{f, g, h\}, u_\varphi\} = \{u_\varphi, \{\Gamma f, \Gamma g, \Gamma h\}, u_\varphi\} \qquad (f, g, h \in C_0(S, \mathbb{R})). \tag{3.16}$$

To achieve this, it suffices to establish (3.16) for the case $f = h$. Since

$$2\{\Gamma f, \Gamma g, \Gamma f\} = (\Gamma f + \Gamma g)^3 + (\Gamma g - \Gamma f)^3 - 2(\Gamma g)^3 - 2\{\Gamma g, \Gamma f, \Gamma f\},$$

we need only show

$$\{u_\varphi, \Gamma\{g, f, f\}, u_\varphi\} = \{u_\varphi, \{\Gamma g, \Gamma f, \Gamma f\}, u_\varphi\}$$

for $f, g \in C_0(S, \mathbb{R})$. Indeed, using (1.18), we have

$$
\begin{aligned}
&\{u_\varphi, \{\Gamma g, \Gamma f, \Gamma f\}, u_\varphi\} \\
&= 2\{u_\varphi, \Gamma g, \{u_\varphi, \Gamma f, \Gamma f\}\} - \{\{u_\varphi, \Gamma g, u_\varphi\}, \Gamma f, \Gamma f\} \\
&= 2f(x_\varphi)^2\{u_\varphi, \Gamma g, u_\varphi\} - \alpha g(x_\varphi)\{u_\varphi, \Gamma f, \Gamma f\} \\
&= 2\alpha f(x_\varphi)^2 g(x_\varphi) - \alpha g(x_\varphi) f(x_\varphi)^2 \\
&= \alpha f(x_\varphi)^2 g(x_\varphi) = \{u_\varphi, \Gamma\{g, f, f\}, u_\varphi\}.
\end{aligned}
$$

We next show that

$$\{u_\varphi, u_\varphi, \Gamma\{f, g, h\}\} = \{u_\varphi, u_\varphi, \{\Gamma f, \Gamma g, \Gamma h\}\} \qquad (f, g, h \in C_0(S, \mathbb{R})). \tag{3.17}$$

We first deduce from (1.18), (3.14) and (3.15) that

$$
\begin{aligned}
\{u_\varphi, \Gamma f, \Gamma g\} &= \{\{u_\varphi, u_\varphi, u_\varphi\}, \Gamma f, \Gamma g\} \\
&= 2\{u_\varphi, u_\varphi, \{u_\varphi, \Gamma f, \Gamma g\}\} - \{u_\varphi, \{u_\varphi, \Gamma g, \Gamma f\}, u_\varphi\} \\
&= 2f(x_\varphi)g(x_\varphi)u_\varphi - g(x_\varphi)f(x_\varphi)u_\varphi \\
&= f(x_\varphi)g(x_\varphi)u_\varphi.
\end{aligned}
$$

Therefore

$$
\begin{aligned}
\{u_\varphi, u_\varphi, \Gamma\{f, g, f\}\} &= \alpha f(x_\varphi)^2 g(x_\varphi)u_\varphi \\
&= \{\Gamma f, \Gamma g, \{u_\varphi, u_\varphi, \Gamma f\}\} \\
&= \{\{\Gamma f, \Gamma g, u_\varphi\}, u_\varphi, \Gamma f\} - \{u_\varphi, \{\Gamma g, \Gamma f, u_\varphi\}, \Gamma f\} + \{u_\varphi, u_\varphi, \{\Gamma f, \Gamma g, \Gamma f\}\} \\
&= \alpha f(x_\varphi)g(x_\varphi)f(x_\varphi) - \alpha g(x_\varphi)f(x_\varphi)^2 + \{u_\varphi, u_\varphi, \{\Gamma f, \Gamma g, \Gamma f\}\} \\
&= \{u_\varphi, u_\varphi, \{\Gamma f, \Gamma g, \Gamma f\}\}.
\end{aligned}
$$

Consider the partial ordering $\leq$ for tripotents in W'', introduced in Definition 1.2.42. We have two cases:

(i) the lattice supremum $u = \bigvee_{\varphi \in Q} u_\varphi$ is a tripotent in W'';

(ii) the closed unit ball W_1' of W' is itself the smallest norm closed face of W_1' containing all $\varphi \in Q$ (cf. [37, theorem 3.7]).

Case (i) Let $W = Z^\mathfrak{t}$ be the real form of a JB*-triple Z. Then W'' is a real form of the JBW*-triple Z^{**}. Consider u as a tripotent in Z^{**}. The u-homotope $(P_2(u)Z^{**}, \circ_u)$ is a JBW*-algebra and $P_2(u)W''$ is a weak* closed *real* *-subalgebra of $P_2(u)Z^{**}$. It follows that the self-adjoint part

$$P_2(u)W_{sa}'' = \{w \in P_2(u)W'' : w = w^* = \{u, w, u\}\}$$

is a JBW-algebra. Each u_φ is a minimal projection in $P_2(u)W_{sa}''$ and we have

$$u \,\square\, u_\varphi = u_\varphi \,\square\, u = u_\varphi \,\square\, u_\varphi.$$

Let $f, g, h \in C_0(S, \mathbb{R})$ and let

$$c = \Gamma\{f, g, h\} - \{\Gamma f, \Gamma g, \Gamma h\} \in W \subset W''.$$

Then we have $\{u, c, u\} \in P_2(u)W''$ and

$$\{u_\varphi, \{u, c, u\}, u_\varphi\} = \{u, \{u_\varphi, \{u, c, u\}, u_\varphi\}, u\} = \{u_\varphi, c, u_\varphi\} = 0.$$

Also, (3.17) implies

$$\begin{aligned} \{u_\varphi, u_\varphi, \{u, c, u\}\} &= \{u, u_\varphi, \{u, c, u\}\} \\ &= \{u, \{u_\varphi, u, c\}, u\} = \{u, \{u_\varphi, u_\varphi, c\}, u\} = 0. \end{aligned}$$

It follows that

$$u_\varphi \circ_u \{u, c, u\} = \{u_\varphi, u, \{u, c, u\}\} = \{u_\varphi, u_\varphi, \{u, c, u\}\} = 0$$

and

$$\{u_\varphi, \{u, c, u\}, u_\varphi\} = 2(u_\varphi \circ_u \{u, c, u\}) \circ_u u_\varphi - u_\varphi^2 \circ_u \{u, c, u\} = 0.$$

By Lemma 1.1.7, u_φ and $\{u, c, u\}$ operator commute in the JBW*-algebra $P_2(u)Z^{**}$. Likewise, one can show that u_φ operator commutes with $\{u, c, u\}^* = \{u, \{u, c, u\}, u\}$. Let

$$\{u, c, u\}_1 = \frac{1}{2}(\{u, c, u\} + \{u, c, u\}^*) \quad \text{and}$$

$$\{u, c, u\}_{-1} = \frac{1}{2i}(\{u, c, u\} - \{u, c, u\}^*).$$

Then $\{u, c, u\}_{\pm 1} \in P_2(u)W''_{sa}$ and

$$\{u, c, u\} = \{u, c, u\}_1 + i\{u, c, u\}_{-1},$$

where $\{u, c, u\}_{\pm 1}$ operator commutes with u_φ

It follows that, for $j = \pm 1$, the element $\{u, c, u\}_j$ operator commutes with $u_\varphi + u_\psi$ for all $\varphi, \psi \in Q$ and hence the JBW-subalgebra $W(\{u, c, u\}_j, u_\varphi + u_\psi)$ generated by $\{u, c, u\}_j$ and $u_\varphi + u_\psi$ in $P_2(u)W''_{sa}$ is associative.

The lattice supremum $u_\varphi \vee u_\psi$ in $P_2(u)W''_{sa}$ is the range projection $r(u_\varphi + u_\psi)$ of $u_\varphi + u_\psi$ and is the strong limit of a sequence $\{p_n(u_\varphi + u_\psi)\}$ of polynomials without constant term [47, lemma 4.2.6]. By associativity of $W(\{u, c, u\}_j, u_\varphi + u_\psi)$ and $\{u, c, u\} \circ_u (u_\varphi + u_\psi) = 0$, we conclude that

$$\{u_\varphi \vee u_\psi, \{u, c, u\}_j, u_\varphi \vee u_\psi\} = \{u, c, u\} \circ_u r(u_\varphi + u_\psi) = 0.$$

By induction, we have

$$\{u_{\varphi_1} \vee \cdots \vee u_{\varphi_k}, \{u, c, u\}_j, u_{\varphi_1} \vee \cdots \vee u_{\varphi_k}\} = 0$$

for any finite collection $\varphi_1, \ldots, \varphi_k$ in Q. By taking finite suprema, we can express u as the supremum of an increasing net $\{v_\alpha\}$ of projections v_α in $P_2(u)W''_{sa}$ with $\{v_\alpha, \{u, c, u\}_j, v_\alpha\} = 0$. Since u is the strong limit of $\{v_\alpha\}$, we obtain

$$\{u, \{u, c, u\}_j, u\} = 0 \qquad (j = \pm 1)$$

and hence $\{u, \{u, c, u\}, u\} = 0$, which enables us to conclude that

$$\{u, \Gamma\{f, g, h\}, u\} = \{u, \{\Gamma f, \Gamma g, \Gamma h\}, u\}.$$

Finally, for any $f \in C_0(S, \mathbb{R})$, pick $x \in S$ with $\|f\| = |f(x)|$. Let $\psi \in \text{ext } E'_1$ with $\Gamma'\psi = \delta_x$, and let $\varphi \in \text{ext } W'_1$ be an extension of ψ. Then $\varphi \in Q$ and $\Gamma'\varphi = \delta_x$. Hence

$$\begin{aligned}
\|\Gamma f\| \geq \|\{u, \Gamma f, u\}\| &\geq \|\{u_\varphi, \{u_\varphi, \{u, \Gamma f, u\}, u_\varphi\}, u_\varphi\}\| \\
&= \|\{u_\varphi, \Gamma f, u_\varphi\}\| \\
&= \|f(x)u_\varphi\| = |f(x)| = \|f\|,
\end{aligned}$$

which gives $\|\{u, \Gamma f, u\}\| = \|f\|$.

Case (ii) By Example 3.3.22, Z^{**} can be embedded as a 1-complemented subtriple of a JBW*-algebra B which has an identity. Let $u' = \bigvee_{\varphi \in Q} u_\varphi$ be the supremum in B. Then

$$F(u') = \{\eta \in B_* : \|\eta\| = 1 = \eta(u')\}$$

is a norm-closed face of the closed unit ball of the predual B_* and $F(u') \cap W_1'$ is a norm-closed face of the closed unit ball W_1', containing all $\varphi \in Q$. Hence $F(u') \cap W_1' = W_1'$ and each extreme point $\rho \in \text{ext } W_1'$ belongs to $F(u')$.

We can repeat the previous arguments in $P_2(u')B$ to show that

$$\{u', \Gamma\{f, g, h\}, u'\} = \{u', \{\Gamma f, \Gamma g, \Gamma h\}, u'\} \qquad (f, g, h \in C_0(S, \mathbb{R})).$$

It follows that, for each $\rho \in \text{ext } W_1'$, its support tripotent u_ρ satisfies

$$
\begin{aligned}
\{u_\rho, \Gamma\{f, g, h\}, u_\rho\} &= \{u', \{u_\rho, \{u', \Gamma\{f, g, h\}, u'\}, u_\rho, \}u'\} \\
&= \{u', \{u_\rho, \{u', \{\Gamma f, \Gamma g, \Gamma h\}, u'\}, u_\rho, \}u'\} \\
&= \{u_\rho, \{\Gamma f, \Gamma g, \Gamma h\}, u_\rho\}.
\end{aligned}
$$

Since $\rho \in \text{ext } W_1'$ is arbitrary, we have

$$\Gamma\{f, g, h\} = \{\Gamma f, \Gamma g, \Gamma h\} \qquad (f, g, h \in C_0(S, \mathbb{R})).$$

$\square$

Remark 3.4.7 If Γ is surjective in Theorem 3.4.6, then it is a triple isomorphism, since surjectivity implies $Q = \text{ext } W_1'$ and $\rho(\Gamma\{f, g, h\}) = \rho\{\Gamma f, \Gamma g, \Gamma h\}$ for all $\rho \in \text{ext } W_1'$.

Theorem 3.4.8 *Let V and W be JB-triples and let $\gamma : V \longrightarrow W$ be a linear isometry which may not be surjective. Then for each $a \in V$, there is a tripotent $u \in W''$ such that*

$$\{u, \gamma\{x, y, z\}, u\} = \{u, \{\gamma(x), \gamma(y), \gamma(z)\}, u\}$$

for all x, y, z in the closed subtriple $R(a)$ generated by a, and $\{u, \gamma(\cdot), u\} : R(a) \longrightarrow W''$ is an isometry.

Proof This follows immediately from Theorem 3.4.6, since $R(a)$ is of the form $C_0(S, \mathbb{R})$, and if γ is a triple monomorphism on $R(a)$, we can take $u \in W''$ to be a complete tripotent satisfying $\{u, \gamma(a), u\} = \gamma(a)$, by Lemma 3.3.7. $\square$

Remark 3.4.9 We stress that the above result is a *local* one in that the tripotent u depends on the element a. This is illustrated by the following example.

Example 3.4.10 Let S be the closed unit disc in the complex plane $\mathbb{C}$ and let $\sigma : z \in S \mapsto \bar{z} \in S$ be the complex conjugation. Consider the real abelian C*-algebra

$$C_\sigma(S) = \{f \in C(S) : f \circ \sigma = \bar{f}\}$$

as in (2.19), which contains complex continuous functions on S. Let $M_2(\mathbb{R})$ be the real C*-algebra of 2×2 real matrices and let $C(S, M_2(\mathbb{R}))$ be the real C*-algebra of $M_2(\mathbb{R})$-valued continuous functions on S. Define a linear isometry

$\gamma : C_\sigma(S) \longrightarrow C(S, M_2(\mathbb{R}))$ by

$$\gamma(f) = \begin{pmatrix} \text{Re } f & \text{Im } f \\ 0 & 0 \end{pmatrix} \qquad (f \in C_\sigma(S)),$$

where Re f and Im f denote the real and imaginary parts of f, respectively. Then γ preserves the cubes f^3, but not the triple product.

Further, there is no tripotent $u \in C(S, M_2(\mathbb{R}))''$ satisfying both conditions

$$\{u, \gamma\{f, g, f\}, u\} = \{u, \{\gamma(f), \gamma(g), \gamma(f)\}, u\} \quad \text{and} \quad \|\{u, \gamma(f), u\}\| = \|f\|$$

for all $f, g \in C_\sigma(S)$. Indeed, if there were such a tripotent u, consider the functions $g = \mathbf{1} \in C_\sigma(S)$ and $f \in C_\sigma(S)$ given by

$$f(z) = i \operatorname{Im} z \qquad (z \in S).$$

Then we would have

$$\{u, \gamma\{f, g, f\}, u\} = \{u, \begin{pmatrix} f^2 & 0 \\ 0 & 0 \end{pmatrix}, u\}$$

$$= \{u, \{\gamma(f), \gamma(g), \gamma(f)\}, u\} = \begin{pmatrix} 0 & 0 \\ 0 & 0 \end{pmatrix}.$$

On the other hand, $f^2 \in C_\sigma(S)$ and we would have

$$\|f^2\| = \|\{u, \gamma(f^2), u\}\| = \|\{u, \begin{pmatrix} f^2 & 0 \\ 0 & 0 \end{pmatrix}, u\}\| = 0,$$

which is impossible.

We also note that the complexification γ_c of γ is not an isometry since $\|g + if\| = 2 \neq \sqrt{2} = \|\gamma_c(g + if)\|$.

3.5 Hilbert spaces

In this section, we study ternary structures in Hilbert spaces. We have already seen that every Hilbert space V has a natural ternary structure given by the Jordan triple product

$$\{x, y, z\} = \frac{1}{2}\langle x, y\rangle z + \frac{1}{2}\langle z, y\rangle x \qquad (x, y, z),$$

which plays an important role in geometry. One can also define other ternary products on a Hilbert space and study the induced ternary structures. Our main concern, however, is the structures of JH-triples, since they correspond to a class of Riemannian symmetric spaces including the Hermitian symmetric spaces.

Given a Hilbert space V which is also a Jordan triple system, we call its inner product $\langle \cdot, \cdot \rangle$ *associative*, as in (2.31), if

$$\langle (a \square b)x, y \rangle = \langle x, (b \square a)y \rangle \qquad (a, b, x, y \in V).$$

Proposition 3.5.1 *Every finite-dimensional JB-triple carries the structure of a continuous JH-triple. Every finite-dimensional JB*-triple carries the structure of a JH*-triple.*

Proof We note that a closed subtriple of a JH-triple is also a JH-triple in the inherited inner product. Since JB-triples are real closed subtriples of JB*-triples, we need only consider finite-dimensional JB*-triples. Let H be a finite-dimensional complex Hilbert space and $L(H)$ the JB*-triple of linear operators on H.

Define the canonical inner product $\langle \cdot, \cdot \rangle$ on $L(H)$ by the trace:

$$\langle x, y \rangle = \mathrm{Trace}\,(xy^*) \qquad (x, y \in L(H)).$$

Then $L(H)$ is a Hilbert space in the inner product norm $\| \cdot \|_2$ and $\langle \cdot, \cdot \rangle$ is associative:

$$
\begin{aligned}
\langle (a \square b)x, y \rangle &= \frac{1}{2}\mathrm{Trace}\,(ab^*xy^* + xb^*ay^*) \\
&= \frac{1}{2}\mathrm{Trace}\,(xy^*ab^* + xb^*ay^*) \\
&= \langle x, (b \square a)y \rangle \qquad\qquad (a, b, x, y \in L(H)).
\end{aligned}
$$

For $a, b, x \in L(H)$, we have $\|a \square b(x)\|_2 \leq \|a \square b\|\|x\|_2 \leq \|a\|\|b\|\|x\|_2 \leq \|a\|_2\|b\|_2\|x\|_2$. It follows that $L(H)$ and all finite-dimensional JC*-triples are JH*-triples.

It remains to consider the two exceptional Cartan factors. As in Example 2.4.32, the JB*-algebra $H_3(\mathcal{O})$ can be equipped with an associative inner product defined by

$$\langle a, b \rangle = \mathrm{Trace}\,(a \square b)$$

and hence $H_3(\mathcal{O})$ is a JH*-triple in the canonical Jordan triple product.

Finally, the JB*-triple $M_{12}(\mathcal{O})$ is a closed subtriple of $H_3(\mathcal{O})$ and is therefore a JH*-triple. $\qquad\square$

Example 3.5.2 Let $M_2(\mathbb{R})$ be the real Hilbert space of 2×2 real matrices, with the usual inner product

$$\langle a, b \rangle = \mathrm{Trace}\,(ab^*) \qquad (a, b \in M_2(\mathbb{R})),$$

where b^* is the transpose of the matrix b. Although $M_2(\mathbb{R})$ is indeed a JH-triple in the triple product

$$\{a, b, c\} = \frac{1}{2}(ab^*c + cb^*a),$$

it is not a JH-triple in the Jordan triple product

$$\{a, b, c\}_1 = \frac{1}{2}(abc + cba).$$

In fact, we have

$$\langle \{a, b, x\}_1, x \rangle = 0 \neq 1 = \langle x, \{b, a, x\}_1 \rangle$$

if

$$x = \begin{pmatrix} 2 & 0 \\ 1 & 0 \end{pmatrix}, \quad a = \begin{pmatrix} 1 & 0 \\ 0 & 0 \end{pmatrix}, \quad b = \begin{pmatrix} 0 & 0 \\ 1 & 1 \end{pmatrix}.$$

Example 3.5.3 Let V and W be Hilbert spaces over the same involutive real division algebra $\mathbb{F} = \mathbb{R}, \mathbb{C}$ or $\mathbb{H}$. With real scalar multiplication, the vector space $L(V, W)$ of continuous $\mathbb{F}$-linear operators between V and W forms a real Banach space in the operator norm $\| \cdot \|$. An operator $T \in L(V, W)$ is said to be of *Hilbert–Schmidt class* if the series

$$\|T\|_2 := \left(\sum_\alpha \|T(e_\alpha)\|^2 \right)^{1/2}$$

is convergent for an orthonormal basis $\{e_\alpha\}$ of V, and hence for any orthonormal basis, in which case the *Hilbert–Schmidt norm* $\|T\|_2$ does not depend on the choice of basis. With this norm, the Hilbert–Schmidt operators form a real Hilbert space $L_2(V, W)$ with inner product

$$\langle T, S \rangle_2 = \operatorname{Re} \operatorname{Trace}(S^*T) \qquad (T, S \in L_2(V, W)).$$

We write $L_2(V)$ for $L_2(V, V)$. Although $L_2(V)$ is not closed in $L(V)$ in the operator-norm topology, it is a weak* dense two-sided ideal in $L(V)$.

With the *canonical Jordan triple product*

$$\{R, S, T\} := \frac{1}{2}(RS^*T + TS^*R)$$

the Hilbert space $L_2(V, W)$ is a continuous JH-triple in which the inner product is associative. In fact, for any $k > 0$, with the inner product

$$k\langle T, S \rangle_2 = k\operatorname{Re} \operatorname{Trace}(S^*T)$$

and the canonical triple product just mentioned, $L_2(V, W)$ is also a JH-triple.

If V is a complex Hilbert space, then $L_2(V)$ is a JH*-triple in the canonical Jordan triple product and the complex inner product $\langle T, S \rangle = \text{Trace}\,(S^*T)$.

The preceding example shows that a triple isomorphism between two JH-triples need not be an isometry.

One can embed a Hilbert space $(V, \langle \cdot, \cdot \rangle)$ over $\mathbb{F}$ into $L_2(V)$ in many ways. We choose to do so in the following way. Given two vectors $u, v \in V$, we define the rank one operator $v \otimes u \in L_2(V)$ by

$$(v \otimes u)(x) = \langle x, u \rangle v \qquad (x \in V).$$

One sees easily that $u \otimes v$ is the adjoint of $v \otimes u$ and $\|v \otimes u\|_2 = \|v\|\|u\|$. Fix a vector e_γ in an orthonormal basis $\{e_\alpha\}$ of V and define the embedding

$$v \in V \mapsto v \otimes e_\gamma \in L_2(V),$$

where

$$\begin{aligned}
\langle u \otimes e_\gamma, v \otimes e_\gamma \rangle_2 &= \text{Re Trace}\,((u \otimes e_\gamma)(v \otimes e_\gamma)^*)\\
&= \text{Re Trace}\,((u \otimes e_\gamma)(e_\gamma \otimes v))\\
&= \text{Re Trace}\,(u \otimes v)\\
&= \text{Re} \sum_\alpha \langle u \otimes v(e_\alpha), e_\alpha \rangle\\
&= \text{Re} \sum_\alpha \langle u, e_\alpha \rangle \langle e_\alpha, v \rangle = \text{Re}\langle u, v \rangle.
\end{aligned}$$

If we regard $v \otimes e_\gamma$ as a matrix represented with respect to the basis $\{e_\alpha\}$, then v occupies the γth column of this matrix which has zero entries elsewhere. In $L_2(V)$, we have

$$(u \otimes e_\gamma)(v \otimes e_\gamma)^*(w \otimes e_\gamma) = \langle w, v \rangle(u \otimes e_\gamma) \qquad (u, v, w \in V).$$

Hence the real Hilbert space $(V, \text{Re}\langle \cdot, \cdot \rangle)$, equipped with the triple product $2\{u, v, w\} = \langle u, v \rangle w + \langle w, v \rangle u$, embeds as a closed subtriple of the JH-triple $L_2(V)$. If V is a complex Hilbert space, then it is a JB*-triple as well as a JH*-triple in the same triple product and the same norm. We note, however, that the JH*-triple $(L_2(V), \| \cdot \|_2)$ is not a JB*-triple since $\|T^3\|_2$ need not equal $\|T\|_2^3$ in $L_2(V)$.

Example 3.5.4 The triple spin factor $Sp_n(H)$ introduced in Theorem 2.5.9 is a JB*-triple which is also a JH*-triple in the same triple product but in a different, albeit equivalent, norm. Indeed, $Sp_n(H)$ is linearly *-isomorphic to a complex Hilbert space $(V, \langle \cdot, \cdot \rangle)$ with a conjugation $* : V \longrightarrow V$ satisfying

$$\langle a^*, b^* \rangle = \langle b, a \rangle \qquad (a, b \in V).$$

Equip V with the Jordan triple product

$$\{a, b, c\} = \frac{1}{2}(\langle a, b \rangle c + \langle c, b \rangle a - \langle a, c^* \rangle b^*).$$

Then V is a JH*-triple and $Sp_n(H)$ is triple isomorphic to V. We call V a *complex triple spin factor*.

We define a *real triple spin factor* to be a real Hilbert space $(V, \langle \cdot, \cdot \rangle)$ equipped with an involutive linear isometry $j : V \longrightarrow V$ and the triple product

$$\{a, b, c\} = \frac{1}{2}(\langle a, jb \rangle c + \langle c, jb \rangle a - \langle a, c \rangle jb).$$

A real triple spin factor is a JH-triple with an associative inner product. In the spin factor $(H \oplus \mathbb{R}, \circ)$ introduced after Theorem 2.3.14, the Hilbert space H, with the identity map as an involutive isometry, is a real triple spin factor in the canonical triple product $\{a, b, a\} = 2(a \circ b) \circ a - a^2 \circ b$, scaled by $1/2$.

We recall that Jordan Hilbert triples are Jordan triple systems *as well as* Hilbert spaces on which the inner derivations

$$d(a, b) = a \,\square\, b - b \,\square\, a$$

are skew-symmetric and, by Lemma 2.4.18, automatically continuous. Jordan Hilbert triples correspond to a class of orthogonal involutive Lie algebras (cf. Lemma 2.4.16). Continuity of the box operators $a \,\square\, b$ on Jordan Hilbert triples follows from *associativity* of the inner product.

Lemma 3.5.5 *Let V be a Jordan Hilbert triple. The following conditions are equivalent:*

(i) *The inner product in V is associative.*
(ii) *For all $a, b \in V$, the box operator $a \,\square\, b$ is continuous with adjoint*
$(a \,\square\, b)^* = b \,\square\, a$.

Further, these conditions imply that the bilinear map

$$(a, b) \in V \times V \mapsto a \,\square\, b \in L(V)$$

is continuous. In other words, there is a constant $c > 0$ such that

$$\|a \,\square\, b\| \leq c \|a\| \|b\| \qquad (a, b \in V).$$

In particular, a real Jordan Hilbert triple with an associative inner product is a continuous JH-triple.

Proof It suffices to show that (i) $\Rightarrow$ (ii). Let $a, b \in V$. Since

$$\langle (a \,\square\, b)x, y \rangle = \langle x, (b \,\square\, a)y \rangle \qquad (x, y \in V),$$

the box operator $a \square b$ is weakly continuous and hence norm continuous. Moreover, $(a \square b)^* = b \square a$.

To show continuity of the bilinear map, we observe that, for $a, x \in V$, the linear map $Q(a, x) : V \longrightarrow V$ is bounded since, as before, it is weakly continuous by associativity of the inner product:

$$\langle Q(a, x)(b), y \rangle = \langle \{a, b, x\}, y \rangle = \langle \{a, y, x\}, b \rangle \qquad (y \in V).$$

Hence, for each $a \in V$, we can define a linear map

$$Q(a, \cdot) : x \in V \mapsto Q(a, x) \in L(V).$$

We show that $Q(a, \cdot)$ is continuous on V. Let (x_n) be a sequence in V converging to 0 and let the sequence $(Q(a, x_n))$ converge to some $T \in L(V)$. For any $u, v \in V$, we have

$$\langle Tu, v \rangle = \lim_n \langle Q(a, x_n)u, v \rangle = \lim_n \langle \{a, u, x_n\}, v \rangle = \lim_n \langle x_n, \{u, a, v\} \rangle = 0,$$

which implies that $T = 0$ and therefore $Q(a, \cdot)$ is continuous by the closed graph theorem. It follows that $\|Q(a, x)\| \leq \|Q(a, \cdot)\| \|x\|$ for all $x \in V$ and

$$\|Q(a, x)\| = \|Q(x, a)\| \leq \|Q(x, \cdot)\| \|a\| \leq \|Q(x, \cdot)\|$$

for all $x \in V$ and $\|a\| \leq 1$.

By the uniform boundedness principle, we have

$$\sup\{\|Q(a, \cdot)\| : \|a\| \leq 1\} \leq c$$

for some $c > 0$ which gives $\|Q(a, x)\| \leq c \|a\| \|x\|$ for all $a, x \in V$, and in turn

$$\|\{a, b, x\}\| = \|Q(a, x)b\| \leq \|Q(a, x)\| \|b\| \leq c \|a\| \|x\| \|b\| \qquad (a, b, x \in V)$$

or equivalently, $\|a \square b\| \leq c \|a\| \|b\|$ for all $a, b \in V$.

$\square$

Lemma 3.5.6 *Let V be a complex Hilbert space as well as a Hermitian Jordan triple. The following conditions are equivalent:*

(i) *The inner product in V is associative; that is, V is a JH*-triple.*
(ii) *The box operator $a \square a : V \longrightarrow V$ is continuous and self-adjoint for each $a \in V$.*

Proof To see (ii) $\Rightarrow$ (i), we observe that

$$2a \square b = (a + b) \square (a + b) + i((a + ib) \square (a + ib))$$

is continuous for all $a, b \in V$. The equation

$$((a + ib) \square (a + ib))^* = (a + ib) \square (a + ib)$$

implies

$$(a \square b)^* - (b \square a)^* = -(a \square b) + (b \square a)$$

which, together with the self-adjointness of $(a + b)\square (a + b)$, yields $(a \square b)^* = b \square a$. $\square$

A Jordan Hilbert triple V is called *Hermitian* if V is a complex Hilbert space and a Hermitian Jordan triple. The inner product in a Hermitian Jordan Hilbert triple is always associative, since we have

$$\langle (a \square b)(x + \alpha y), x + \alpha y \rangle = \langle x + \alpha y, (b \square a)(x + \alpha y) \rangle$$

for $\alpha = 1, i$. Hence the Hermitian Jordan Hilbert triples are exactly the JH*-triples. This may suggest that the hermitification of a JH-triple V would have an associative inner product and hence would be a JH*-triple. That, in fact, is not the case unless the inner product of V is already associative.

Lemma 3.5.7 *Let V be a JH-triple. The following conditions are equivalent:*

(i) *The inner product in V is associative.*
(ii) *The box operator $a \square a : V \longrightarrow V$ is symmetric for each $a \in V$.*
(iii) *The inner product in the hemitification $(V_c, \{\cdot, \cdot, \cdot\}_h)$ of V is associative.*
(iv) *The hermitification $(V_c, \{\cdot, \cdot, \cdot\}_h)$ of V is a JH*-triple.*
(v) *The inner product in the complexification $(V_c, \{\cdot, \cdot, \cdot\}_c)$ of V is associative.*
(vi) *The real restriction of the complexification $(V_c, \{\cdot, \cdot, \cdot\}_c)$ is a JH-triple.*

Proof (ii) $\Rightarrow$ (i). Let $a, b \in V$. Since the box operators $a \square a$, $b \square b$ and $(a + b)\square (a + b)$ are symmetric, the operator $a \square b + b \square a$ is symmetric. It follows that $2(a \square b)^* = (a \square b + b \square a + d(a, b))^* = a \square b + b \square a + d(b, a) = 2b \square a$.

(i) $\Rightarrow$ (iii). Given $a, x, y \in V_c$, with the Hermitian triple product $\{\cdot, \cdot, \cdot\}_h$ defined in (1.24), it is straightforward to verify that

$$\langle \{a, a, x\}_h, y \rangle_c = \langle x, \{a, a, y\}_h \rangle_c$$

by associativity of the inner product $\langle \cdot, \cdot \rangle$.

(iv) $\Rightarrow$ (i). Let $a, b, x, y \in V$. The identities

$$\mathrm{Re}\langle \{a, b, x + y\}_h, x + y \rangle_c = \mathrm{Re}\langle x + y, \{b, a, x + y\}_h \rangle_c$$

and

$$\mathrm{Re}\langle \{a, ib, x + iy\}_h, x + iy \rangle_c = \mathrm{Re}\langle x + iy, \{ib, a, x + iy\}_h \rangle_c$$

together yield

$$\langle \{a, b, x\}, y \rangle = \langle x, \{b, a, y\} \rangle.$$

(i) $\Rightarrow$ (v). Straightforward.

(vi) $\Rightarrow$ (i). Let $a, b, x, y \in V$. The identity

$$\mathrm{Re}\langle \{a \oplus ia, b \oplus ib, x \oplus iy\}_c, \ x \oplus iy \rangle_c$$
$$= \mathrm{Re}\langle x \oplus iy, \ \{b \oplus ib, a \oplus ia, x \oplus iy\}_c \rangle_c$$

yields

$$\langle \{a, b, x\}, y \rangle - \langle \{a, b, y\}, x \rangle = \langle y, \{b, a, x\} \rangle - \langle x, \{b, a, y\} \rangle$$

which, together with the identity

$$\langle d(a, b)x, y \rangle = \langle x, d(b, a)y \rangle,$$

yields

$$\langle \{a, b, x\}, y \rangle = \langle y, \{b, a, x\} \rangle.$$

$\square$

Example 3.5.8 Let V be a Hilbert space of at least two dimensions over $\mathbb{F} = \mathbb{R}, \mathbb{C}$ or $\mathbb{H}$. The real Hilbert space $L_2(V)$ is a Jordan triple in the triple product

$$\{a, b, c\} = \frac{1}{2}(abc + cba)$$

but is not a JH-triple since $M_2(\mathbb{R})$ in Example 3.5.2 can be embedded in $L_2(V)$ as a closed subtriple and $M_2(\mathbb{R})$ is not a JH-triple in this triple product.

However, any closed commutative subalgebra $\mathcal{A}$ of $L_2(V)$ is a continuous JH-triple in the triple product, but the inner product in $\mathcal{A}$ need not be associative, where

$$\langle \{a, b, x\}, y \rangle = \mathrm{Re}\,\mathrm{Trace}\,(y^*\{a, b, x\}) \quad \text{and}$$
$$\langle x, \{b, a, y\} \rangle = \mathrm{Re}\,\mathrm{Trace}\,(x^*\{a, b, y\})$$

for $a, b, x, y \in \mathcal{A}$. Consider, for instance, the commutative algebra

$$\mathcal{A} = \left\{ \begin{pmatrix} \alpha & \beta \\ 0 & \alpha \end{pmatrix} : \alpha, \beta \in \mathbb{R} \right\}.$$

We have $\langle a \square a(a), a^3 \rangle = 11 \neq 7 = \langle a, a \square a(a^3) \rangle$ for $a = \begin{pmatrix} 1 & 1 \\ 0 & 1 \end{pmatrix}$.

To avoid confusion, two elements a and b in a Jordan Hilbert triple V such that $a \square b = b \square a = 0$ will be called *triple orthogonal* to each other. If the inner product in V is associative, it is readily seen that two triple orthogonal tripotents are also orthogonal with respect to the inner product. Given a subset S of a Jordan Hilbert triple, we denote its complements with respect to the two notions of orthogonality by

$$S^\perp = \{v \in V : \langle v, S \rangle = \{0\}\}$$
$$S^\square = \{v \in V : v \,\square\, S = \{0\}\},$$

which will be called respectively the *orthogonal complement* and the (*triple*) *annihilator* of S. We note that the annihilator $S^\square$ is a subtriple of V by the triple identity

$$\{a, b, a\} \,\square\, v(x) = 2\{a, b, \{a, v, x\}\} - \{a, \{b, v, x\}, a\} = 0$$
$$(a, b \in S^\square, v \in S, x \in V).$$

Lemma 3.5.9 *Let V be a Jordan Hilbert triple with an associative inner product and let I be a triple ideal in V. Then its orthogonal complement $I^\perp$ is a closed ideal of V and $I^\perp \subset I^\square$. If the annihilator $V^\square$ vanishes, then $\overline{I}$ is the closed linear span of $\{I, I, I\}$ and $I^\perp = I^\square$.*

Proof The first assertion follows from

$$\langle \{I^\perp, V, V\} + \{V, I^\perp, V\}, I \rangle$$
$$= \langle I^\perp, \{I, V, V\}\rangle + \langle\{V, I, V\}, I^\perp\rangle$$
$$\subset \langle I^\perp, I\rangle + \langle I, I^\perp\rangle = \{0\}$$

by associativity of the inner product.

Let $a \in I^\perp$ and $b \in I$. Then

$$\langle\{a, b, x\}, y\rangle = \langle a, \{y, x, b\}\rangle \in \langle a, I\rangle = \{0\} \qquad (x, y \in V)$$

implies $a \in I^\square$.

Let $V^\square = \{0\}$ for the rest of the proof. To show that $\overline{I}$ is the closed linear span of $\{I, I, I\}$, it suffices to establish

$$\{I, I, I\}^\perp \subset I^\perp.$$

We note that $\{I^\perp, I, I\} = \{0\}$ since

$$\langle\{I^\perp, I, I\}, V\rangle = \langle I^\perp, \{V, I, I\}\rangle \subset \langle I^\perp, I\rangle = \{0\}.$$

Similarly, $\{I^\perp, I, I^\perp\} = 0$. It follows from $V = \overline{I} \oplus I^\perp$ that

$$\{V, I, V\} = \{\overline{I}, I, \overline{I}\} + \{I^\perp, I, \overline{I}\} + \{\overline{I}, I, I^\perp\} + \{I^\perp, I, I^\perp\} \subset \{\overline{I}, \overline{I}, \overline{I}\} \subset \overline{I}.$$

Let $a \in \{I, I, I\}^{\perp}$ and write $a = b + c$ with $b \in \overline{I}$ and $c \in I^{\perp}$. Then

$$\langle \{b, V, V\}, V \rangle = \langle \{b, V, \overline{I}\}, V \rangle + \langle \{b, V, I^{\perp}\}, V \rangle$$
$$= \langle a - c, \{V, \overline{I}, V\} \rangle + \langle b, \{V, I^{\perp}, V\} \rangle$$
$$\subset \langle a, \{\overline{I}, \overline{I}, \overline{I}\} \rangle + \langle c, \overline{I} \rangle + \langle b, I^{\perp} \rangle = \{0\}.$$

Hence we have $b \in V^{\square}$ and $b = 0$, that is, $a \in I^{\perp}$.

Finally, if $a \in I^{\square}$, then

$$\langle a, \{I, I, I\} \rangle = \langle \{a, I, I\}, I \rangle = \{0\}$$

and therefore $a \in I^{\perp}$. $\square$

Let V be a Jordan triple system. An element $a \in V$ is called an *absolute zero divisor* if the quadratic operator $Q_a = 0$. Let

$$V^0 = \{a \in V : Q_a = 0\}$$

denote the set of absolute zero divisors of V and let

$$V^{\square} = \{a \in V : a \square x = 0, \forall x \in V\}$$

be the annihilator of V. Given $a \in V^0$, we have

$$(a \square x)^2 = (x \square a)^2 = 0$$

for all $x \in V$. This follows from the identity (1.18):

$$2(a \square x)^2 = Q_a(x) \square x + Q_a Q_x$$

and also $2(x \square a)^2 = x \square Q_a(x) + Q_x Q_a$ by (1.17). The annihilator $V^{\square}$ of a JH-triple V is always contained in V^0, as will be shown.

By (2.30), a JH-triple V has an orthogonal decomposition

$$V = Z(V) \oplus Z(V)^{\perp},$$

where $Z(V) = \{a \in V : a \square x = x \square a, \forall x \in V\}$ is a closed abelian subtriple of V.

Lemma 3.5.10 *For a JH-triple V, we have $V^{\square} \subset V^0 \cap Z(V)$ and $V^{\square}$ is a closed triple ideal in V.*

Proof Let $a \in V^{\square}$ and $x \in V$. Then $Q_a = 0$ since $Q_a(v) = a \square v(a)$. We have $(x \square a)^2 = 0$ by the previous remark. Hence $d(a, x)^2 = (a \square x - x \square a)^2 = 0$. Since the inner derivation $d(a, x)$ is a normal operator, we have $d(a, x) = 0$, that is, $a \square x = x \square a$. This proves $a \in Z(V)$. The last assertion follows from $\{V, V^{\square}, V\} = \{V^{\square}, V, V\} = \{0\}$. $\square$

We now derive some useful characterisations of absolute zero divisors in Jordan Hilbert triples.

Lemma 3.5.11 *Let V be a Jordan Hilbert triple with an associative inner product. Let $a \in V$. The following conditions are equivalent:*

(i) $Q_a = 0$.
(ii) $(a \,\square\, x)^2 = 0$ *for all* $x \in V$.
(iii) $(x \,\square\, a)^2 = 0$ *for all* $x \in V$.
(iv) $a \,\square\, a = 0$.
(v) $a^3 = 0$.

Proof (ii) $\Rightarrow$ (iv) and (iii) $\Rightarrow$ (iv). Since the inner product is associative, the box operator $a \,\square\, a$ is symmetric and therefore $(a \,\square\, a)^2 = 0$ implies $a \,\square\, a = 0$.

(iv) $\Leftrightarrow$ (v). We need only show that $a^3 = 0$ implies $a \,\square\, a = 0$. Observe first that $a^3 = 0$ implies $Q_a(a \,\square\, a) = 0$, from the identity

$$\{a, x, \{a, a, a\}\} = \{a, \{x, a, a\}, a\} \qquad (x \in V).$$

The identity (1.17) gives

$$2\{a, a, \{x, a, a\}\} = \{a, \{a, x, a\}, a\} + \{a, \{a, a, a\}, x\} \qquad (x \in V)$$

and hence $2(a \,\square\, a)^2 = Q_a^2$. It follows that $2(a \,\square\, a)^3 = Q_a^2(a \,\square\, a) = 0$. Therefore $a \,\square\, a = 0$ since $a \,\square\, a$ is symmetric by associativity of the inner product.

(iv) $\Rightarrow$ (i). Let $x \in V$. We have, from the Jordan triple identity,

$$\{x, a, \{a, x, a\}\} = \{\{x, a, a\}, x, a\} - \{a, \{a, x, x\}, a\} + \{a, x, \{x, a, a\}\}$$
$$= -\{a, \{a, x, x\}, a\}.$$

Therefore

$$\|\{a, x, a\}\|^2 = \langle \{a, x, a\}, \{a, x, a\}\rangle = \langle \{x, a\{a, x, a\}\}, a\rangle$$
$$= -\langle \{a, \{a, x, x\}, a\}, a\rangle = -\langle a, \{\{a, x, x\}, a, a\}\rangle = 0.$$ $\square$

Proposition 3.5.12 *Let V be a Jordan Hilbert triple with an associative inner product. Then we have $V^0 = V^\square$.*

Proof We first show that the absolute zero divisors V^0 form a closed subspace of V. Let $a, b \in V^0$. Since the inner product is associative, we have $(a \,\square\, b)^* = b \,\square\, a$. From Lemma 3.5.11 and the triple identity (1.26), we obtain

$$[a \,\square\, b, b \,\square\, a] = \{a, b, b\} \,\square\, a - b \,\square\, \{a, a, b\} = 0.$$

It follows that $a \,\square\, b$ is a normal operator on the Hilbert space V. Since $(a \,\square\, b)^2 = 0$ by the remark before Lemma 3.5.10, we have $a \,\square\, b = 0 = b \,\square\, a$

and hence $(a + b) \Box (a + b) = 0$. This implies $a + b \in V^0$ by Lemma 3.5.11. Hence V^0 is a closed subspace of V.

We now show that V^0 is a triple ideal of V, using an argument in Neher [92, p. 155]. Let $a \in V^0$ and $x, y \in V$. Then $\{z, a, z\} \in V^0$ for all $z \in V$, and hence $\{x, a, y\} \in V^0$. Also, the identity

$$Q(B(x, y)a, B(x, y)a) = B(x, y)Q_a B(y, x)$$

for the Bergmann operator $B(x, y)$ in Theorem 1.2.18 implies $B(x, y)a \in V^0$. It follows from the fact that V^0 is a subspace of V and

$$B(x, y)a = a - 2\{x, y, a\} + Q_x Q_y(a)$$

that $\{x, y, a\} \in V^0$, proving that V^0 is a triple ideal of V.

Finally, we prove that $a \Box x = 0$ for $a \in V^0$ and all $x \in V$. Since $a \Box b = 0$ for all $b \in V^0$ as shown before, we need only show that $a \Box x = 0$ for all x in the orthogonal complement $(V^0)^\perp$ of V^0. For any $y, z \in V$, the associativity of the inner product gives

$$\langle \{a, x, y\}, z \rangle = \langle \{y, \{x, a, z\}\} = \langle \{a, z, y\}, x \rangle = 0$$

since V^0 is an ideal and $x \in (V^0)^\perp$. Hence $a \Box x = 0$. $\qquad\square$

Let V be a Jordan triple as well as a Hilbert space. If $a, b \in V$ satisfies

$$\langle (a \Box b)x, y \rangle = \langle x, (b \Box a)y \rangle$$

for all $x, y \in V$, then $a \Box b$ is weakly continuous and hence $a \Box b \in L(V)$.

Lemma 3.5.13 *Let V be a Jordan triple and a Hilbert space. Let*

$$V_\sigma = \{a \in V : a \Box x \in L(V) \text{ and } (a \Box x)^* = x \Box a \text{ for all } x \in V\}$$
$$= \{a \in V : \langle (a \Box x)y, z \rangle = \langle y, (x \Box a)z \rangle \text{ for all } x, y, z \in V\}.$$

Then V_σ is a subtriple of V and the inherited inner product on V_σ is associative.

Proof Let $a, x \in V_\sigma$ and $b, z, w \in V$. Using the identity

$$2\{a, \{x, a, b\}, z\} = \{\{a, x, a\}, b, z\} + \{\{a, b, a\}, x, z\}$$

we deduce that

$$\langle (\{a, x, a\} \Box b)z, w \rangle = 2\langle (a \Box \{x, a, b\})z, w \rangle - \langle (\{a, b, a\} \Box x)z, w \rangle$$
$$= 2\langle z, (\{x, a, b\} \Box a)w \rangle - \langle z, (x \Box \{a, b, a\})w \rangle$$
$$= \langle z, (b \Box \{a, x, a\})w \rangle$$

by the identity (1.17). Hence $\{a, x, a\} \in V_\sigma$ and V_σ is a subtriple of V. The last assertion is obvious. $\qquad\square$

If V is a continuous JH-triple in Lemma 3.5.13, then

$$V_\sigma = \{a \in V : (a \,\square\, x)^* = x \,\square\, a \text{ for all } x \in V\}$$

is closed by continuity of the triple product and is the largest subtriple of V in which every element a induces a symmetric operator $a \,\square\, b + b \,\square\, a$ for each $b \in V$. We call V_σ the *associative part* of V.

Example 3.5.14 The JH-triple $\mathcal{A}$ in Example 3.5.8 has the orthogonal decomposition

$$\mathcal{A} = \left\{ \begin{pmatrix} \alpha & 0 \\ 0 & \alpha \end{pmatrix} : \alpha \in \mathbb{R} \right\} \oplus \left\{ \begin{pmatrix} 0 & \beta \\ 0 & 0 \end{pmatrix} : \beta \in \mathbb{R} \right\},$$

where the first summand is the associative part $\mathcal{A}_\sigma$ of $\mathcal{A}$.

We have a precise relationship between the annihilator $V^\square$ and the zero divisors V^0 of a JH-triple V.

Proposition 3.5.15 *Let V be a JH-triple. Then we have* $V^\square = V^0 \cap V_\sigma \cap Z(V)$.

Proof By Lemma 3.5.10, we have $V^\square \subset V^0 \cap V_\sigma \cap Z(V)$. If $a \in V^0 \cap V_\sigma \cap Z(V)$, then $(a \,\square\, x)^* = x \,\square\, a = a \,\square\, x$ for all $x \in V$ and therefore $a \,\square\, x = 0$, since $(a \,\square\, x)^2 = 0$ by the remark before Lemma 3.5.10. $\square$

If V is a Hermitian Jordan Hilbert triple, then $Z(V)$ is contained in the annihilator $V^\square$ since $a \,\square\, v = 0$ for $a \in Z(V)$ and $v \in V$. Indeed, we have $-ia \,\square\, v = a \,\square\, iv = iv \,\square\, a = ia \,\square\, v$. It follows that $V^\square = V^0 = Z(V)$.

The arguments in the proof of Lemma 3.5.11 imply that for $a \in V_\sigma$, the condition $a^3 = 0$ is equivalent to $Q_a(V) = \{0\}$ which, in turn, is equivalent to $Q_a|_{V_\sigma} = 0$. The following lemma shows that non-degeneracy of V_σ is equivalent to anisotropy as well as vanishing of the annihilator $V_\sigma^\square = \{a \in V_\sigma : a \,\square\, x|_{V_\sigma} = 0, \forall x \in V_\sigma\}$.

Lemma 3.5.16 *Let V be a JH-triple. Then*

$$V_\sigma^\square = V^0 \cap V_\sigma = \{a \in V_\sigma : Q_a = 0\} = \{a \in V_\sigma : a^3 = 0\},$$

which is a triple ideal of V_σ.

Proof We have already established the last equality. Let $a \in V_\sigma^\square$. Then $a^3 = 0$ and hence $a \in V^0$ by the earlier remark. Since the inner product of V_σ is associative, the arguments in the proof of Proposition 3.5.12 can be used to show that $V^0 \cap V_\sigma$ is a triple ideal in V_σ. These arguments can also be used to show $V^0 \cap V_\sigma \subset V_\sigma^\square$ by considering

$$V = \overline{V^0 \cap V_\sigma} \oplus (V^0 \cap V_\sigma)^\perp.$$

The only additional argument is that showing $(V^0 \cap V_\sigma) \square \overline{V^0 \cap V_\sigma} = \{0\}$ because continuity of the triple product is not assumed and hence the closure $\overline{V^0 \cap V_\sigma}$ is involved. Nevertheless, given $a \in V^0 \cap V_\sigma$ and $b \in \overline{V^0 \cap V_\sigma}$ with $b = \lim_n b_n$ and $b_n \in V^0 \cap V_\sigma$, we have

$$\langle a \square b(x), y \rangle = \langle \{a, y, x\}, b \rangle = \lim_n \langle \{a, y, x\}, b_n \rangle = \lim_n \langle a \square b_n(x), y \rangle = 0$$

$$(x, y \in V)$$

since $a \square b_n = 0$, as in the proof of Proposition 3.5.12. One can show, as before, $a \square z|_{\overline{V}_\sigma} = 0$ for $z \in (V^0 \cap V_\sigma)^\perp$. $\qquad\square$

Let V be a continuous JH-triple. Then its associative part V_σ is a continuous JH-triple with an associative inner product. The triple product vanishes on its annihilator $V^\square$ and therefore the nontrivial part of V is the orthogonal complement $(V^\square)^\perp$. The annihilator $V^\square$ is a closed triple ideal in V_σ and induces the orthogonal decomposition

$$V_\sigma = V^\square \oplus (V_\sigma \cap (V^\square)^\perp), \tag{3.18}$$

where, by Lemma 3.5.9, $V_\sigma \cap (V^\square)^\perp$ is a closed triple ideal of V_σ and in particular a subtriple of V.

Hence we have the orthogonal decomposition

$$V = V^\square \oplus (V_\sigma \cap (V^\square)^\perp) \oplus V_\sigma^\perp$$

and we reduce the study of V to that of the latter two nontrivial summands.

The JH-triple $V_\sigma \cap (V^\square)^\perp$ is weakly semisimple. It inherits an associative inner product and therefore its structure can be described completely by the results in Kaup [71] and Neher [92]. The inner product in the orthogonal complement $V_\sigma^\perp$ is not associative. We will discuss the ternary structure of $V_\sigma^\perp$ later. Let us first extend the spectral theory for finite-dimensional Jordan triples in Theorem 1.2.34 to the associative part V_σ. The spectral decomposition of elements in V_σ involves *signed* tripotents.

Definition 3.5.17 Let V be a Jordan triple system. A *signed tripotent* in V is an element $e \in V$ such that

$$\{e, e, e\} = e \quad \text{or} \quad \{e, e, e\} = -e.$$

If $e \neq 0$, then e is also called a *positive* tripotent in the first case, whereas in the second case, e is called a *negative* tripotent.

Lemma 3.5.18 *Let V be a Jordan Hilbert triple with an associative inner product. Let $e, f \in V$ be two signed tripotents of opposite sign. Then $e \,\square\, f = 0$.*

Proof Say f is a negative tripotent. Then it induces a Peirce decomposition

$$V = V_0(f) \oplus V_1(f) \oplus V_2(f)$$

where the Peirce k-space

$$V_k(f) = \left\{ v \in v : (f \,\square\, f)(v) = -\frac{k}{2} v \right\}$$

is the range of the Peirce k-projection $P_k(f)$ which has the same form as that for a positive tripotent.

By associativity of the inner product, we see readily that

$$\langle P_k(u)(x), y \rangle = \langle x, P_k(u)(y) \rangle \qquad (x, y \in V)$$

for each Peirce k-projection $P_k(u)$ of a signed tripotent u.

Let $e_k = P_k(f)e$ and $f_k = P_k(e)f$ for $k = 0, 1, 2$. Then we have

$$\langle f, \{e, e, f\} \rangle = \left\langle f_0 + f_1 + f_2, \frac{1}{2} f_1 + f_2 \right\rangle$$

$$= \frac{1}{2} \langle f_1, f_1 \rangle + \langle f_2, f_2 \rangle \geq 0$$

and also

$$\langle f, \{e, e, f\} \rangle = \langle \{f, f, e\}, e \rangle$$

$$= \left\langle -\frac{1}{2} e_1 - e_2, e_0 + e_1 + e_2 \right\rangle$$

$$= -\frac{1}{2} \langle e_1, e_1 \rangle - \langle e_2, e_2 \rangle \leq 0.$$

Hence $f_1 = f_2 = 0$ and $f = f_0 \in P_0(e)(V)$, which gives $e \,\square\, f = 0$ by Theorem 1.2.44. $\qquad\qquad\square$

Let V_σ be the associative part of a continuous JH-triple V and let $a \in V_\sigma$ with $a^3 \neq 0$. As in the case of JB*-triples, we consider the closed subtriple $R(a)$ of V_σ generated by a. Let $\mathcal{R}$ be the real closed linear span of $R(a) \,\square\, R(a)$ in $L(V)$. Then $\mathcal{R}|_{R(a)}$ is a real Banach subalgebra of $L(R(a))$ and is abelian by power associativity. Let

$$\mathcal{A} = R(a) \oplus \mathcal{R}|_{R(a)}$$

and define a product in $\mathcal{A}$ by

$$(u \oplus h)(v \oplus k) = (h(v) + k(u)) \oplus (u \,\square\, v|_{V(a)} + hk).$$

By Lemma 3.5.5, there is a constant $c > 0$ such that $\|u \,\square\, v\| \le c\|u\|\|v\|$ for all $u, v \in V_\sigma$. By scaling the inner product of V_σ, which would not change $R(a)$ nor orthogonality in V_σ, we may assume $c = 1$, so that $\mathcal{A}$ is a real abelian Banach algebra. As before (cf. proof for Theorem 3.1.12), there is a real algebra homomorphism $\chi : \mathcal{A} \longrightarrow \mathbb{C}$ such that $\chi(a) \neq 0$. For each $b \in R(a)$, the box operator $b \,\square\, b : V \longrightarrow V$ is symmetric and therefore has real spectrum. We infer that the quasi-spectrum of $b^2 = b \,\square\, b|_{R(a)}$ in $\mathcal{A}$ is real, for the same reason as given in the proof of Lemma 2.5.20 and Theorem 3.1.12. It follows from $\chi(b^2) = \chi(b)^2$ that $\chi(b) \in \mathbb{R} \cup i\mathbb{R}$. Hence we have $\chi(R(a)) = \mathbb{R}$ or $\chi(R(a)) = i\mathbb{R}$, which implies $\dim R(a)/\chi|_{R(a)}^{-1}(0) = 1$. Therefore we have the orthogonal decomposition

$$R(a) = \chi|_{R(a)}^{-1}(0) \oplus \mathbb{R}z$$

for some $z \in R(a)$ with $\chi(z) = 1$. Since χ is an algebra homomorphism on $\mathcal{A}$, the kernel $\chi|_{R(a)}^{-1}(0)$ is a triple ideal in $R(a)$, and it follows from Lemma 3.5.9 that $\mathbb{R}z$ is a triple ideal of $R(a)$ and in particular, $\{z, z, z\} \in \mathbb{R}z$. Hence $R(a)$ contains a signed tripotent.

Proposition 3.5.19 *Let V_σ be the associative part of a continuous JH-triple of V. Then for each $a \in V_\sigma$ with $a^3 \neq 0$, we have an orthogonal decomposition*

$$R(a) = \bigoplus_{n=1}^{\infty} \mathbb{R}e_n$$

where each e_n is a signed tripotent.

Proof We first observe that $z^3 \neq 0$ for all $z \in R(a)$. Otherwise we would have $z \,\square\, z = 0$ for some $z \in R(a)$, by Lemma 3.5.11, and power associativity would imply nilpotency of $a \,\square\, a$, which is impossible, since $a \,\square\, a$ is symmetric and $a^3 \neq 0$.

By the above remark, $R(a)$ contains a signed tripotent e such that $\mathbb{R}e$ is a triple ideal in $R(a)$. Let $\{e_\alpha\}$ be a maximal orthogonal family of such signed tripotents. We must have $R(a) = \bigoplus_\alpha \mathbb{R}e_\alpha$. Indeed, if there is an element z in the orthogonal complement $(\bigoplus_\alpha \mathbb{R}e_\alpha)^\perp \cap R(a)$ which is a subtriple of $R(a)$, then repeating the argument in the remark yields a signed tripotent in the orthogonal complement, contradicting maximality.

It follows that there is an orthogonal sequence (e_n) of signed tripotents in $R(a)$ such that $a = \sum_n \lambda_n e_n$ with $\lambda_n \in \mathbb{R}$. Since $R(a)$ is generated by odd powers of a, we conclude that

$$R(a) = \bigoplus_{n=1}^{\infty} \mathbb{R}e_n.$$

$\square$

The lemma that follows implies that the signed tripotents e_n in Proposition 3.5.19 can be chosen to be *primitive*. A signed tripotent in a Jordan Hilbert triple is called *primitive* if it cannot be decomposed into a sum of two triple orthogonal signed tripotents.

Lemma 3.5.20 *Let e be a signed tripotent in a Jordan Hilbert triple with an associative inner product. Then e is the sum of at most a finite number of primitive signed tripotents.*

Proof If e is not primitive, then it can be decomposed as a sum of two triple orthogonal signed tripotents, e_1 and e_2. By orthogonality, we have

$$\|e\|^2 = \|e_1\|^2 + \|e_2\|^2.$$

By Lemma 3.5.5, we have

$$\|e\|^2 = \|\{e, e, e\}\|^2 \le c\|e\|^6$$

for some $c > 0$. Hence e can be decomposed into a sum of at most finitely many mutually triple orthogonal signed primitive tripotents. $\qquad\square$

To study $V_\sigma^\perp$, we begin by showing that the associative part V_σ is invariant under inner derivations. To see this, we first note that

$$\langle \{c, x, y\}, z \rangle = \langle x, \{c, z, y\} \rangle \qquad (c \in V_\sigma \text{ and } x, y, z \in V),$$

which follows from $\langle \{c, x, y\}, z \rangle = \langle y, \{x, c, z\} \rangle = \langle \{c, z, y\}, x \rangle$.

Lemma 3.5.21 *Let V be a JH-triple with the associative part V_σ. Then $d(a, b)V_\sigma \subset V_\sigma$ for all $a, b \in V$.*

Proof Let $c \in V_\sigma$. We need to show $d(a, b)c \in V_\sigma$. Let $y, z \in V$. Then we have

$$\{d(a, b)c, x, y\} = d(a, b)\{c, x, y\} - \{c, d(a, b)x, y\} - \{c, x, d(a, b)y\}$$

for each $x \in V$. Hence

$$
\begin{aligned}
\langle d(a, b)c &\,\square\, x(y), z \rangle \\
&= \langle d(a, b)\{c, x, y\}, z \rangle - \langle \{c, d(a, b)x, y\}, z \rangle - \langle \{c, x, d(a, b)y\}, z \rangle \\
&= \langle \{c, x, y\}, d(b, a)z \rangle - \langle \{c, z, y\}, d(a, b)x \rangle - \langle d(a, b)y, \{x, c, z\} \rangle \\
&= -\langle y, \{x, c, d(a, b)z\} \rangle - \langle y, \{d(a, b)x, c, z\} \rangle + \langle y, d(a, b)\{x, c, z\} \rangle \\
&= \langle y, x \,\square\, d(a, b)c(z) \rangle.
\end{aligned}
$$

This proves $d(a, b)c \in V_\sigma$. $\qquad\square$

To discuss the structure of the complement $V_\sigma^\perp$ of V_σ, we introduce another ternary structure related to JH-triples.

A *Hilbert ternary algebra* is a real Hilbert space $(V, \langle \cdot, \cdot \rangle)$ equipped with a trilinear map $[\cdot, \cdot, \cdot] : V^3 \longrightarrow V$, called a *ternary product*, for which the ternary and inner products are related by

$$\langle [a, b, x], y \rangle = \langle x, [b, a, y] \rangle = \langle a, [y, x, b] \rangle \qquad (a, b, x, y \in V).$$

In other words, the inner product is *associative* with respect to the ternary product $[\cdot, \cdot, \cdot]$.

Given a Hilbert ternary algebra V with ternary product $[\cdot, \cdot, \cdot]$ and $a, b \in V$, we can also define the box operator $a \,\square\, b : x \in V \mapsto [a, b, x] \in V$. Applying the arguments in the proof of Lemma 3.5.5, one can show that the ternary product is continuous.

Given subsets X, Y and Z of a Hilbert ternary algebra V, we define

$$[X, Y, Z] = \{[x, y, z] : x \in X, y \in Y, z \in Z\}.$$

A subspace J of V is called a *ternary subalgebra* if $[J, J, J] \subset J$, and is called a *ternary ideal* if $[J, V, V] + [V, J, V] + [V, V, J] \subset J$. A closed ternary subalgebra is also called a *Hilbert ternary subalgebra*.

Analogous to Lemma 3.5.9, the orthogonal complement $J^\perp$ of a ternary ideal J in a Hilbert ternary algebra is also a ternary ideal (cf. Lemma 3.5.27).

A Hilbert ternary algebra need not be a Jordan triple, but a JH-triple with an associative inner product is a Hilbert ternary algebra in the given triple product.

Example 3.5.22 On the real Hilbert space $(L_2(V, W), \langle \cdot, \cdot \rangle_2)$ of Hilbert–Schmidt operators, we define a ternary product by

$$[a, b, c]_2 = ab^*c \qquad (a, b, c \in L_2(V, W)),$$

which is called the *canonical ternary product*. We define the *symmetrized ternary product* by

$$\{a, b, c\}_2 = \frac{1}{2}([a, b, c]_2 + [c, b, a]_2) \qquad (a, b, c \in L_2(V, W)).$$

With each of these two ternary products, $L_2(V, W)$ is a Hilbert ternary algebra. We note that $\{\cdot, \cdot, \cdot\}_2$ is the canonical triple product in $L_2(V, W)$ introduced before and $(L_2(V, W), \{\cdot, \cdot, \cdot\}_2)$ is a JH-triple with an associative inner product.

Definition 3.5.23 A Hilbert ternary algebra $(V, \{\cdot, \cdot, \cdot\})$ is called *abelian* if

$$[a, b, [x, y, z]] = [[a, b, x], y, z]$$

for all $a, b, x, y, z \in V$.

The ternary algebra $(L_2(V, W), [\cdot, \cdot, \cdot]_2)$ in Example 3.5.22 is abelian. In an abelian Hilbert ternary algebra V, we have

$$
\begin{aligned}
\langle [a, b, [x, y, z]], w \rangle &= \langle [x, y, z], [b, a, w] \rangle \\
&= \langle z, [y, x, [b, a, w]] \rangle \\
&= \langle z, [[y, x, b], a, w]] \rangle \\
&= \langle \{a, [y, x, b], z], w \rangle,
\end{aligned}
$$

which gives

$$
[a, b, [x, y, z]] = [a, [y, x, b], z], \tag{3.19}
$$

but it is not always true that $[a, b, [x, y, z]] = [a, [b, x, y], z]$! For instance, if $\quad a = \begin{pmatrix} 1 & 0 \\ 0 & 0 \end{pmatrix}$ and $\quad b = \begin{pmatrix} 0 & 1 \\ 0 & 0 \end{pmatrix}$, then $\quad [b, b, [a, a, b]_2]_2 \neq [b, [b, a, a]_2, b]_2$ in $L_2(\mathbb{R}^2)$.

In an abelian Hilbert ternary algebra V, we have $[a, a, [a, a, x]] = [[a, a, a], a, x]$ for $a, x \in V$. Hence $[a, a, a] = 0$ implies $(a \,\square\, a)^2 = 0$ and therefore $a \,\square\, a = 0$, since $a \,\square\, a$ is a symmetric operator on V. Extending the construction of the symmetrised product $\{\cdot, \cdot, \cdot\}_2$ in Example 3.5.22, we show that every abelian Hilbert ternary algebra carries a natural JH-triple structure.

Proposition 3.5.24 *Let* $(V, \langle \cdot, \cdot \rangle, [\cdot, \cdot, \cdot])$ *be an abelian Hilbert ternary algebra. Then* $(V, \langle \cdot, \cdot \rangle)$ *is a JH-triple in the symmetrized ternary product*

$$
\{x, y, z\}_s := \frac{1}{2}([x, y, z] + [z, y, x])
$$

for which the inner product $\langle \cdot, \cdot \rangle$ *is associative.*

Proof It is clear that the triple product $\{\cdot, \cdot, \cdot\}_s$ is symmetric in the outer variables and that the inner product is associative with respect to $\{\cdot, \cdot, \cdot\}_s$. The main triple identity for $\{\cdot, \cdot, \cdot\}_s$ follows from the abelian condition and (3.19). □

In the previous construction, we have $\{a, \cdot, a\}_s = [a, \cdot, a]$, and one infers readily from Lemma 3.5.11 that, in an abelian Hilbert ternary algebra V, $[a, a, a] = 0$ if and only if $[a, x, a] = 0$ for all $x \in V$. This fact can also be easily established directly. If we define the concepts of *non-degeneracy* and *anisotropy* in terms of the ternary product for Hilbert ternary algebras as we did for Jordan triple systems, then the previous remark implies that these two notions are identical for *abelian* Hilbert ternary algebras.

As a converse to Proposition 3.5.24, we show that every JH-triple also has an inherent Hilbert ternary algebraic structure.

Proposition 3.5.25 *Let* $(V, \langle \cdot, \cdot \rangle, \{\cdot, \cdot, \cdot\})$ *be a JH-triple. Then* $(V, \langle \cdot, \cdot \rangle)$ *is a Hilbert ternary algebra in the ternary product* $[\cdot, \cdot, \cdot]_d$ *defined by*

$$[a, b, x]_d = d(a, b)x \qquad (a, b, x \in V).$$

Proof We need to show that the inner product is associative with respect to the ternary product $[\cdot, \cdot, \cdot]_d$. By skew-symmetry of the inner derivation $d(a, b)$, we have

$$\langle [a, b, x]_d, y \rangle = \langle x, [b, a, y]_d \rangle$$

for all $a, b, x, y \in V$. Using the identity

$$d(a, b)x = d(a, x)b + d(x, b)a$$

we deduce that

$$
\begin{aligned}
\langle a, [y, x, b]_d \rangle &= \langle a, d(y, x)b \rangle \\
&= \langle a, d(y, b)x \rangle + \langle a, d(b, x)y \rangle \\
&= \langle d(b, y)a, x \rangle + \langle d(x, b)a, y \rangle \\
&= \langle d(b, a)y, x \rangle + \langle d(a, y)b, x \rangle + \langle d(x, a)b, y \rangle + \langle d(a, b)x, y \rangle \\
&= 2\langle d(b, a)y, x \rangle + \langle b, d(y, a)x \rangle + \langle b, d(a, x)y \rangle \\
&= 2\langle d(b, a)y, x \rangle + \langle b, d(y, x)a \rangle \\
&= 2\langle d(b, a)y, x \rangle - \langle d(y, x)b, a \rangle,
\end{aligned}
$$

which gives $\langle a, [y, x, b]_d \rangle = \langle [b, a, y]_d, x \rangle$. $\qquad\qquad\square$

Definition 3.5.26 Given a JH-triple $(V, \langle \cdot, \cdot \rangle, \{\cdot, \cdot, \cdot\})$, the Hilbert ternary algebra $(V, \langle \cdot, \cdot \rangle, [\cdot, \cdot, \cdot]_d)$ defined in Proposition 3.5.25 is called the *derived Hilbert ternary algebra* of V and is denoted by $[V]$.

We note that the Hilbert ternary algebra $[V]$ is never abelian unless V is flat or $V = \{0\}$. We also note that $[V]$ is *isotropic* in the sense that $[a, a, a] = 0$ for all $a \in V$. One can strengthen Lemma 3.5.21 to assert that the associative part V_σ is a ternary ideal of $[V]$ for a continuous JH-triple V.

Lemma 3.5.27 *Let V be a JH-triple. Then the closure $\overline{V_\sigma}$ of the associative part V_σ is a ternary ideal of the derived Hilbert ternary algebra $[V]$. The orthogonal complement $V_\sigma^\perp$ is also a ternary ideal of $[V]$.*

Proof We note that the inner derivation $d(a, b) = -d(b, a)$ is continuous. In view of Lemma 3.5.21, it suffices to show $[V, V_\sigma, V]_d \subset \overline{V_\sigma}$, since

$$[V_\sigma, V, V]_d = -[V, V_\sigma, V]_d.$$

Let $a \in V_\sigma$ and $b, c \in V$. We need to show $[b, a, c]_d = d(b, a)c \in \overline{V_\sigma}$. Assume first that $b = b_1 + b_2$ with $b_1 \in V_\sigma$ and $b_2 \in V_\sigma^\perp$. By Lemma 3.5.21, we have $d(b_1, a)c \in V_\sigma$, since

$$d(b_1, a)c = d(b_1, c)a + d(c, a)b_1.$$

We show $d(b_2, a)c = 0$. For each $v \in V$, we have

$$\langle d(b_2, a)c, v \rangle = \langle [b_2, a, c]_d, v \rangle = \langle b_2, [v, c, a]_d \rangle = 0$$

since $[v, c, a]_d \in V_\sigma$ by Lemma 3.5.21. Hence $d(b_2, a)c = 0$ and $d(b, a)c \in V_\sigma$.

Now $V = \overline{V_\sigma} \oplus V_\sigma^\perp$ and continuity of the inner derivation implies $d(b, a)c \in \overline{V_\sigma}$ for any $b \in V$.

The fact that $V_\sigma^\perp$ is also a ternary ideal of $[V]$ follows from

$$\langle [V_\sigma^\perp, V, V]_d + [V, V_\sigma^\perp, V]_d + [V, V, V_\sigma^\perp]_d, \; V_\sigma \rangle$$
$$= \langle V_\sigma^\perp, [V_\sigma, V, V]_d + [V, V_\sigma, V]_d + [V, V, V_\sigma]_d \rangle$$
$$\subset \langle V_\sigma^\perp, \overline{V_\sigma} \rangle = \{0\}.$$

$\square$

Lemma 3.5.28 *Let V be a JH-triple and let $[V]$ be the derived Hilbert ternary algebra. If J is a triple ideal in V, then J is a ternary ideal in $[V]$.*

Proof Let $a \in J$ and $x, y \in V$. Then we have $[a, x, y]_d = \{a, x, y\} - \{x, a, y\} \in J$ and likewise $[x, y, a]_d \in J$. Also, $[x, a, y]_d = \{x, a, y\} - \{a, x, y\} \in J$. $\square$

Let W be a Hilbert ternary algebra. The *ternary annihilator* of W is defined to be the closed ternary ideal

$$\mathrm{Ann}\, W = \{a \in W : [a, W, W] = \{0\}\} = \{a \in W : [W, W, a] = \{0\}\}.$$

Lemma 3.5.29 *Let V be a JH-triple and $[V]$ the derived Hilbert ternary algebra. Then we have $\mathrm{Ann}\,[V] = Z(V)$.*

Proof Indeed, $a \in Z(V)$ if and only if the inner derivation $d(a, x) = 0$ for all $x \in V$. $\square$

We now describe the structure of $V_\sigma^\perp$, which is a Hilbert ternary algebra in the derived ternary product $[a, b, c]_d = d(a, b)c$.

Lemma 3.5.30 *Let V be a JH-triple. Then $V_\sigma^\perp \cap Z(V)$ is the ternary annihilator of the Hilbert ternary algebra $(V_\sigma^\perp, [\cdot, \cdot, \cdot]_d)$, that is,*

$$V_\sigma^\perp \cap Z(V) = \{a \in V_\sigma^\perp : [a, V_\sigma^\perp, V_\sigma^\perp]_d = \{0\}\}.$$

Proof It suffices to show that the two conditions $a \in V_\sigma^\perp$ and $[a, V_\sigma^\perp, V_\sigma^\perp]_d = \{0\}$ imply $a \in Z(V)$. Assume the former conditions. We show $[a, V, V]_d = \{0\}$. We have

$$
\begin{aligned}
\langle [a, V, V]_d, V \rangle &= \langle [a, V_\sigma^\perp, V_\sigma^\perp]_d, V \rangle + \langle [a, V_\sigma^\perp, V_\sigma^{\perp\perp}]_d, V \rangle + \langle [a, V_\sigma^{\perp\perp}, V_\sigma^\perp]_d, V \rangle \\
&\quad + \langle [a, V_\sigma^{\perp\perp}, V_\sigma^{\perp\perp}]_d, V \rangle \\
&= \langle [a, V_\sigma^\perp, \overline{V_s}]_d, V \rangle + \langle [a, \overline{V_\sigma}, V_\sigma^\perp]_d, V \rangle + \langle [a, \overline{V_s}, \overline{V_s}]_d, V \rangle \\
&= \langle a, [V, \overline{V_\sigma}, V_\sigma^\perp]_d \rangle + \langle a, [V, V_\sigma^\perp, \overline{V_s}]_d \rangle + \langle a, [V, \overline{V_s}, \overline{V_s}]_d \rangle \\
&\subset \langle a, \overline{V_\sigma} \rangle = \{0\}
\end{aligned}
$$

which completes the proof.   □

We have the decomposition

$$
V_\sigma^\perp = (V_\sigma^\perp \cap Z(V)) \oplus (V_\sigma^\perp \cap Z(V)^\perp),
$$

where $[a, b, c]_d = 0$ for all $a, b, c \in V_\sigma^\perp \cap Z(V)$ and the ternary annihilator of of the Hilbert ternary algebra $(V_\sigma^\perp \cap Z(V)^\perp, [\cdot, \cdot, \cdot]_d)$ is zero.

In the terminology of Castellon and Cuenca [19], a Hilbert ternary algebra is a *real H*-triple system with identity involution* and contains a nonzero minimal closed ternary ideal. Using the fact that the orthogonal complement of a ternary ideal is also a ternary ideal, it is not difficult to see that a Hilbert ternary algebra with zero annihilator; in particular, the summand $V_\sigma^\perp \cap Z(V)^\perp$, is an orthogonal sum of minimal closed ternary ideals (cf. [19, theorem 1.6]).

Notes

We have only included the most basic properties of JB*-triples in this chapter. A useful survey, albeit without proofs, of recent results in JB*-triples, among other applications of Jordan theory, can be found in Iordănescu [60]. Earlier surveys have been given in Chu and Mellon [29] and Russo [100]. A survey including JB*-algebras can be found in the article '*Jordan structures in analysis*' by Rodriguez-Palacios [Jordan Algebras (Oberwolfach 1992) 97–186, de Gruyter, Berlin, 1994]. Real forms of JB*-triples were introduced and studied in Isidrc *et al.* [61] and Kaup [74]. Other than this, the results in the first section can be found in Kaup [69] and [70] in some form or another.

Contractivity of Peirce projections has been shown in Friedman and Russo [40]. Theorem 3.2.3 has been proved in Braun, Kaup, and Upmeier [14]. The proof of Corollary 3.2.4 follows the arguments in Harris [49, corollary 8]. In the case of JC*-triples, the formula for the Möbius transformation in Example 3.2.11 has been given in Harris [49] and Potapov [95].

The estimate of the norm of the positive and negative square roots of the Bergmann operator $B(a, a)$ is important in applications. The norm of the negative square root was obtained in Kaup [73], as well as the formula (3.3) and an explicit expression for the integral there. The alternative computation of the norm $\| B(a, a)^{-1/2} \|$ in Proposition 3.2.13 is an extension of a computation in Hamada *et al.* [45] for JC*-triples. Lemma 3.2.8 was first observed in Chu *et al.* [24], as well as other estimates of the norm $\| B(a, a)^{1/2} \|$. The distortion results given in Theorem 3.2.16 were obtained in Chu *et al.* [24]. Iteration of holomorphic maps has been widely studied. The iteration results in Theorem 3.2.19 and Lemma 3.2.21 for compact holomorphic maps on a Hilbert ball have been shown in Chu and Mellon [28]. Mellon [89] considered compact holomorphic maps on open unit balls of JB*-triples and obtained the results in Lemma 3.2.17 and Theorem 3.2.18. The holomorphic invariance of ellipsoids in a Hilbert ball was shown in Goeble [44].

Peirce projections play a fundamental role in the strucutre theory of JB*-triples. Contractive projections are important in the study of operator algebras and Banach spaces. Ranges of contractive projections on C*-algebras have been studied by Friedman and Russo, among others. They proved in [41] a special case of Theorem 3.3.1, namely, the same result for JC*-triples, using a functional-analytic method. A result relevant to Theorem 3.3.1 can also be found in Stacho [106]. The application of Theorem 3.3.1 to showing that the second dual of a JB*-triple is also a JB*-triple is due to Dineen [32]. The terminology of a structural projection originates from the notion of a *structural transformation* for Jordan pairs given in Loos [86].

The notion of a type I JBW*-triple, as well as the classification of type I JBW*-triples given in Theorem 3.3.12 and Theorem 3.3.13, is due to Horn [56]. Lemma 3.3.8 and Lemma 3.3.16, as well as Theorem 3.3.9, are proved in Horn [57]. Lemma 3.3.14 has been proved in Friedman and Russo [40], where the decomposition of a JBW*-triple into the direct sum of the atomic part and a weak* closed triple ideal was also shown. Theorem 3.3.19 has been proved in Friedman and Russo [42].

The Cauchy–Schwarz inequality for JB*-triples has been proved in Barton and Friedman [6]. The results in Theorem 3.4.6 and Theorem 3.4.8 describing the structure of non-surjective linear isometries on real JB*-triples are new. They extend the results in Chu and Mackey [27] for JB*-triples. Example 3.4.10 has been given in Apazoglou [4].

The concept of a real or complex ternary algebra was first introduced by Hestenes [53] to develop a spectral theory for non-hermitian matrices and operators. An isotropic abelian Hilbert ternary algebra is, in the definition of Hestenes, a *real ternary algebra*. Abelian Hilbert ternary algebras are exactly the *associative real H*-triple systems with identity involution* defined in Castellon and Cuenca [19]. It has been shown in Castellon and Cuenca [19] that a *simple* abelian Hilbert ternary algebra V, where V is said to be *simple* if it has a nontrivial ternary product and does not contain a nonzero proper closed ternary ideal, is ternary isomorphic to the ternary algebra of Hilbert–Schmidt operators between Hilbert spaces over $\mathbb{R}$, $\mathbb{C}$ or $\mathbb{H}$. The results on JH-triples in the last section are new.

Bibliography

[1] C.A. Akemann, The dual space of an operator algebra, *Trans. Amer. Math. Soc.* **126** (1967) 286–302.

[2] A.A. Albert, On a certain algebra of quantum mechanics, *Annals of Math.* **35** (1934) 65–73.

[3] A.A. Albert, On Jordan algebras of linear transformations, *Tran. Amer. Math. Soc.* **59** (1946) 524–555.

[4] M. Apazoglou, *Linear maps on real C*-algebras and related structures*, Ph.D. Thesis (University of London, 2010).

[5] J. Arazy, Isometries of Banach algebras satisfying the von Neumann inequality, *Math. Scand.* **74** (1994) 137–151.

[6] T. Barton and Y. Friedman, Grothendieck's inequality for JB*-triples and applications, *J. London Math. Soc.* **36** (1987) 513–523.

[7] T. Barton and R.M. Timoney, Weak*-continuity of Jordan triple products and its applications, *Math. Scand.* **59** (1986) 177–191.

[8] G. Birkhoff, Analytic groups, *Trans. Amer. Math. Soc.* **43** (1938) 61–101.

[9] E. Bishop and R.R. Phelps, A proof that every Banach space is subreflexive, *Bull. Amer. Math. Soc.* **67** (1961) 97–98.

[10] F.F. Bonsall and J. Duncan, *Numerical Range of Operators on Normed Spaces and Elements of Normed Algebras* (Cambridge: Cambridge Univ. Press, 1971).

[11] A. Borel, *Semisimple groups and Riemannian symmetric spaces*, Texts & Readings in Math. **16** (New Delhi: Hindustan Book Agency, 1998).

[12] N. Bourbaki, *Lie Groups and Lie Algebras*, Chapters 1–3 (Berlin: Springer-Verlag, 1989).

[13] H. Braun and M. Koecher, *Jordan-Algebren* (Berlin: Springer-Verlag, 1966).

[14] R. Braun, W. Kaup and H. Upmeier, A holomorphic characterization of Jordan C*-algebras, *Math. Z.* **161** (1978) 277–290.

[15] L.J. Bunce and C.-H. Chu, Compact operations, multipliers and Radon–Nikodym property in JB*-triples, *Pacific J. Math.* **153** (1992) 249–265.

[16] É. Cartan, Sur les domaines bornés homogènes de l'espace de n variables complexes, *Abh. Math. Semin. Univ. Hamburg* **11** (1935) 116–162.

[17] H. Cartan, Les fonctions de deux variables complexes et le problème de la représentation analytique, *J. Math. Pures Appl.* **10** (1931) 1–114.

[18] H. Cartan, *Sur les Groupes de Transformations Analytiques*, Act. Sci. Ind. **198** (Paris: Hermann, 1935).

[19] A. Castellon and J.A. Cuenca, Associative H*-triple systems, in *Nonassociative Algebraic Models*, Zaragoza 1989 (New York: Nova Sci. Publ. 1982), pp. 45–67.

[20] C. Chevalley, *Theory of Lie Groups* (Princeton: Princeton Univ. Press, 1946).

[21] C.-H. Chu, Grassmann manifolds of Jordan algebras, *Archiv Math.* **87** (2006) 179–192.

[22] C.-H. Chu, Jordan triples and Riemannian symmetric spaces, *Adv. Math.* **219** (2008) 2029–2057.

[23] C.-H. Chu, T. Dang, B. Russo and B. Ventura, Surjective isometries of real C*-algebras, *J. London Math. Soc.* **47** (1993) 97–118.

[24] C.-H. Chu, H. Hamada, T. Honda and G. Kohr, Distorsion theorems for convex mappings on homogeneous balls, *J. Math. Anal. Appl.* **369** (2010) 437–442.

[25] C.-H. Chu and B. Iochum, Complementation of Jordan triples in von Neumann algebras, *Proc. Amer. Math. Soc.* **108** (1990) 19–24.

[26] C.-H. Chu and J.M. Isidro, Manifolds of tripotents in JB*-triples, *Math. Z.* **233** (2000) 741–754.

[27] C.-H. Chu and M. Mackey, Isometries between JB*-triples, *Math. Z.* **251** (2005) 615–633.

[28] C.-H. Chu and P. Mellon, Iteration of compact holomorphic maps on a Hilbert ball, *Proc. Amer. Math. Soc.* **125** (1997) 1771–1777.

[29] C.-H. Chu and P. Mellon, Jordan structures in Banach spaces and symmetric manifolds, *Expos. Math.* **16** (1998) 157–180.

[30] T. Dang, Real isometries between JB*-triples, *Proc. Amer. Math. Soc.* **114** (1992) 971–980.

[31] S. Dineen, *The Schwarz Lemma* (Oxford: Oxford Univ. Press, 1989).

[32] S. Dineen, The second dual of a JB* triple system, in *Complex Analysis, Functional Analysis and Approximation Theory*, ed. J. Mujica (Amsterdam: North-Holland, 1986), pp. 67–69.

[33] J. Dieudonné, *Foundations of Modern Analysis* (London: Academic Press, 1969).

[34] C.J. Earle and R.S. Hamilton, A fixed point theorem for holomorphic mappings, *Proc. Symp. Pure Math.* **16** (1969) 61–65.

[35] C.M. Edwards and G. Rüttimann, On the facial structure of the unit balls in a JBW*-triple and its predual, *J. London Math. Soc.* **38** (1988) 317–332.

[36] C.M. Edwards, K. McCrimmon and G. Rüttimann, The range of a structural projection, *J. Funct. Analysis* **139** (1996) 196–224.

[37] C.M. Edwards and G. Rüttimann, The facial and inner ideal structures of a real JBW*-triple, *Math. Nachr.* **222** (2001) 159–184.

[38] J. Faraut and A. Koranyi, *Analysis on Symmetric Cones* (Oxford: Clarendon Press, 1994).

[39] Y. Friedman, *Physical Applications of Homogeneous Balls* (Boston: Birkhäuser, 2005).

[40] Y. Friedman and B. Russo, Structures of the predual of a JBW*-triple, *J. Reine Angew. Math.* **356** (1985) 67–89.

[41] Y. Friedman and B. Russo, Solution of the contractive projection problem, *J. Funct. Analy.* **60** (1985) 56–79.

[42] Y. Friedman and B. Russo, The Gelfand Naimark theorem for JB*-triples, *Duke Math. J.* **53** (1986) 139–148.

[43] C.M. Glennie, Some identities valid in special Jordan algebras but not valid in all Jordan algebras, *Pacific J. Math.* **16** (1966) 47–59.

[44] K. Goeble, Fixed points and invariant domains of holomophic mappings of the Hilbert ball, *Nonlinear Analysis* **6** (1982) 1327–1334.

[45] H. Hamada, T. Honda and G. Kohr, Bohr's theorem for holomorphic mappings with values in homogeneous balls, *Israel J. Math.* **173** (2009) 177–187.

[46] H. Hamada and G. Kohr, Φ-like and convex mappings in infinite dimensional spaces, *Rev. Roum. Math. Pures Appl.* **47** (2002) 315–328.

[47] H. Hanche-Olsen and E. Størmer, *Jordan Operator Algebras* (London: Pitman, 1984).

[48] Harish-Chandra, Representations of semi-simple Lie groups VI, *Amer. J. Math.* **78** (1956) 564–628.

[49] L.A. Harris, Bounded symmetric domains in infinite dimensional spaces, in *Lecture Notes in Math.* 364 (Berlin: Springer-Verlag 1974), pp. 13–40.

[50] S. Heinrich, Ultraproducts in Banach space theory, *J. Reine Angew. Math.* **313** (1979) 72–104.

[51] S. Helgason, *Differential Geometry, Lie Groups and Symmetric Spaces* (London: Academic Press, 1980).

[52] K.H. Helwig, Jordan-Algebren und Symmetrische Rume I, *Math. Z.* **115** (1970) 315–349.

[53] M.R. Hestenes, A ternary algebra with applications to matrices and linear transformations, *Arch. Rational Mech. Analysis* **11** (1962) 138–194.

[54] E. Hille and R.S. Phillips, *Functional Analysis and Semi-groups*, AMS Colloquium Publ. **31** (Providence: Amer. Math. Soc. 1957).

[55] U. Hirzebruch, Über Jordan-Algebren und kompakte Riemannsche symmetrische Räume von Rang 1, *Math. Z.* **90** (1965) 339–354

[56] G. Horn, Classification of JBW*-triples of type I, *Math. Z.* **196** (1987) 271–291.

[57] G. Horn, Characterization of the predual and ideal structure of a JBW*-triple, *Math. Scand.* **61** (1987) 117–133.

[58] G. Horn and E. Neher, Classification of continuous JBW*-triples, *Trans. Amer. Math. Soc.* **306** (1988) 553–578.

[59] L.K. Hua, *Harmonic Analysis of Functions of Several Complex Variables in the Classical Domains*, Translations of Math. Monographs 6 (Providence: Amer. Math. Soc. 1963).

[60] R. Iordănescu, *Jordan Structures in Analysis, Geometry and Physics* (Bucharest: Editura Academiei Române, 2009).

[61] J.M. Isidro, W. Kaup and A. Rodríguez Palacios, On real forms of JB*-triples, *Manuscripta Math.* **86** (1995) 311–335.

[62] N. Jacobson, *Structure and Representations of Jordan Algebras*, Amer. Math. Soc. Colloq. Publ. **39** (Providence: Amer. Math. Soc. 1968).

[63] L. Ji, Introduction to symmetric spaces and their compactifications, In *Lie Theory*, ed. J.-P. Anker and B. Orsted, Progress in Math. **229** (Boston: Birkhauser, 2005) pp. 1–67.

[64] P. Jordan, J. von Neumann and E. Wigner, On an algebraic generalisation of the quantum mechanical formalism, *Ann. Math.* **36** (1934) 29–64.

[65] R.V. Kadison, Isometries of operator algebras, *Ann. of Math.* **54** (1951) 325–338.

[66] I.L. Kantor, Classification of irreducible transitive differential groups, *Dokl. Akad. Nauk SSSR* **158** (1964) 1271–1274.

[67] I.L. Kantor, Transitive differential groups and invariant connections on homogeneous spaces, *Trudy Sem. Vecktor. Tenzor. Anal.* **13** (1966) 310–398.

[68] I. Kaplansky, *Lie Algebras and Locally Compact Groups*, Chicago Lectures in Math. (Chicago: University of Chicago Press, 1971).

[69] W. Kaup, *Algebraic Characterization of Symmetric Complex Banach Manifolds*, Math. Ann. **228** (1977) 39–64.

[70] W. Kaup, A Riemann mapping theorem for bounded symmetric domains in complex Banach spaces, *Math. Z.* **183** (1983) 503–529.

[71] W. Kaup, Über die Klassifikation der symmetrischen hermiteschen Mannigaltigkeiten unendlicher Dimension II, *Math. Ann.* **262** (1983) 57–75.

[72] W. Kaup, Contractive projections on Jordan C*-algebras and generalizations, *Math. Scand.* **54** (1984) 95–100.

[73] W. Kaup, Hermitian Jordan triple systems and their automorphisms of bounded symmetric domains. In *Non-associative Algebra and its Applications*, Oviedo 1993, ed. S. González (Dordrecht: Kluwer Acad. 1994), pp. 204–214.

[74] W. Kaup, On real Cartan factors, *Manuscripta Math.* **92** (1997) 191–222.

[75] W. Klingenberg, *Riemannian Geometry* (Berlin: Walter der Gruyter, 1982).

[76] M. Koecher, *Jordan Algebras and their Applications*, University of Minnesota, Minneapolis, 1962, (Lecture Notes in Math. **1710**, Heidelberg: Springer-Verlag, 1999).

[77] M. Koecher, Imbedding of Jordan algebras into Lie algebras I, *Bull. Amer. J. Math.* **89** (1967) 787–816.

[78] M. Koecher, *An Elementary Approach to Bounded Symmetric Domains*, Lecture Notes (Rice University, 1969).

[79] M. Koecher, Jordan algebras and differential geometry. In *Actes du Congrès International des Mathématiciens*, Nice (1970), pp. 279–283.

[80] S. Kobayashi, *Transformation Groups in Differential Geometry* (Heidelberg: Springer-Verlag, 1972).

[81] S. Lang, *Differential and Riemannian Manifolds* (Heidelberg: Springer-Verlag, 1995).

[82] O. Loos, Jordan triple systems, R-spaces, and bounded symmetric domains, *Bull. Amer. Math. Soc.* **77** (1971) 558–561.

[83] O. Loos, A structure theory of Jordan pairs, *Bull. Amer. Math. Soc.* **80** (1974) 67–71.

[84] O. Loos, *Jordan Pairs*, Lecture Notes in Math. **460** (Heidelberg: Springer-Verlag, 1975).

[85] O. Loos, *Bounded Symmetric Domains and Jordan Pairs*, Mathematical Lectures (University of California, Irvine, 1977).

[86] O. Loos, On the socle of a Jordan pair, *Collect. Math.* **40** (1989) 109–125.

[87] K. McCrimmon, Jordan algebras and their applications, *Bull. Amer. Math. Soc.* **84** (1978) 612–627.

[88] K. McCrimmon, *A Taste of Jordan Algebras*, Universitext (Heidelberg: Springer-Verlag, 2004).

[89] P. Mellon, Holomorphic invariance on bounded symmetric domains, *J. Reine Angew. Math.* **523** (2000) 199–223.

[90] K. Meyberg, Jordan-Tripelsysteme und die Koecher-Konstruktion von Lie-Algebren, *Math. Z.* **115** (1970) 58–78.

[91] S.B. Myers and N. Steenrod, The group of isometries of a Riemannian manifold, *Ann. of Math.* **40** (1939) 400–416.

[92] E. Neher, *Jordan Triple Systems by the Grid Approach*, Lecture Notes in Math. **1280** (Berlin: Springer-Verlag, 1987).

[93] T. Nomura, Grassmann manifold of a JH-algebra, *Annals of Global Analysis and Geometry* **12** (1994) 237–260.

[94] H. Omori, *Infinite-dimensional Lie Groups*, Transl. Math. Monographs **158** (Providence: Amer. Math. Soc., 1997).

[95] V.P. Potapov, The multiplicative structure of J-contractive matrix functions, *Amer. Math. Soc. Transl.* **15** (1960) 131–243.

[96] I. Pyatetzki-Shapiro, On a problem posed by É. Cartan, *Dokl. Akad. Nauk. SSSR* **124** (1959) 272–273.

[97] T. Robart, Sur l'intégrabilité des sous-algèbres de Lie en dimension infinie, *Can. J. Math.* **49** (1997) 820–839.

[98] W. Rudin, *Real and Complex Analysis* (New York: McGraw-Hill, 1966).

[99] W. Rudin, *Functional Analysis* (New York: McGraw-Hill, 1973).

[100] B. Russo, Structure of JB*-triples, in *Proceedings of Oberwolfach Conference on Jordan Algebras*, 1992, eds. W. Kaup, K. McCrimmon, H.P. Petersson (Berlin: Walter de Gruyter, 1994), pp. 208–280.

[101] S. Sakai, A characterization of W*-algebras, *Pacific J. Math.* **6** (1956) 763–773.

[102] J. Sauter, *Randstrukturen beschränkter symmetrischer Gebiete*, Dr. der Nat. Dissertation (Universität Tübingen, 1995).

[103] I. Satake, *Algebraic Structures of Symmetric Domains* (Princeton: Princeton Univ. Press, 1980).

[104] R.D. Schafer, *An Introduction to Nonassociative Algebras* (London: Academic Press, 1966).

[105] T.A. Springer, *Jordan Algebras and Algebraic Groups*, Ergebnisse der Math. und ihrer Grenzgebiete **75** (Heidelberg: Springer-Verlag, 1973).

[106] L.L. Stacho, A projection principle concerning biholomorphic automorphisms, *Acta. Sci. Math.* **44** (1982) 99–124.

[107] A. Stachura, Iterates of holomorphic self-maps of the unit ball in Hilbert space, *Proc. Amer. Math. Soc.* **93** (1985) 88–90.

[108] J. Tits, Une classe d'algèbres de Lie en relation avec les algèbres de Jordan, *Indag. Math.* **24** (1962) 530–535.

[109] D. Topping, *Jordan Algebras of Self-adjoint Operators*, Mem. Amer. Math. Soc. **53** (Providence: Amer. Math. Soc., 1965).

[110] H. Upmeier, Über die Automorphismengruppe von Banach-Mannigfaltigkeiten mit invarianter Metrik, *Math. Ann.* **223** (1976) 279–288.

[111] H. Upmeier, *Symmetric Banach Manifolds and Jordan C*-algebras*, Math. Studies **104** (Amsterdam: North Holland, 1985).

[112] V.S. Varadarajan, *Lie Groups, Lie Algebras and their Representations*, Grodnote Texts in Math. **102**, (Heidelberg: Springer-Verlag, 1984).

[113] E.B. Vinberg, The theory of convex homogeneous cones, *Trudy Moskov. Mat. Obsc.* **12** (1963) 303–358.

[114] J.P. Vigué, Le groupe des automorphismes analytiques d'un domaine borné d'un espace de Banach complexe. Application aux domaines bornés symétriques, *Ann. Sc. Ecob. Norm. Sup.* **9** (1976) 203–282.

[115] W. van Est and T. Korthagen, Nonenlargeable Lie algebras, *Indag. Math.* **26** (1964) 15–31.

[116] J. Wolf, Fine structure of Hermitian symmetric spaces, in *Symmetric Spaces* (Marcel Dekker, 1972), pp. 271–357.

[117] J.D.M. Wright, Jordan C*-algebras, *Michigan Math. J.* **24** (1977) 291–302.

[118] J.D.M. Wright and M. Youngson, On isometries of Jordan algebras, *J. London Math. Soc.* **17** (1978) 339–344.

[119] M. Youngson, Hermitian operators on Banach Jordan algebras, *Proc. Edinburgh Math. Soc.* **22** (1979) 169–180.

[120] E.I. Zelmanov, On prime Jordan algebras, *Algebra i Logika* **8** (1979) 162–175, English transl. *Algebra and Logic* **18** 1979.

[121] E.I. Zelmanov, On prime Jordan algebras II, *Sibirsk Mat. Zh.* **24** (1983) 89–104. English Transl. *Siberian Math. J.* **24** 1993.

[122] E.I. Zelmanov, Lie algebras with a finite grading, *Math. USSR Sbornik* **52** (1985) 347–385.

[123] K.A. Zhevlakov, A.M. Slinko, I.P. Shestakov and A.I. Shirshov, *Rings that are Nearly Associative* (London: Academic Press, 1982) [Russian original (Moscow: Nauka, 1978)].

Index